ROUTLEDGE LIBRARY EDITIONS: GEOLOGY

Volume 1

ADJUSTMENTS OF THE FLUVIAL SYSTEM

ADJUSTMENTS OF THE FLUVIAL SYSTEM

Binghamton Geomorphology Symposium 10

Edited by
DALLAS D. RHODES AND
GARNETT P. WILLIAMS

LONDON AND NEW YORK

First published in 1979 by Kendall/Hunt
Reissued in 1982 by George Allen & Unwin

This edition first published in 2020
by Routledge
2 Park Square, Milton Park, Abingdon, Oxon OX14 4RN

and by Routledge
52 Vanderbilt Avenue, New York, NY 10017

Routledge is an imprint of the Taylor & Francis Group, an informa business

© 1979, 1982 D.D. Rhodes, G.P. Williams and Contributors

British Library Cataloguing in Publication Data
A catalogue record for this book is available from the British Library

ISBN: 978-0-367-18559-6 (Set)
ISBN: 978-0-429-19681-2 (Set) (ebk)
ISBN: 978-0-367-46057-0 (Volume 1) (hbk)
ISBN: 978-0-367-46058-7 (Volume 1) (pbk)
ISBN: 978-1-00-302670-9 (Volume 1) (ebk)

Publisher's Note
The publisher has gone to great lengths to ensure the quality of this reprint but
points out that some imperfections in the original copies may be apparent.

Disclaimer
The publisher has made every effort to trace copyright holders and would welcome
correspondence from those they have been unable to trace.

ADJUSTMENTS OF THE FLUVIAL SYSTEM

Binghamton Geomorphology Symposium 10

Edited by
DALLAS D. RHODES AND
GARNETT P. WILLIAMS

Routledge
Taylor & Francis Group

LONDON AND NEW YORK

First published in 1979 by Kendall/Hunt
Reissued in 1982 by George Allen & Unwin

This edition first published in 2020
by Routledge
2 Park Square, Milton Park, Abingdon, Oxon OX14 4RN

and by Routledge
52 Vanderbilt Avenue, New York, NY 10017

Routledge is an imprint of the Taylor & Francis Group, an informa business

British Library Cataloguing in Publication Data
A catalogue record for this book is available from the British Library

ISBN: 978-0-367-18559-6 (Set)
ISBN: 978-0-429-19681-2 (Set) (ebk)
ISBN: 978-0-367-46057-0 (Volume 1) (hbk)
ISBN: 978-0-367-46058-7 (Volume 1) (pbk)
ISBN: 978-1-00-302670-9 (Volume 1) (ebk)

Publisher's Note
The publisher has gone to great lengths to ensure the quality of this reprint but points out that some imperfections in the original copies may be apparent.

Disclaimer
The publisher has made every effort to trace copyright holders and would welcome correspondence from those they have been unable to trace.

ADJUSTMENTS OF THE FLUVIAL SYSTEM

DALLAS D. RHODES
GARNETT P. WILLIAMS
editors

A Proceedings Volume of the
Tenth Annual Geomorphology Symposia Series
held at Binghamton, New York
September 21-22, 1979

London
GEORGE ALLEN & UNWIN
Boston   Sydney

George Allen & Unwin (Publishers) Ltd,
40 Museum Street, London WC1A 1LU, UK

George Allen & Unwin (Publishers) Ltd,
Park Lane, Hemel Hempstead, Herts. HP2 4TE, UK

Allen & Unwin Inc.
9 Winchester Terrace, Winchester, Mass 01890, USA

George Allen & Unwin Australia Pty Ltd,
8 Napier Street, North Sydney, NSW 2060, Australia

First published in 1979 by Kendall/Hunt
Re-issued in 1982 by George Allen & Unwin

British Library Cataloguing in Publication Data

Adjustments of the fluvial system. — (Annual geomorphology symposia
series; 10th)
   1. Watersheds — Congresses
   2. Sediments (Geology) — Congresses
   3. Rivers — Congresses
   I. Rhodes, Dallas D.
   II. Williams, Garnett P.
   III. Series
   551.3'5   GB561

   ISBN 0−04−551059−8
   ISBN 0−04−551060−1 Pbk

Library of Congress Catalog Card Number: 79−89524

B   402108 01

Printed in the United States of America

To Marie Morisawa and Don Coates, who
initiated these annual geomorphology
symposia ten years ago. They have
nurtured and developed the series
over the past decade, providing
leadership, inspiration, and an
important stimulus to geomorphology.

CONTENTS

Introduction   vii

J. Hoover Mackin (1905-1968) as remembered by his friends
and students 1
       Assembled by R. Ken Fahnestock

Part 1   FLUVIAL THEORY

1   Catastrophe Theory as a Model for Change in Fluvial Systems 13
       William L. Graf

2   Invariant Power Functions as Applied to Fluvial Geomorphology 33
       Waite R. Osterkamp

3   Dynamic Adjustments of Alluvial Channels. 55
       Chih Ted Yang and Charles C.S. Song

Part 2   CHANNEL PROCESSES

4   Hydraulic Adjustment of the East Fork River, Wyoming to the
       Supply of Sediment 69
       Edmund D. Andrews

5   Distribution of Boundary Shear Stress in Rivers 95
       James C. Bathurst

6   Bank Processes, Bed Material Movement, and Planform Development
       in a Meandering River 117
       Colin R. Thorne and John Lewin

Part 3   ADJUSTMENTS TO NATURAL EVENTS

7   Event Frequency and Morphological Adjustment of Fluvial Systems
       in Upland Britain. 139
       A.M. Harvey, D.H. Hitchcock, and D.J. Hughes

8   Effects of Large Organic Debris on Channel Form and Fluvial
       Processes in the Coastal Redwood Environment 169
       Edward A. Keller and Taz Tally

9   Forest-Fire Devegetation and Drainage Basin Adjustments in
       Mountainous Terrain 199
       William D. White and Steven G. Wells

Part 4   INTERPRETATION OF PALEO-ADJUSTMENTS

10   Slack-Water Deposits: A Geomorphic Technique for the
     Interpretation of Fluvial Paleohydrology 225
          Peter C. Patton, Victor R. Baker, and R. Craig Kochel

11   River Channel and Sediment Responses to Bedrock Lithology
     and Stream Capture, Sandy Creek Drainage, Central Texas 255
          Russell G. Shepherd

12   Quaternary Fluvial Geomorphic Adjustments in Chaco Canyon,
     New Mexico 277
          David W. Love

Part 5   ADJUSTMENTS TO MAN-MADE CHANGES

13   Channel Adjustment to Sediment Pollution by the China-Clay
     Industry in Cornwall, England 309
          Keith S. Richards

14   Hydraulic Geometry, Stream Equilibrium, and Urbanization 333
          Marie Morisawa and Ernest LaFlure

15   Some Canadian Examples of the Response of Rivers to
     Man-Made Changes 351
          Dale I. Bray and Rolf Kellerhals

Introduction

This is the tenth meeting in the series of annual geomorphology symposia sponsored by the Department of Geological Sciences and Environmental Studies, State University of New York at Binghamton. After its inception ten years ago, the symposium quickly became a highlight for geomorphologists of all backgrounds. At these meetings such outstanding studies as Geoffrey Boulton's Kirk Bryan Award-winning paper on glacial erosion (1974) and Stanley Schumm's (1973) study of geomorphic thresholds have been presented. Schumm's paper provided the seminal idea for the 1978 symposium. Nearly 150 other papers (counting those in this volume) have been presented during the symposia. Many are frequently cited in geomorphic literature.

Each symposium has been organized around a central theme. Past themes have been environmental (1970), quantitative (1971), coastal (1972), fluvial (1973), and glacial geomorphology (1974), theories of landform development (1975), geomorphology and engineering (1976), and geomorphic thresholds (1978). This year we have come back to the very popular fluvial theme.

Occasionally a symposium is dedicated to an individual who, during his career, made outstanding contributions to a certain discipline. The present symposium honors one such person--J. Hoover Mackin. His impact on geology in general and geomorphology in particular was considerable. Mackin's contribution derives as much from his teaching and inspiration as from his lengthy bibliography. Other authors have adequately recounted the many honors, societies, publications, and related aspects of Mackin's career. In this symposium volume Ken Fahnestock, one of Hoover's former students, presents the fascinating story of Hoover the man.

The title of this year's meeting, "Adjustments of the Fluvial System", also honors Mackin. It comes from his classic paper on the concept of the graded river (1948) in which Mackin repeatedly refers to "adjustments" of various aspects of river systems. The interrelationships among water discharge, sediment load, slope, and channel form were central to Mackin's work. During the 31 years since Mackin's paper was published, the study of these interrelationships has continued to occupy geomorphologists. This symposium brings together several recent studies which advance our understanding of river systems.

The first group of papers presented deals with fluvial theory. William Graf discusses the possible applicability of catastrophe theory to fluvial geomorphology. Chih Ted Yang and Charles Song relate the theory of minimum rate of energy expenditure to observed channel adjustments. Waite Osterkamp suggests invariant power functions as a fresh approach to the relations between hydraulic variables.

Channel processes are the second general topic. Ned Andrews relates measured bedload-transport rates to channel morphology and hydraulic geometry on a Wyoming river. James Bathurst presents his detailed measurements of shear stress on the channel boundary of a meander bend. Colin Thorne and John Lewin, working on the same reach of river as Bathurst, discuss the bank processes, bed-material movement, and planform development of that reach.

The adjustment of streams to some natural events is the subject of the third group of papers. Adrian Harvey and co-workers discuss event frequency as it influences process thresholds in upland Britain. Two aspects of fluvial adjustment to vegetation are presented. The effects of downed redwood trees and other organic debris on channel form in coastal northern California are analyzed by Edward Keller and Taz Tally. William White and Steve Wells demonstrate how forest-fire devegetation influenced drainage-basin adjustments in a mountainous area of New Mexico.

Three papers emphasize the interpretation of paleo-adjustments. Peter Patton and colleagues explore the possible use of overbank slackwater deposits as indicators of the magnitude of ancient floods. Russ Shepherd examines channel adjustments to stream capture and bedrock type in central Texas. The fluvial geomorphic history of Chaco Canyon, New Mexico is analyzed by Dave Love.

Channel adjustments to the effects of man are the subject of the final group of papers. Keith Richards shows how some British stream channels have reacted to increased suspended loads of kaolin introduced by the china-clay industry. Marie Morisawa and Ernest LaFlure trace the changes in channel size and hydraulic geometry that accompany the urbanization of areas in Pennsylvania and New York. The adjustments of several Canadian rivers to the works of man is treated by Dale Bray and Rolf Kellerhals.

Organizing the symposium and preparing the proceedings volume has required the help of many individuals. Marie Morisawa has been a source of encouragement and guidance. The authors have been models of patience, cooperation, and understanding. We are grateful to the session chairmen-- Ian Campbell, Michael Foley, James Knox, and Neil Salisbury--for insuring the smooth presentation of the papers at the symposium. Barbara Hostettler's proficiency as our typist and Tom Wisz' (Kendall/Hunt Publishing Co.) helpful suggestions contributed considerably toward a prompt and attractive published volume. Lisa Rossbacher ably assisted in proof reading. We sincerely thank all of these people.

For us, this symposium is the culmination of nearly two years effort. Performing this service has given us an appreciation of the time and energy that have gone into the nine previous meetings. We also learned the value of teamwork. It's rare in any joint enterprise that the responsibilities are equally shared, but somehow we managed, finished the job, and still speak to each other. We enjoyed the opportunity to contribute to this important series. We hope that the annual symposia will continue as a stimulating forum for new ideas in geomorphology.

Dallas D. Rhodes

Garnett P. Williams

Whittier, California

Denver, Colorado

June 1979

REFERENCES

Boulton, G.S., 1974, Processes and patterns of glacial erosion; in Coates, D.R., ed., Glacial geomorphology: Binghamton, State University of New York, Publications in Geomorphology, p. 41-87.

Mackin, J.H., 1948, Concept of the graded river: Geological Society of America Bulletin, v. 59, p. 463-511.

Schumm, S.A., 1973, Geomorphic thresholds and complex response of drainage systems; in Morisawa, M., ed., Fluvial geomorphology: Binghamton, State University of New York, Publications in Geomorphology, p. 299-310.

J. HOOVER MACKIN

(1905-1968)

As remembered by his friends and students

Assembled by

R. K. FAHNESTOCK

Department of Geology
New York State University College
Fredonia, New York

INTRODUCTION

Much of this brief introduction to the man we honor with this symposium has been taken from the eloquent memoirs by James Gilluly (1971) and Harold James (1974). For a bibliography and a list of honors that Hoover Mackin received during his lifetime, the reader is referred to these memoirs. The remainder of the present paper contains stories and anecdotes from Hoover's friends and students. I am truly grateful to these and other individuals whose stories and comments have been used in this compilation, either as direct quotes or as reinforcement of the ideas describing the dimensions of this man.

Mackin's early work was in geomorphology in the Appalachians and in the Big Horn Basin of Wyoming, and he always regarded himself as primarily a geomorphologist. His interests in geology carried him into any area that challenged him. He became successively an engineering geologist, working on dam and reservoir problems; a structural geologist, skilled in the preparation and analysis of complex maps; an economic geologist, successful in ore finding; a field petrologist, expert in the problems of ignimbrites and flood basalts; and finally a student of the moon. Surely he was one of the most versatile geologists of our time.

Mackin's Ph.D. dissertation was a study of the erosional history of the Big Horn Basin, Wyoming. In this he was the first to apply the idea of pedimentation to the northern Rockies. In this work, too, Mackin showed as early as 1936 that a low-gradient stream of slight competence, if it is working on weak rocks, may be able to capture a much more powerful stream working on steep alluvial slopes. How many doctoral theses have launched two such seminal ideas?

During World War II, Mackin joined the Strategic Minerals Program of the United States Geological Survey, with his major project concerning the iron ores of the Iron Springs District, Utah.

During the Iron Springs work Mackin became interested in the problems of ignimbrites. He demonstrated the persistence of recognizable sequences of ash flows over areas of thousands of square miles and emphasized that they had been emplaced on very low gradients. He further pointed out the need

Joe and Esther Mackin
on Pole Cat Bench, 1933

(photography by John Lucke)

to study the ignimbrite sequences in order to determine the nature and amount of Cenozoic deformation in the Basin and Range province. The graduate theses of several students of this "geomorphologist" were devoted to further studies in field petrology and structural geology inspired by his work.

In 1948, after years of study, Mackin finally published his analytical discussion of graded streams (Mackin, 1948). This model of scientific logic clarified and pointed up many of the principal factors involved in stream evolution.

An eloquent advocate of the scientific method, Hoover was never better than in his 1963 paper "Rational and Empirical Methods of Investigation in Geology." This paper should be carefully read by all students of geology and reread at intervals by all professionals. The following extracts from this paper reiterate a standard by which our efforts in this symposium and all our other work may be judged.

Hoover maintained that time would soon take care of the older generation. His paper was written . . .

"for the youngsters--the graduate students--and its purpose is to show that as they quantify, which they are bound to do, it is neither necessary nor wise to cut loose from the classical geologic method. . .
. . .those who use the method all the time never follow the steps in the order stated; the method has become a habit of thought that checks reasoning against other lines of reasoning, evidence against other kinds of evidence, reasoning against evidence, and evidence against reasoning, thus testing both the evidence and the reasoning for relevancy and accuracy at every stage of the inquiry. . .
. . .This means that the investigator admits to his graphs, so to speak, only items of evidence that are relevant to the particular matter under investigation. . .And once an item of information has been admitted to the graph, it cannot be disregarded; as a rule, the items that lie outside the clusters of points are at least as significant, and usually much more interesting, than those that lie within the clusters. It is from inquiry as to why these strays are where they are that most new ideas--most breakthroughs in science-- develop. . .
. . .The best and highest use of the brains of our youngsters is the working out of cause and effect relations in geologic systems, with all the help they can get from the other sciences and engineering, and mechanical devices of all kinds, but with basic reliance on the complex reasoning processes described by Gilbert, Chamberlin, and Johnson."[1]

I would like to add the name of Mackin to this list.

---

[1]Quoted with the permission of the Geological Society of America.

PERSONALITY

Harold James states: "Mackin was a gregarious man and typically was the center of discussion groups in the field or at meetings. His views always were expressed crisply, concisely, and with humor. He loved a good argument and he started many; in one of his papers he remarks that it is more important that a working hypothesis be provocative than it be right. He was often on the attack, but he attacked ideas, not people, and his vigorous and sometimes earthy remarks never left a residue of ill will."

"As Rubey (personal communication) pointed out, the fact that successive terrace levels in a single river valley have gradients which are so nearly alike suggests that slope may actually be less adaptable than cross-sectional form." (Hoover--"while I love Bill, I hate this kind of meaningless generalization.")

Hoover in letter to M.G. Wolman, October 1953:

"I like very much the kind of work indicated in Fig. 28. I suppose I like it because I was weaned on that sort of approach--the testing of each brick before it is built into the structure (Johnson--Role of Analysis and Criteria for Marine Terrace correlation, for example) and I get no glow at all from the use of v^2/d in the pool-riffle problem. So you see, if you read the earlier correspondence with Luna, that I'm still me, unreclaimed."

"If there is any doubt that I'm still me, here are a few more of my reactions: Manning's n is a great deal more and a great deal less than roughness in the plain English meaning of the word. No matter now clearly you say this, the word still conveys the wrong meaning.

Arriving at what you call roughness by elimination seems to me equivalent to the worst way of bringing a chemical analysis to exactly 100 %. "Easy is the descent to Hell" by this route. . .

. . .As to whether the Brandywine is graded in the steep parts, it seems to me that it is simply a question whether the resistant rocks present to the river a problem of abrasion or transportation.

All the same, I think that your paper is a fine job and an important contribution."

John Anderson reveals Hoover's attitude toward geomorphology that should be considered seriously by all. When John entered Hoover's office at Texas with another new student, he welcomed them with his usual warmth and asked what they wanted to work on. The other student immediately replied that he wanted to do geomorphology. Hoover thought this over for a moment and then gently asked, "Don't you think you had better learn some geology first?"

Hoover set high personal standards for research and morality. He encouraged his students and colleagues by word and example. Don Mullineaux recalls, "Many of us will remember with some regret his brusqueness, and that many times he didn't recognize that his barks could inflict wounds. But we will also remember that he hurt no one by design, and avoided disparaging remarks about us or others to third parties, even though he might disagree with approaches or results. He also buffered and defused, when possible, the kinds of conflicts and antagonisms that inevitably arise during geological investigations. He would not accept poor standards of either competence or integrity as objectives, but did recognize and accept the fact that we could not match his own standards. In short, he offered knowledge, methodology, enthusiasm, support, and friendship, and we couldn't have asked for more."

Pete Rowley writes, "Hoover had a most engaging personality. He was totally natural, however; none of these adventures were ever contrived or manufactured. He was never stuffy or haughty with students, but treated us as he would the most distinguished visitor. He was intensely loyal to his students. He criticized ideas, but not people, and was sensitive to the feelings of others. Nonetheless, a person developed a thick skin working with him because he was such a perfectionist."

If we chided him about one or another eccentricity, he claimed that he had worked hard at cultivating a reputation of a character for more than 20 years (when Don Mullineaux first heard the claim), and he wasn't about to give it up then.

Hoover was an insomniac and a noisy sleeper; on trips, nobody who knew better ever shared a room with him. But even more legendary was his restlessness in bed; pictures on the wall and lamps on bedside tables had to be moved to safer places, far from the bed. His wife Esther marveled at the time when Hoover woke up in the morning on a bare mattress. The bottom sheet was missing and could not be found; it wasn't until months later, when they turned the mattress, that the crumpled, errant sheet was found--under the mattress. Thus, during that memorable night, Hoover had pulled off the sheet, crumpled it up, and stuffed it under the mattress--all in his sleep!

Howard Coombs recalls, "In his home Hoover, together with Esther, surrounded their guests with warm hospitality and interesting conversation. Discussions about geology were the order of the day, or far into the night, with delightful comments from Esther, a gracious lady with a great sense of humor. No time ever passed more quickly or more enjoyably than in the Mackin home. I feel most fortunate in having had this opportunity to be with Hoover and Esther for so many years--no one can forget them."

Pete Rowley writes, "His wife, Esther, is also a joy to all those who know her, and has an excellent sense of humor. At parties, people would gather around her also; she was especially loved by the grad students. After Hoover's death, Willie Nelson remembers a comment by her in which she marveled at the outpouring of letters and love for Hoover.

Hoover enjoyed geology, people, ideas, travel; in short, life. His bibliography, like his life, was far too short. There were too few hours in the day for his family, students, friends and profession to write all of the papers that he had within. His standards, both in clarity of thought and exposition, required that each paper be written, reviewed, and rewritten many times. Yet he once wrote in a letter to Wolman,

"The 1936 article covers most of the points we discussed in Denver--
I'd like to rewrite it, but have yet to reread a paper (of mine)
that did not give the same feeling."

TEACHING

Many of his students, including James, echo the idea that "as a teacher of
earth science, Mackin was almost without a peer during his lifetime. His lectures were models of clarity, and they were delivered with a completely in-
fectious intensity and enthusiasm, whether given to the beginning freshman
class or to a group of advanced graduate students. . . . .He encouraged divergent views--provided they were based on good, logical thinking--and de-
tested the mere parroting of textbook or classroom notes. In his. . .course
in map interpretation, he surprised students continually with an A grade for
the wrong answer reached by careful analysis and reasoning, and a C or worse
for the right answer based on an inadequate or improper approach. Logical
thinking was paramount."

Don Mullineaux remembers that Hoover taught as much by example as well
as words. "He demanded that students think and question as they learned,
taught that the process of attacking a problem was important as well as the
answer. He also required that students for their own research truly consider
alternate hypotheses and pry into the meaning of their results beyond the
first or most evident answer. Perhaps most of all, he emitted ideas and questions like a radioactive body, and bombarded students and associated professionals alike with them."

Howard Coombs recalls, "The Geology Department during his stay at the
University of Washington was a close-knit group that provided an opportunity
to discuss geologic and other problems with complete freedom. Not only was
I a colleague of Hoover's but I sat through several of his classes and the
class on map interpretation several times. It was most stimulating to follow
his analysis of the problem from the evidence on maps, air photos and remote
sensing data. Hoover was quite impatient with illogical thinking but at the
same time gave full credit for imaginative solutions if tempered with a solid
data base. The students soon learned about logical thought processes as
applied to geology."

Gerald Parker remembers, "All the while Hoover talked he was rapidly
smoking one cigarette after another and illustrating his lecture on the structure and stratigraphy of Long Island, N.Y. with rapidly- and accurately-
drawn geologic cross sections. These he developed as he explained, chalk
in one hand and a cigarette in the other. So wrapped up in his lectures did
he become that he sometimes unconsciously tried to write with a cigarette or
smoke his piece of chalk! He was like an artistic magician with his colored
chalk diagrams and so thoroughly organized in his thinking, intent in his
purpose, and intense in his concentration that his explanations left no doubt
whatever of what he was describing. Cause and effect were the two sides
of the same coin; form and process went hand-in-hand; if we were to understand the how, when, where, and why of the earth's construction and confi-

guration, we would need to understand geologic and geomorphic processes."

OFFICE

Hoover's office was a classic of apparent disarray, and he made the customary claim of a "place for everything, and everything in its place." When Don Mullineaux finished a Seattle thesis project under him, the geologic maps were too large for the library, and by agreement with the library were left with Hoover. "Some years later I returned to Seattle to work there with the USGS, and went to Hoover's office to use those maps. But he remembered nothing about the situation and couldn't find them, even though I swore that he had to have them. Once he realized, however, that I claimed to have personally left them there, he pointed out that if I had really put them somewhere in his office, they would still be in that place. And so they were, dusty but undisturbed."

Pete Rowley remembers, "His first office at Texas, in the old geology building, was tiny and always a complete mess; it was stacked high with teetering piles of books and papers; every so often students or the head librarian would have to sort through the piles to find lost homework papers or borrowed books. When the department moved (1967) into its present large impressive building, Hoover was given probably the largest office--the size of a small classroom--that was dominated in the center by a huge conference table, which soon after similarly became piled high with Mackin's projects. Any student or visitor to the office was required to have his head pressed against Hoover's chest in order to listen to the "click" of his artificial heart valve."

Paul Williams tells the story--also repeated by Gilluly--about Mackin continually misplacing the key to his ground-floor office in the Geology Department at the University of Washington. So numerous mornings he entered his office by means of his window, until he decided to simply leave the office door unlocked. Hoover's office always was a mess. Warren Hobbs remembers, after the 1936 earthquake in Seattle, Hoover reminded him that an earthquake could never adversely affect his office because "everything already is at the angle of repose."

FIELD TRIPS

Grant Heiken describes one field trip "During Easter vacation, 1966. Hoover wanted several of his students to see the remarkable sequence of ignimbrites exposed in the Sierra Madre Occidental of Western Mexico. So, four of us piled into a U.T. carryall and drove all day and most of one night to Torreon. The next day we collected Hoover and Dick Blank at the Durango airport where we also met two geologists from the Instituto de Geologia. During the next 36 hours, guided by Mackin's remarkable ability to observe and interpret, we were exposed to more ignimbrites than any of us had ever imagined could be present in any single volcanic field. Since that trip, students from U.T. and U.C.-Santa Cruz have been working in this area; it is, indeed, one of the world's largest silicic volcanic fields.

The first field day was spent looking at the section between Durango and Mazatlan. After an all-too-short night in a hotel at Mazatlan, we hit the same section again, beginning at sunrise and finishing after dark, near Durango. Since it was Easter week, our two colleagues from the Instituto had stayed out all night with the celebration in the streets. Hoover, not knowing this, was somewhat upset when the celebrants kept dozing off on outcrops the next day! We also, in our enthusiasm for the geology, forgot that there weren't any gas stations along the Durango highway and literally coasted the last 20 miles, down dip slopes of tuff units dipping toward Durango.

After dinner in Durango, our colleagues from Mexico City headed for the nearest hotel in a state of collapse. No such luxury for us! Mackin had to be at a meeting in the U.S. the next day--so we drove through the night toward Monterrey where he was to catch a plane. Along the way we witnessed the Mackin "lead foot"; I looked over his shoulder, at 4 am, and noticed that the U.T. carryall was being stretched to 85 mph down a shoulderless Mexican highway."

DRIVING

While Hoover told many stories on himself with great glee, he never told any about his driving. In fact he loved to drive anything he could get his hands on from a truck to a Honda trail machine and prided himself on his skill. He was just a bit hurt when his students and colleagues insisted on driving.

Bates McKee certainly had an early introduction to Hoover's driving habits. "Over one of the first weekends I was at UW Hoover was leading a field trip to Mount Rainier, and kindly asked me to go along, and to ride with him in his car. I was unprepared, and by the time we had cleared the Seattle city limits, we had gone through 5 red lights. I knew then that this was a blessed man, and God must have been saving him for something beautiful."

Paul Williams remembers in 1955 when he and Hoover decided to drive non-stop from Seattle to Salt Lake City in Paul's 1940 Plymouth, which was on its last legs. He woke up with dawn breaking, and Hoover talking to a highway patrolman. He remembers the patrolman say "Do you realize that you were driving 70 on the wrong side of the road?" As luck would have it, the patrolman was a rockhound and amateur geologist, and for 20 minutes Hoover--dripping with Irish charm--regaled the man with a thumbnail sketch of the geology of southern Idaho. Finally the patrolman retired, without issuing a ticket! When he had left, Hoover drove off--at 70 mph on the wrong side of the road!

Thirty miles from nowhere on a trip to Tatman Mountain where Bill Bradley and Ken Fahnestock wanted to walk to the mountain top, Hoover, because of his heart condition, volunteered to drive the Survey jeep along the base of the mountain to pick us up at the other end. As soon as we were out of earshot, Bill asked why on earth I would let him do that? I could see little harm in letting Hoover drive a mile on the dirt road and besides, what was I supposed to say under the circumstances? As we neared the end of the mountain, we could see the jeep leave the road below and start across boulder strewn flats through waist high sagebrush. We left the mountain top immediately to try to head off a very long walk home, but before we had gotten far, the jeep came to a halt perched on a boulder with three of four wheels in the air.

ABSENT-MINDEDNESS

Hoover was notoriously absent-minded. Alan Cary recalls that "After making a speaking tour of several U.S. universities, Hoover flew to Iceland for some consulting work. He arrived in a pouring rain, going directly from the plane into the field to examine dam sites. After several hours of tramping through wet brush and the downpour he arrived at his hotel totally drenched. Even though it was after hours, the Hotel roused the cleaner from his evening paper to accommodate the consulting geologist from the United States. When his clothes were returned, on very short notice, they consisted of slacks and a sport jacket, not the suit which Hoover insisted he had sent out.

A thorough search of the cleaning establishment revealed no suit, so the local tailor was routed out and Hoover fitted with a new suit.

Upon his return home Esther remarked "Hoover, you have a new suit." An explanation from Hoover followed, to which Esther replied "But Hoover, you wore your slacks and sport jacket. Your suit is here at home in the closet."

Paul Williams tells the story of the day that Hoover captured a small bull snake and put it in his jacket pocket, then buttoned down the flap on the pocket. "He intended to save it for a friend who collected snakes. A day or two later, people started avoiding Hoover. Another day went by during this field trip, and it became clear that there was a strange aroma to Hoover. Finally, wearing the same jacket, Hoover began an active search for the cause of the smell and--you guessed it--pulled out of his pocket a dead bull snake."

Esther tells that on a field trip with students to Mud Mountain Dam on White River in Washington, the day warmed up and Hoover discarded his jacket into "one of his student's" cars. Missing the jacket the next day, he quizzed his students with little success. To this day, some Mud Mountain visitor must be wondering where the jacket came from.

His adventures with hats were mentioned by many. Bates McKee remembers Hoover hunting for his hat when it was on his head. Pete Rowley topped that one by recalling Hoover walking down the hall at Texas with two hats on his head, one perched on top of the other.

If that story can be topped it is by one told with glee by Hoover himself and recalled by Dan Barker.

"One of the last to leave a panel meeting in Washington, D.C., Hoover reached for his new hat in the anteroom of the Cosmos Club and found that the only remaining hat was not his, and further it was not new. Irate, he took the hat as hostage and, after returning to Austin, immediately wrote to the owner whose name was in the hatband, offering to return the hat upon receipt of his own. Shortly after he mailed the letter, his telephone rang:

"Professor Mackin, this is Braniff Airlines. You left your hat on the plane when you flew to Washington last Thursday. Should we deliver it to your home or office?" A letter from the Cosmos Club member denied the theft of Hoover's hat but suggested that Hoover could keep the hat if he needed it so badly."

Hoover provided another explanation:

"Dear Dick and John [Flint and Rogers of Yale]:

Several years ago after a couple of days of meetings in Washington, I came out of a late evening committee session at the Cosmos Club to find in the cloak room only one hat, brown like mine and my size, but way beyond its prime; evidently this beaten-up wreck had been exchanged for my crisp, new $22 Stetson. In the next few days I inveighed with righteous indignation to the Manager and friends who were members about the scoundrel, probably a member, who would make such a switch. I then discovered the name and address of the original owner of the old hat on a card inside the sweatband; I promptly wrote him, taking the charitable position that the switch was accidental, and suggesting (rather firmly) that we exchange hats.

About a week later my hat was returned by Braniff, with a note to the effect that it was found by the cleaners at the end of the Austin-Washington-New York flight, and was finally traced through my seat number, initials in the hat, etc. The same mail brought a letter from the aforementioned miscreant to the effect that he was never near the Cosmos Club, could not find the hat when he left the plane in New York, and would like to have it back. I was of course not taken in by this cock-and-bull story, but still wonder how he forced Braniff to support it.

For some reason--perhaps the sins of an ancestor--I have been repeatedly the victim of this sort of villany;--in Iceland, the pants to a new suit, lost (Ha!) by the drycleaner; in a Mexico City hotel, a pair of shoes; in Idaho, a camera and a wide scattering of shirts, ties, etc. And almost always, when I get to the bottom of it, there is the same preposterous suggestion that the missing articles were mislaid by me. This is duplicity on top of wickedness, and it shakes my sunny regard for the essential goodness of my fellow man.

Now it's an entire car. [Author: When visiting the Kline Laboratory on one occasion, Hoover came out a different door than the one he had entered, couldn't see his rented car and reported it stolen.] If it had happened at Cal Tech or if slippery Dick Jahns were in the offing, I would consider the possibility of a practical joke, but under the circumstances--in the shadow of the Hallowed Halls--that is out of the question. Anyway, as usual, I lucked out.

Thanks to you for finding the car and sending the stuff. And to John and Alan Bateman for their hospitality. The enclosed check, a contribution to the Kline Liquor Fund, is in your names."

Sincerely,

Hoover

CONCLUSIONS

Hoover's presence is still felt wherever his students and colleagues are working. Whenever I am on a gravel bar, I remember the care with which this member of the National Academy placed his carefully-folded pants on a small ant-free island in mid-channel before wading out in his shorts to pebble-count a Greybull River gravel sample by the "Wolman" method. Each pebble had to be located using a measuring tape, because "after 50 years of watching where I step, I cannot step without looking."

Pete Rowley said it well: "In my opinion, he is the standard of excellence, both as a scientist and as a person, for our profession. My memories of how he did things in geology will be an important influence throughout my career."

REFERENCES

Gilluly, James, 1971, Memorial to Joseph Hoover Mackin (1905-1968): Geological Society of America Proceedings for 1968, p. 206-211.

James, H.L., 1974, Joseph Hoover Mackin Nov. 16, 1905 - Aug. 12, 1968: National Academy of Sciences Biographical Memoirs, v. 45, p. 249-262.

Mackin, J.H., 1948, Concept of the graded river: Geological Society of America Bulletin, v. 59, p. 463-511.

_____, 1963, Rational and empirical methods of investigation in geology, in Albritton, C.C., Jr., ed., The fabric of geology: Stanford, California, Freeman, Cooper & Co., p. 135-163.

CATASTROPHE THEORY AS A MODEL FOR

CHANGE IN FLUVIAL SYSTEMS

WILLIAM L. GRAF

Department of Geography
Arizona State University
Tempe, Arizona 85281

ABSTRACT

Catastrophe theory provides a language for describing space-time changes in systems and is potentially applicable to fluvial processes. The theory, which is based on topology, indicates that all changes in the four dimensional natural universe proceed according to one of seven different singularities, or catastrophes. The cusp catastrophe seems likely to be useful to geomorphologists: it is characterized by abrupt and smooth changes, divergent and bimodal behavior, hysteresis, and stability of structure. The utility of catastrophe theory in geomorphology is limited as indicated by tests using data from arroyo systems and rapids in canyon rivers. Disadvantages of the theory include difficulty in identification of system control factors, definition of energy functions, and the generality of the theory. Advantages include a marriage of concepts of equilibrium and change, the stability of the change structure, and its perspective, which is unlike previous models. An evaluation of catastrophe theory in fluvial applications suggests that it has limited usefulness for the description of change, but in some situations it can provide a unifying mechanism for general concepts and specific observations.

INTRODUCTION

Perhaps the most significant social and economic contributions of the science of geomorphology lie in the analysis of change in fluvial systems. Politically and economically important changes in the courses of major rivers such as the Mississippi, destruction of valuable range and irrigated lands by channel entrenchment, catastrophic erosion of soils in agricultural lands and sedimentation problems in channels, reservoirs, and harbors indicate the close association between geomorphic adjustment and societal response. The first step toward meaningful explanation and accurate prediction of fluvial geomorphic changes is to describe them. The objectives of this paper are 1) to introduce catastrophe theory as a means of describing change in fluvial systems, 2) to test this approach using data from entrenched stream channels and rapids in canyon rivers, and 3) to evaluate the potential utility of catastrophe theory in fluvial geomorphology.

Nature of Change in Fluvial Systems

Since the introduction of the hydro-physical approach to fluvial processes by Horton (1932, 1945) and Strahler (1952), the majority of research efforts have been built on the concept of equilibrium. The emphasis on those time periods during which fluvial systems are relatively unchanging or when the systems change in a systematic fashion has been at once rewarding and limiting. The development of systems of explanation that include hydraulic geometry (Leopold and Maddock, 1953) and dynamic equilibrium (Hack, 1960) has advanced our understanding of rivers and their landscapes.

The emphasis on stable time periods has led the science away from analysis of change or adjustment, and only very recently have geomorphologists turned their attention to those periods between the equilibrium states. Schumm (1973), for example, has identified the importance of thresholds and complex responses intrinsic in some fluvial systems. Bull (1975) has shown that some systems do not trend toward a steady state, and Graf (1977) has analyzed response times during periods of fluvial adjustment. A recent major symposium addressed the question of change in fluvial systems rather than their stability (Gregory, 1977). This new direction of research, with its emphasis on change rather than stability, has its roots in, and is a logical outgrowth of, the previous work dealing with equilibrium. The two threads of work are complementary rather than opposed.

The adjustments from one stable state to another may be gradual, and if some geomorphic variables are plotted against time, a smooth curve results (for example, the amount of fill in a sedimenting reservoir). Other geomorphic systems operate with sudden changes from one stable state to another, so that when variables are plotted against time, a step function results (e.g., Knox, 1972). In many cases, the temporal scale of analysis is an important factor (Schumm and Lichty, 1965), and some changes that appear as smooth transitions over one period of observation appear as step functions when viewed in a longer or shorter perspective.

The mathematical language used by geomorphologists to describe changes observed in geomorphic variables has been based on calculus, a body of mathematics that is extremely flexible and adaptable, especially for continuous processes and those that have smooth transitions from one state to another. Geomorphic research concerned with equilibrium has made effective use of calculus-based mathematics, which have yet to be fully exploited for this purpose. The flexibility of calculus is limited, however, when discontinuous operations or abrupt transitions are considered (Zeeman, 1976). Catastrophe theory was designed as a language to replace calculus, especially for those situations where discontinuous processes or abrupt changes are common.

The importance of the artificial language used in scientific research is clear. The language chosen determines what is possible to express and what is to be condemned as an illusion (Boulding, 1956, p. 71). Just as our theories determine how we view the world and its evidence, our language determines the construction of theories. If geomorphic theories are based only on calculus, those theories may be unnecessarily restricted. Catastrophe theory has been employed with varying success by social scientists, engineers, physicists, biologists and paleontologists. Its utility for geomorphology remains untested, but the similarity between the general concepts of catastrophe theory and the changes in some fluvial systems suggests that an unbiased examination of the theory's potential is in order.

Catastrophe Theory

As an alternative to calculus, catastrophe theory is not a theory in the sense that geomorphologists usually employ the term, and it is not necessarily appropriate only for catastrophic changes. The misnomer has led to considerable controversy in other fields where catastrophe theory has been suggested as a useful tool. Because catastrophe theory is a language, it does not explain anything. It may be useful for describing the behavior of geomorphic systems but it does not answer the basic question of "why?"

Catastrophe theory was created by Professor Rene Thom of the Institute for Advanced Scientific Studies in France. His major exposition of the theory is an exercise in differential topology, which provides the derivation of the theory (Thom, 1975). Topology is the study of those properties of geometric figures that are unaffected by deformation. Interpretations of the theory for non-mathematicians have been presented by Zeeman (1976), and Woodcock and Poston (1974) have provided useful background information. A highly readable introductory discussion of the theory and its history is provided by Woodcock and Davis (1978). Thom envisioned the theory as a way to characterize change in systems, reasoning that although natural processes are quantitatively complex, they are qualitatively simple and stable. For example, despite a great variation in climate, lithology, and geologic structure, stream channels have remarkable similarities throughout the world. Their dimensions may be different but their basic characteristics are similar. Thom proposes that the quantitative differences are insignificant in comparison with the qualitative similarities. This emphasis on the qualitative aspects of the real world is a product of the evolution of catastrophe theory from topology, where arrangement rather than magnitude is the essence.

At the heart of Thom's theory is the idea that topological singularities can describe changes through time as well as through space. In calculus and analysis, singularities are the points on a graphic curve where the direction or quality of curvature changes. In topology, singularities are phenomena that occur when points are projected from one surface to another while the surfaces are distorted (Woodcock and Davis, 1978). Through these distortions, the singularity may change in size or magnitude, but it retains its basic form. For example, in space, a topologic singularity called the triangle is always a three sided plane figure defined by straight lines. According to Thom's theory, the change experienced by systems through time can also be characterized by singularities (or forms of space/time), which he called catastrophes. The shape of the triangle, no matter what its size, is determined by three lines. The catastrophe (or type of change in a system), no matter what its rate, is determined by the number of system control factors. Thom (1975) provides a topological proof of this general concept just as Newton once provided proofs of the precepts of calculus.

Objections to catastrophe theory are not a product of its development, but rather result from attempts at application. Thom's initial application of his abstract catastrophe theory was in the analysis of visual caustics, patterns of light resulting from imperfect reflections and refraction. The forms of the caustics successfully represented visible evidence of the predicted catastrophe forms (Woodcock and Davis, 1978). Thom also found that catastrophe theory provided useful approaches to evolutionary and morphogenetic problems in embryology. Other researchers have used catastrophe theory in much closer relationships with quantitative theorizing and prediction, and it is here that controversy develops. Applications in the realms of nerve responses, social behavior, and paleontology have generated more questions than answers, and have not been as successful as Thom's original examples. Critical discussions of the theory are provided in general by Croll (1976), Wagstaff (1976), and Zahler and Sussman (1977), and geological applications were discussed by Cubitt and Shaw (1976) and Henley (1976).

CATASTROPHES AND THE NATURE OF CHANGE

Background

The most profound implication of catastrophe theory for geomorphology is that geomorphic changes proceed according to topologic singularities just like all other changes in the four-dimension universe. These topologic singularities include abrupt as well as smooth changes, and if the appropriate singularity can be identified, the researcher can look beyond the observed data with a great deal of confidence. Precise numerical description may be possible since according to the theory, change through space/time proceeds in only a limited number of ways. By analogy, a plane may be completely subdivided into equal-sized units by only three types of figures: squares, triangles, and hexagons. Thom (1975) conclusively demonstrates that the space-time continuum can also be subdivided by only a limited number of forms: in the case of the natural universe there are only seven singularities or catastrophes. With the plane, the form is determined by the number of sides: with the space/time continuum the form is determined by the number of control and behavior factors. Thus, changes in fluvial systems can be characterized by only one of a few singularities or catastrophes.

A series of concepts, each considered in turn below, can lead to an understanding of basic catastrophe theory. The interpretation taken here is essentially an algebraic one for the sake of simplicity, and has been more fully developed by Stewart (1975) and Wilson (1976). First, consider a system that is controlled by two variables, a and b. System behavior is measured by a third variable, x. These three variables are related to each other by a dynamic or potential energy function, $E(a, b, x)$. For each combination of (a, b) there is a corresponding value of x that minimizes E, representing an equilibrium state as shown in figure 1 (Wilson, 1976, p. 351). These equilibrium states are defined by the solutions of $\frac{E}{x} = 0$. The covariations of a, b, and x that satisfy the equation commonly form a plane, with each combination of (a, b) corresponding to one value of x. But in some instances, one combination of (a, b) might produce two or more corresponding values of x, resulting in a surface that is convoluted by folds. These surfaces, as defined by the equation, are the singularities that characterize system change: they are called catastrophes because a small adjustment in one variable may result in a catastrophic change in the behavior variable as its value changes radically across a fold.

In summary, a catastrophe is a surface that is a graph of all (a, b, x) points where the first derivative of the energy function is zero. Each system

16

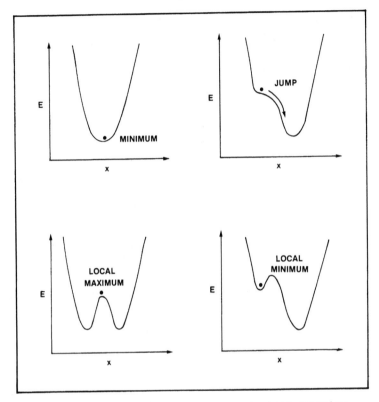

Figure 1.   *Versions of an energy or potential function.
The most likely state for each is where the
first derivative, or slope of the line representing the function, is zero. Local
maxima are unstable.*

state plots as a point on that surface, and as the system changes through time
the point moves about on the surface. If the system is not in equilibrium, the
point does not plot on the surface. In this scheme, system change is defined
by movement from one equilibrium condition to another. Also, there may be
multiple control or behavior variables adding dimensions and complexity to the
catastrophe surface.

In fluvial geomorphic applications, the system concept is readily acceptable
and has been explicitly employed by researchers for nearly two decades
(Chorley, 1962; Chorley and Kennedy, 1972). Control variables could be defined in a number of ways, but the most basic are force and resistance.
Possible behavior variables include channel characteristics such as width,
depth, shape, and roughness. It is now recognized that most fluvial systems
tend toward some kind of equilibrium, dynamic or steady state, so that the

general concept of an energy function is familiar to geomorphologists, though the exact forms of these functions have not been deduced. The definition of the equilibrium states as those where the potential energy is equal to zero is not common in geomorphology, but since this is only one of several possible interpretations of the equilibrium that is associated with the catastrophe surface, the discrepancy is not a serious one. The major consideration is that there are tendencies toward stable or dynamic equilibriums.

Basic Catastrophes

Thom's derivation of catastrophe theory indicates that the form of the energy functions that ultimately define the shapes of the singularities are catastrophe surfaces. All systems in the natural, four-dimensional (length, width, height, and time) universe have energy functions that depend solely on the number of control factors. These functions, along with their first derivatives that are used to define the corresponding catastrophe surfaces, are given in table 1.

The simplest system has one control factor (such as force) and one behavior variable (such as channel depth) and is represented by a two-dimensional graph that is a fold (fig. 2). A fold is a catastrophe defined in two dimensions and that has the following properties. As the control variable changes, behavior changes to reflect only one equilibrium state or minimum value for the energy function given in line 1 of table 1. Beyond a certain range of values for the control variable, the system breaks down and is undefined (fig. 2). Few, if any, geomorphic systems are controlled by a single variable, so the fold catastrophe is not particularly useful.

A system with two control factors (such as force and resistance) and a single behavior variable (such as channel width), more similar to geomorphic systems, produces a complex three-dimensional surface that is characterized by a tuck or cusp (fig. 2). Unlike the simpler fold, the cusp is a surface with smooth sections separated in places by a reverse slope. As the behavior variable ($\underline{x}$) responds to changes in the control variables ($\underline{a}$ and $\underline{b}$) a point representing an equilibrium system state moves about the surface. In some areas of the surface, such movements may be smooth, while in other areas drastic falls or leaps may be required to maintain equilibrium.

Systems of more complexity than the cusp catastrophe are definable in a topologic and algebraic sense (table 1), but they cannot be drawn in three dimensions unless one or more dimensions are held constant. In all cases, the surfaces are characterized by multiple sheets and folds. Because the simple fluvial systems analyzed here can be defined by two control variables, we do not consider the higher order surfaces further. If we could define the control factors for such complex systems, catastrophe theory supplies a conceptual framework.

The Cusp Catastrophe

The cusp catastrophe appears to be a useful representation of change in those geomorphic systems definable by two control factors and a behavior dimension. Strahler (1952) argued that geomorphologists ought to consider the most basic aspects of applied physics in the analysis of surficial processes, but

Table 1

The Seven Basic Catastrophes

Singularity	Control Factors	Behavior Factors	E, Energy Function	Derivative: when equal to zero defines singularity
Fold	1	1	$\frac{1}{3}x^3 - ax$	$x^2 - a$
Cusp	2	1	$\frac{1}{4}x^4 - ax - \frac{1}{2}bx^2$	$x^3 - a - bx$
Swallowtail	3	1	$\frac{1}{5}x^5 - ax - \frac{1}{2}bx^2 - \frac{1}{3}cx^3$	$x^4 - a - bx - cx^2$
Butterfly	4	1	$\frac{1}{6}x^6 - ax - \frac{1}{2}bx^2 - \frac{1}{3}cx^3 - \frac{1}{4}dx4$	$x^5 - a - bx - cx^2 - dx^3$
Hyperbolic	3	2	$x^3 + y^3 + ax + by + cxy$	$3x^2 + a + cy$ $3y^2 + b + cx$
Elliptic	3	2	$x^3 - xy^2 + ax + by + cx^2 + cy^2$	$3x^2 - y^2 + a + 2cx$ $-2xy + b + 2cy$
Parabolic	4	2	$x^2y + y^4 + ax + by + cx^2 + dy^2$	$2xy + a + 2cx$ $x^2 + 4y^3 + b + 2dy$

it has only been recently that researchers have taken this direction
(e.g., Birkeland, 1967; Baker,
1974; Bull, 1979). Although there
are many secondary environmental
factors that are brought to bear
in surficial processes, all ultimately have their effect by influencing force or resistance, or both.
Force and resistance (an opposite
force) might be replaced by forcing and resisting power or other
surrogates such as momentum and
friction.

The behavior variable to be
considered is largely a product
of the convenience or necessity
of the researcher, but it is
usually some physical dimension
or measurable characteristics of
the system. Examples include
width in the case of a channel,
diameter in the case of a particle,
or mass in the case of a landslide.

If a geomorphic system can
be adequately described by measures of force (a), resistance
(b), and response (x), Thom's
work indicates that the changes
during the system's operation
can be described by a cusp catastrophe. The surface, defined
by the relationships given in
table 1, line 2, has six significant characteristics that have

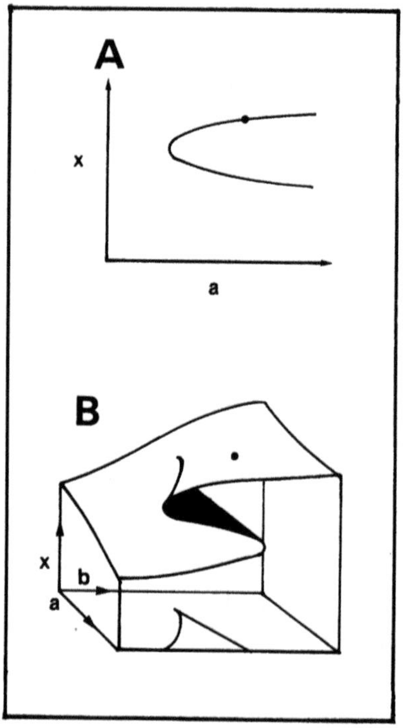

Figure 2. *Two basic catastrophes: A,
the fold defined by a single
control factor and behavior
variable; and B, the cusp
defined by two control factors and a behavior vari-
able. Example system
states shown by points.*

geomorphic implications: abrupt and smooth changes, divergence, hysteresis,
bimodal behavior, an unstable or inaccessible region, and structural stability.

First, the cusp catastrophe suggests that a system may change abruptly or
gradually depending on the antecedent conditions. If a point (1A in fig. 3)
representing the system state is located away from the fold, a small change
in (a, b) variables will result in a relatively small adjustment in x, the behavior.
On the other hand, if the point (1B in fig. 3) is located at the fold, a small
change in the controls will cause the transgression of a threshold and will result in a drastic change in the value of x as the point jumps to a different part
of the surface. This characteristic of the cusp accommodates the concepts of
geomorphic thresholds as expressed by Schumm (1973, 1978), especially the
view that such thresholds are intrinsic and not necessarily the product of
outside factors. The acceptance of the cusp as a model, however, indicates
that some thresholds may be circumvented so that rapid adjustments do not

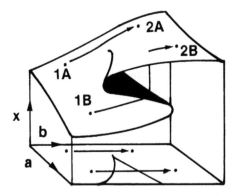

Figure 3.   Smooth (A) and abrupt (B)
            changes.

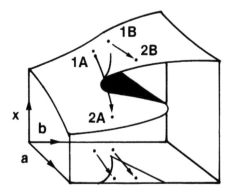

Figure 4.   Divergent behavior.

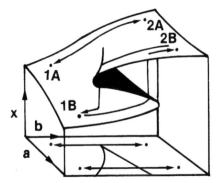

Figure 5.   Hysteresis.

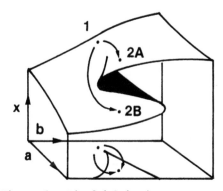

Figure 6.   Bimodal behavior.

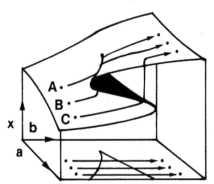

Figure 7.   Jump conventions: A, random; B, Maxwell
            convention; C, perfect delay.

In figures 3-7, the axis across the diagram is control variable _b_, the depth
axis is control variable _a_, and the vertical axis is the behavior variable _x_.

21

occur. That is, the point representing the system state may move from beginning condition to final condition by going around instead of over the fold.
For example, if a meandering stream experiences a change in sediment supply
that requires an increase in channel gradient, the channel may change rapidly
by meander cut-offs, or it may change more slowly by filling in the headwaters and cutting in the lower reaches. The result in either case is the
same. The nature of change is determined by the sequence and magnitude
of changes in the control factors. Human intervention in natural systems
might guide changes of this type or they might occur on a haphazard basis
naturally.

The second characteristic of the cusp is divergence. Given nearly the
same starting point, the system behavior may proceed in different directions
and hence may conclude a change at two radically different values. In
figure 4, for example, the system beginning at point 1A changes to point
2A in response to a change in one of the control factors. If the system had
point 1B as a starting point, however, and experienced precisely the same
change in the control factor, it would conclude its change at point 2B, quite
different from the first example. Divergence suggests that antecedent conditions are highly significant in the operation of some geomorphic systems.
Artificial loading of a potential landslide, for example, may result in slow
adjustment or a rapid mass movement, depending on the antecedent moisture
conditions, which in certain ranges of resistance would have to vary only
slightly to produce radically different responses.

The possibility of hysteresis, the third characteristic in the cusp catastrophe, is significant for geomorphic applications because cyclic activity is
commonly noted in surficial processes. Hysteresis occurs when a single control factor regularly increases and decreases while the other control remains
constant (or in a more complicated instance when several variables increase
and decrease regularly in concert with each other). As shown in figure 5,
the point representing system behavior may respond with smooth adjustments
(1A to 2A) or it may experience smooth transitions interrupted regularly by
major jumps (1B to 2B). Cyclic climatic changes in semi-arid regions produce alternating cutting and filling of channels that can be represented by
the hysteresis properties of the cusp catastrophe.

The fourth characteristic of the cusp is bimodal behavior. For each combination of (a, b) that occurs in the fold, called the bifurcation set, two
stable values of the behavior variable are possible (fig. 6). The underside
of the fold is a local maximum of the potential function and therefore is an
unstable state. The exact state is determined by the course of system change
as the moving point enters the fold, so that antecedent conditions as well as
the sequence of change are important considerations. An example of bimodal
behavior familiar to geomorphologists is the flow behavior of water in channels.
At certain combinations of depth and velocity, those that produce Reynolds
numbers between 500 and 750, flow may be either laminar or turbulent, depending in part on the sequence of adjustment (Simons, 1969). For combinations of depth and velocity outside the fold, the behavior of flow is either
of one type or the other.

The process of jumping from one sheet to another in the fold, or having
the behavior variable change drastically with a minor change in the control

variables, may be governed by one of three rules: the perfect delay (C in fig. 7), the Maxwell convention (B in fig. 7) or a random approach (A in fig. 7). The appropriate rule is determined by the nature of the system. With perfect delay, the point representing system conditions moves into the fold (representing changes through time) and jumps to the other surface at the farthest edge of the fold (fig. 7, point C). With the Maxwell convention (B), the jump occurs as soon as possible, and the system adjusts to the global maximum or minimum whenever possible (Poston and Wilson, 1977). With a random approach (A), the jump may occur at any time within the fold.

In the fold, between the two sheets representing equilibrium starts, is a third sheet that is the underside of the fold, (the dark area in fig. 2B). In terms of the energy function, this third surface represents a local maximum of the energy or potential function (see fig. 1). The surface has no significance in interpreting the change processes, and in some representations of the cusp it is left out of the graph entirely.

The final and most important characteristic of the cusp catastrophe is a property shared with all other catastrophes: it is a stable structure. Unlike other models for change, specifically mathematical models, the cusp is unchanging from one situation to another. If the behavior of a geomorphic system is controlled by two factors, Thom's work shows its change will always be governed by a cusp catastrophe. There will always be a possibility of smooth change, abrupt adjustments, divergence, hysteresis, and bimodal behavior.

APPLICATIONS

The utility of catastrophe theory in geomorphic applications depends on four criteria: 1) geomorphic systems must be capable of definition within the control behavior framework, with defined control and behavior variables, 2) the catastrophe representation must be capable of accommodating the forms and types of commonly-available geomorphic data, 3) the catastrophe surfaces representing geomorphic systems must be able to describe observed behavior, and 4) catastrophe theory must add some interpretive value beyond models presently in use. Tests of the applicability of the theory are afforded by data from arroyo development and rapids in canyon rivers.

Arroyo Development

The process of arroyo development has been a long-standing question in geomorphic research. Although the majority of studies have concentrated on the factors of arroyo initiation, recent efforts have shown that the entrenched channels are examples of equifinality (Cooke and Reeves, 1976). That is, the same forms have resulted from a wide variety of causes, some of which are detectable after their operation and some of which are not. Emphasis in arroyo-related research is now turning to the processes of channel change as the fluvial system adjusts to internal (Schumm, 1977) or external (Cooke and Reeves, 1976) stimuli.

Arroyo development suggests itself as a candidate for the application of catastrophe theory because the process includes abrupt changes as thresholds are crossed and because the system controls are readily identified as (a) the force of flowing water operating against (b) the resistance of earth materials. The behavior variable reflecting this operation is the area of the channel cross section at given points in the stream network.

Summaries of the data collection and reduction for a series of arroyos in montane Colorado are being published elsewhere (Graf, 1979a; 1979b), so that detailed reviews are not presented here. Field data provided measurements of channel characteristics at channel cross sections in Jefferson and Gilpin Counties in the central Front Range of Colorado. Maps and air photographs provided data on drainage basin characteristics upstream from each site as well as information on vegetation density and distribution. Hydrologic equations converted the field- and remotely-sensed data into representations of the force of flowing water at each site. The force of flowing water was expressed as unit stream power (gm/cm-sec). Because particle size and texture were relatively invariate from one site to the next, the material resistance was expressed as a function of the density of vegetation on the valley floor at each site. Vegetation density was represented by biomass, measured in kg of vegetation per m^2 of ground area. The behavior variable was the amount of material excavated at each cross section by the process of entrenchment.

Previous analysis of the data showed little systematic covariation among force, resistance, and amount of material excavated (Graf, 1979a). A logical expectation would be that as force increased and resistance decreased, the amount of material excavated would systematically become greater, but a three dimensional plot of the variables showed that this did not occur. A reinterpretation of the data within the framework of catastrophe theory provides a useful test of the applicability of the theory.

Figures 8A and 8B show a cusp catastrophe with quantitative characteristics as defined by the data from the 67 sampled sites. The data were smoothed by a spatial averaging process because the number of points was not large enough to specify the surface equally well throughout. The left-hand edge of the fold represents a geomorphic threshold (Graf, 1979a) corresponding to the relationship $w = 0.45 \, B_v^{2.23}$, where w = tractive force of the 10-year flood in gm/cm-sec and B_v = the biomass on the valley floor in kg/m^2 (fig. 8A). This equality is the threshold: if the observed conditions indicate an inequality and the left side of the function is greater, channel entrenchment results. If the right side is greater, stability and no entrenchment are observed. This scenario provides an explanation of the upper surface. However, the entire catastrophe indicates the presence of a fold, a second surface beneath the first, whereby some points (in the bifurcation set) representing measured sites plot so low on the scale of excavation that they are not entrenched, even though force and resistance relationships would suggest great excavation. The presence of the fold, or bifurcation set, explains why the trends in the earlier study were not clear--values from both the upper and lower surfaces were included without recognition of the fold. The catastrophe also suggests a second threshold, the right side of the fold in figure 8B which represents the boundary of the bifurcation set. In the area of the surface to the left of this second threshold, channels may be either entrenched or not, but to the right of this threshold all channels are entrenched and plot as points on the upper surface.

In the case of arroyo development it appears that catastrophe theory is a useful model for change in fluvial systems. The system operation is easily defined by control and behavior factors, the graphic representation accommodates the data, observed behavior is included in the catastrophe surface, and the fold with its two thresholds rather than one adds interpretive value beyond other general models.

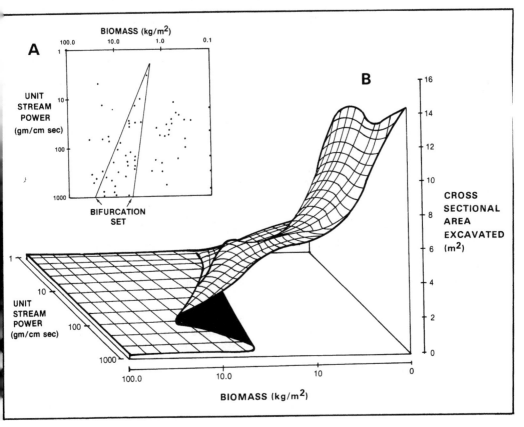

Figure 8. *The cusp catastrophe of arroyo development in the Colorado Front Range as defined by empirical data. A, the distribution of sampled sites; B, the catastrophe surface. Major trenching occurs in and to the right of the bifurcation set. Note that the ordinate of figure 8A increases downward so that the two dimensional representation is oriented in the same way as the perspective view of figure 8B.*

River Rapids

The behavior variable described above was a continuous one: amount of material excavated. Some applications of catastrophe theory have used binary behavior variables to accommodate either-or situations. In some social science
applications, for example, the behavior of commuters is modeled by a binary
choice of behavior: to ride the bus or drive a car (Wilson, 1976). Some potential geomorphic applications also are of a binary type. For example, in
canyon rivers, numerous tributary streams join the main stream where both
channels have steep gradients and carry large particles. At some junctions,
boulder rapids occur, but at other junctions, no rapids develop. The role of
transportation of large boulders by tributaries has long been understood
(Powell, 1875), and recent work has suggested that the distribution of rapids
in the Grand Canyon is related to the fault-controlled tributary streams (Do-
land, Howard, and Trimble, 1978). In some canyons of the Colorado Plateau,
however, only a portion of the tributaries create rapids.

Catastrophe theory might be useful since there are two obvious controlling
factors at the junctions: the transportation capability of the main channel and
the transportation capability of the tributary. The behavior variable is binary: the presence or absence of a boulder rapid.

Data from the main channel of the Colorado River and its tributaries in
eastern Utah between Dewey Bridge and the town of Moab provide the input
for a test of the utility of catastrophe theory. In this 53 km canyon reach
the Colorado River is joined by 17 significant tributaries, 10 of which have
caused rapids in the main channel. Based on stream-gage data for the main
stream and drainage area data from maps for the tributaries, the largest
floods of record (about the 100-year flood) were determined. Field data were
used with the DuBoys Equation to determine the probable tractive force in
main and tributary channels during these events. Generally, it is logical to
expect that when the force in the main channel is greater than the force in
the tributary channels at the junction of the two, no rapids would occur.
All the debris deposited by the tributary would be swept away by the main
stream. If the force was greater in the tributary, rapids in the main channel
would be unable to transport the larger particles deposited by the more powerful tributary.

A graph showing tributary force on one axis and main-channel force on the
other represents the junctions as points in the field of the graph, with the
third dimension as a measure of the probability of the occurrence of a rapid.
Theoretically, the geomorphic threshold would be that set of points where the
two forces are equal: on one side, all system states or points would represent
junctions without rapids because that would be the part of the graph where
the main channel is more powerful than the tributaries. On the other side of
the threshold, all the points or system states would represent junctions with
rapids where the main channel is less powerful than the tributaries. Figure 9
shows that this expected situation does not develop. However, if a third axis
(probability of a rapid) is added to represent binary behavior, a catastrophe
surface appears, and the presence of a fold provides a framework for interpreting the bifurcation set.

The main concept added by catastrophe theory in the case of rapids is
that of bimodal behavior. Instead of a simple threshold defined by equal

forces, it appears that at force levels within the bifurcation set, either mode of behavior (rapids or no rapids) is possible. The switching convention between the two states is apparently random in this middle area of the graph. Besides this additional perspective, however, the application of catastrophe theory is only partly successful. Although the geomorphic system is capable of definition by control and behavior factors, the binary variable (probability of a rapid) is not especially successful. The undulations of the catastrophe surface have no meaning except when the surface crosses the horizontal plane above which rapids occur and below which they do not. Also, data on many geomorphic systems frequently represent observation of a limited number of systems states. Clearly, with only seventeen points much of the catastrophe surface is not supported by observations. A catastrophe surface may be so poorly perceived as to be nearly useless, depending on the nature of the phenomena, the variables, and the availability of data.

DISADVANTAGES

The geomorphic researcher potentially interested in the use of catastrophe theory to describe a system in change must weigh several advantages and disadvantages, many of which are in evidence in the two test cases cited above but not specifically described.

Because of the rapid and sometimes uncritical acceptance of catastrophe theory for application by some researchers, one commentator refers to the theory as an "emperor with no clothes" (Kolata, 1977). Because of its name and sweeping implications, the theory and its potential applications have been described and supported in popular unrefereed publications. The development of the theory and its framework have not been seriously questioned, but applications, especially in social sciences and paleontology, have been controversial (Woodcock and Davis, 1978). Any geomorphologist using the theory as a perspective must defend not only the data and conclusions, but also the use of catastrophe theory to link the two.

There are many disadvantages to use of the theory depending on the application, but four problem areas are most commonly encountered: designation of control factors, definition of potential, qualitative nature of the theory, and its generality. The choice of the control factors is important because it determines the definition of the catastrophe surface, yet that choice depends on the judgment and experience of the investigator. A related problem is the orientation of the axis and the direction of increase for the control factors. The axis of each control factor may be at an angle to the rectangular base lines of the figures, or each axis may parallel the sides of the figure (compare the bases of figures 8B and 9).

The existence of an energy system or potential is critical in all applications of catastrophe theory because it is this potential that defines equilibrium or most likely states for each combination of control variables. In many fluvial systems a tendency toward a steady state (Leopold and Maddock, 1953) or dynamic equilibrium (Bull, 1975) has been identified and at least partly verified, but this is not necessarily the case for all potential applications.

At a time when geomorphology is completing a quantitative revolution, catastrophe theory strikes a discordant note because of its qualitative nature.

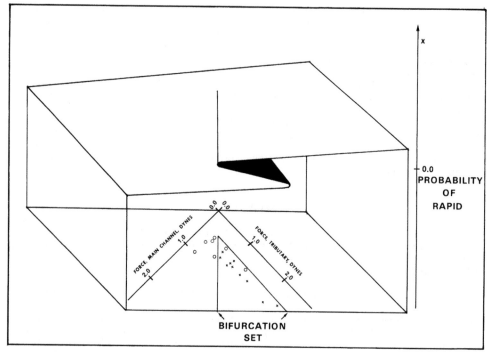

Figure 9. The cusp catastrophe of rapids in the Colorado River above Moab,
          Utah, and the distribution of data points used to construct it.
          Exact form of surface is unknown except for the limits of the bi-
          furcation set. The "o" symbols indicate sites without rapids,
          "x" symbols indicate sites with rapids.

The theory cannot predict when a sudden adjustment will occur; all it can do
is indicate the presence of a fold and the direction in which it lies. It can be
argued that this characteristic complements quantitative approaches when numerical data are used to define the catastrophe surface, but if enough is known
about the process in question to define the surface completely, the theory may
not be needed at all. The data and deterministic models will be the most effective representations of change.

Finally, the wide applicability of catastrophe theory may be a corollary of
shallowness. The theory may describe change in such broad terms that it con-
tributes little in the search for explanation. The ultimate condemnation of the
theory may be that it adds nothing to our understanding of fluvial processes.

ADVANTAGES

Despite the reservations outlined above, catastrophe theory appears to have value in fluvial geomorphology. It provides a useful marriage of the concepts of equilibrium and change. Since many fluvial systems have identifiable equilibrium states, and since internal adjustments or changes from one state to another occur intrinsically or in response to external interference by climate or man's activities, catastrophe theory seems well suited for some research designs.

As discussed above, the theory is qualitative, and though at times this may be a disadvantage, it is a useful tool for generalization. In the cusp catastrophe, for example, there is always a fold, an upper surface and a lower surface. The size of the fold and the magnitudes of the surfaces are not so important as the fact of their existence in any system controlled by two variables. According to catastrophe theory all such systems have the possibility of smooth and abrupt change, divergence, hysteresis, bimodal behavior and possess stable structures of change.

The partial definition of a catastrophe surface can be a valuable input to research design, because if part of the surface is specified, the general direction of changes in variables can be identified for the search for the fold, or for highest or lowest values for the behavior variable.

The most important advantage of catastrophe theory is that it provides a new way for describing the natural world: it is a new language that has characteristics not previously available. The theory allows us to conveniently picture, in a graphic fashion, space-time changes that are representative of the system under study. If all change takes place according to the specified singularities, an infinite number of other possibilities has been eliminated and our research efforts in fluvial geomorphology can be more efficient, more focused, and hopefully more productive. Human interactions with geomorphic processes can be placed on a stronger theoretical footing.

CONCLUSIONS

The tests of the applicability of catastrophe theory to the problems of arroyo development and river rapids showed that the theory has limited utility in the study of fluvial change. If the behavior variable is a continuous one rather than a binary variable, interpretation and theoretical support of the resulting surface is likely to be stronger. If the choice of control variables can be specified with confidence, as was the case with the arroyo problem, the concepts of catastrophe theory can be imparted into geomorphology successfully. Less success attends problems where uncertainty surrounds the choice of variables, as in the case of river rapids.

Catastrophe theory may be applicable to geomorphic processes other than fluvial. Mass movement processes that experience sudden motion alternating with periods of relative stability (e.g., Schumm and Chorley, 1964) appear to be possible examples of hysteresis. The lack of smooth transitions between various dune forms resulting from eolian processes as reported by Wilson (1972) suggests the presence of a fold in a catastrophe surface. Glacial surges, first detailed by Tarr (1909), seem susceptible to description by catastrophe theory. The processes of breaking waves operating on the material of beaches involves specific thresholds of energy and wave steepness (e.g., Davies, 1973, p. 118) that might be viewed from the catastrophic perspective.

Catastrophe theory will not replace previous generalizations in fluvial geomorphology such as dynamic equilibrium, hydraulics, hydraulic geometry, complex responses, thresholds, allometry, or rate laws, but it can provide a unifying mechanism by which these diverse concepts can be bound together. Such an array will provide a better and more complete understanding of change in fluvial systems than reliance on any single alternative.

ACKNOWLEDGEMENTS

Michael Woldenberg (Department of Geography, SUNY Buffalo) provided thought-provoking comments on an initial draft of this paper. Preliminary discussions with John E. Costa (Department of Geography, University of Denver) and Stanley A. Schumm (Department of Earth Resources, Colorado State University) were helpful in the refinement of applications. Financial support from the National Science Foundation for the montane arroyo research and from the National Geographic Society for the canyon rapids research permitted the collection of field data. Supplemental funds from a Faculty Development Grant at the University of Iowa and from the Research Incentives Fund of the Department of Geography at Arizona State University aided in completion of the paper.

REFERENCES

Baker, V.R., 1974, Paleohydraulic interpretation of Quaternary alluvium near Golden, Colorado: Quaternary Research, v. 4, p. 94-112.

Birkeland, P.W., 1968, Mean velocities and boulder transport during Tahoe-age floods of the Truckee River, California-Nevada: Geological Society of America Bulletin, v. 79, p. 137-141.

Boulding, K.E., 1956, The image: knowledge in life and society: Ann Arbor, University of Michigan Press, 175 p.

Bull, W.B., 1975a, Allometric change of landforms: Geological Society of America Bulletin, v. 86, p. 1489-1498.

_____, 1975b, Landforms that do not tend toward a steady state; in Melhorn, W.N., and Flemal, R.C., eds., Theories of Landform Development: Binghamton, State University of New York, Publications in Geomorphology, p. 111-128.

_____, 1979, The threshold of critical power in streams: Geological Society of America Bulletin, v. 90, in press.

Chorley, R.J., 1962, Geomorphology and general systems theory: U.S. Geological Survey Professional Paper 500-B, 10 p.

Chorley, R.J., and Kennedy, B.A., 1971, Physical geography: a systems approach: London, Prentice Hall, 370 p.

Croll, J., 1976, Is catastrophe theory dangerous?: New Scientist, v. 70, p. 630-632.

Cubitt, J.M., and Shaw, B., 1976, The geological implications of steady-state mechanisms in catastrophe theory: Mathematical Geology, v. 8, p. 657-662.

Davies, J.L., 1973, Geographical variation in coastal development: New York, Hafner Publishing Co., 204 p.

Dolan, R., Howard, A., and Trimble, D., 1978, Structural control of rapids and pools of the Colorado River in the Grand Canyon: Science, v. 202, p. 629-631.

Graf, W.L., 1977, The rate law in fluvial geomorphology: American Journal of Science, v. 277, p. 178-191.

_____, 1979a, The development of montane arroyos and gullies: Earth Surface Processes, v. 4, p. 1-14.

_____, 1979b, Mining and channel response: Association of American Geographers Annals, in press, June issue.

Gregory, K.J., ed., 1977, River channel changes: New York, Wiley-Interscience, 448 p.

Hack, J.T., 1960, Interpretation of erosional topography in humid temperature regions: American Journal of Science, v. 258, p. 80-97.

Henley, S., 1976, Catastrophe theory models in geology: Mathematical Geology, v. 8, p. 649-655.

Horton, R.E., 1932, Drainage basin characteristics: American Geophysical Union Transactions, v. 13, p. 350-361.

_____, Erosional development of streams and their drainage basins: hydrophysical approach to quantitative morphology: Geological Society of America Bulletin, v. 56, p. 275-370.

Knox, J.C., 1972, Valley alluviation in southwestern Wisconsin: Association of American Geographers Annals, v. 62, p. 401-410.

Kolata, G.B., 1977, Catastrophe theory: the emperor has no clothes: Science, v. 196, p. 287, 350-351.

Leopold, L.B., and Maddock, T. Jr., 1953, The hydraulic geometry of stream channels and some physiographic implications: U.S. Geological Survey Professional Paper 252, 57 p.

Powell, J.W., 1875, Exploration of the Colorado River of the West and its tributaries: Washington, U.S. Government Printing Office, 175 p.

Schumm, S.A., 1973, Geomorphic thresholds and complex response of drainage systems; in M. Morisawa, ed., Fluvial geomorphology: Binghamton, State University of New York, Publications in Geomorphology, p. 299-310.

_____, 1977, The fluvial system: New York, Wiley-Interscience, 338 p.

Schumm, S.A., and Chorley, R.J., 1964, The fall of Threatening Rock: American Journal of Science, v. 262, p. 1041-1054.

Schumm, S.A., and Lichty, R.W., 1965, Time, space, and causality in geomorphology: American Journal of Science, v. 263, p. 110-119.

Simons, D.B., 1969, Open channel flow; in R.J. Chorley, ed., Water, earth, and man: London, Methuen and Co. Ltd., p. 297-318.

Stewart, I., 1975, The seven elementary catastrophes: New Scientist, v. 68, p. 477-454.

Strahler, A.N., 1932, Dynamic basis of geomorphology: Geological Society of America Bulletin, v. 63, p. 923-938.

Tarr, R.S., 1909, The Yakutat Bay Region, Alaska: U.S. Geological Survey Professional Paper 64, 183 p.

Thom, R., 1975, Structural stability and morphogenesis: an outline of a general theory of models: Reading, W.A. Benjamin, 348 p.

Thornes, J.B., 1978, The character and problems in contemporary geomorphology: in C. Embleton, D. Brunsden, and D.K.C. Jones, eds., Geomorphology, present problems and future prospects: Oxford, Oxford University Press, p. 14-24.

Wagstaff, J.M., 1976, Some thoughts about geography and catastrophe theory: Area, v. 8, p. 319-320.

Wilson, A.G., 1976, Catastrophe theory and urban modeling: an application to modal choice: Environment and Planning, v. 8, p. 351-356.

Wilson, I., 1972, Sand waves: New Scientist, v. 53, p. 634-637.

Woodcock, A.E.R., and Davis, M., 1978, Catastrophe theory: New York, E.P. Dutton, 152 p.

Woodcock, A.E.R., and Poston, T., 1974, A geometrical study of the elementary catastrophes: New York, Springer-Verlag, 257 p.

Zahler, R.S., and Sussmann, H.J., 1977, Claims and accomplishments of applied catastrophe theory: Nature, v. 269, p. 759-763.

Zeeman, E.C., 1976, Catastrophe theory: Scientific American, v. 234, p. 65-83.

INVARIANT POWER FUNCTIONS AS APPLIED TO

FLUVIAL GEOMORPHOLOGY

WAITE R. OSTERKAMP

Water Resources Division
U.S. Geological Survey
Lawrence, Kansas

ABSTRACT

In the latter part of the past century, geomorphic thought was first dominated by the closed-system, cycle-of-erosion model of W. M. Davis. About 25 years ago, the popularity of the Davisian approach was largely replaced by an open-system, dynamic-equilibrium model. An alternative model, that of allometric change, was first described explicitly for geomorphic purposes by W. B. Bull, although the technique had been used previously by many researchers.

Allometric analysis is the development of simple or multiple power-function equations that express the relative rates of change among the variables of a system. A principal geomorphic utility of the method is to show adjustment between two variables. For many geomorphic systems, however, and for fluvial systems in particular, an allometric relation unaffected by other variables rarely can be identified. Unless the effects of complicating variables are held constant and thereby eliminated from consideration, a bivariant allometric relation is likely to be in error. To avoid this difficulty and to provide for the development of reliable multivariant power functions, a simple modification to the allometric-change model is advocated for many studies of fluvial systems. This modification is the determination of a fixed or constant exponent for a bivariant power-function relation by holding the effects of other variables constant. Having established an exponent, that value is imposed on subsequent power-function equations, whether bivariant or multivariant. The coefficient must be evaluated accordingly. The technique seems especially applicable to fluvial systems owing to their complexity and lack of distinction between dependent and independent variables.

The determination of fixed exponents for width-discharge and gradient-discharge relations of alluvial stream channels serves to illustrate the use, advantages, and limitations of the invariant power-function technique. Among the advantages suggested by the examples are:

1. The method results in increased accuracy and sophistication of the adjustment between two variables for empirical studies.
2. When employing multiple regression (or a similar curve-fitting technique), a specified exponent for an independent variable avoids error that would otherwise be inherent in the computation owing to non-linear effects by other independent variables.
3. Conflict caused by defining separate regional relations between two variables is eliminated.
4. Preestablished exponents, based on numerous data, provide a measure of safety when relating and extrapolating very limited data.

5. Invariant power functions provide a uniformity that permits the comparison of results within a study, or with other studies.
6. The method helps focus attention on geomorphic and hydrologic processes, whereas free bivariant analysis ignores process.

Limitations of the invariant power-function technique include its largely empirical approach and the assumption that a power function adequately describes the relation between two variables.

INTRODUCTION

Within the earth sciences, if not the physical sciences in general, models of thought tend to be initiated on a purely empirical framework. Gradual maturing of a model is accompanied by the accumulation of ever greater volumes of data, some perhaps anomalous to the model, and an ever greater awareness of process. The effects of voluminous data and an orientation to process often combine to make continued use of a model awkward and occasionally unmanageable. Inevitably, such awkwardness results in the abandoning of the model or its demotion to a more limited role than it had held previously. The process of model, or paradigm, replacement is, as proposed by Kuhn (1962), fundamental to the continued health and progress of a scientific discipline.

Two models have dominated thought in geomorphology since its emergence as an identifiable discipline within geography or geology. The first defined the cycle of erosion, a general explanation of landform development initially espoused in the latter part of the last century (Davis, 1889). The model was basically descriptive and treated all geomorphic features as the result of three variables--geologic structure, process, and stage. Consideration of structure, which included rock type, jointing, and spatial arrangement, and process (referring to denudation and deposition) was partially neglected at the expense of stage (a description of the extent to which a landscape had matured following a relatively rapid uplift of the surface). Although process was a defined part of the scheme, the model was empiric in the respect that landscapes were regarded as subject to an evolutionary and irreversible sequence, the stage of which was readily observable. The processes that had caused a landscape to achieve a degree of maturity were largely inconsequential.

The cycle-of-erosion model is representative of a closed system (Chorley, 1962). It requires a defined set of initial conditions, progresses irreversibly with an increase in entropy (unusable energy), and is time dependent. The cycle is a process of continual wearing down of upland areas until landscape planation has been achieved. At first the undeniable tendency toward formation of peneplanes was readily accepted, partly because comparisons of landscapes appeared to prove the general validity of the cycle-of-erosion model. As data and understanding of specific landforms increased, however, it became apparent that the model is an oversimplification and that modern examples of peneplanes are rare or lacking.

These deficiencies resulted in a partial shift to an open-system model, that of dynamic equilibrium. Although this model (Gilbert, 1877) had been proposed earlier than the cycle of erosion, it did not gain general acceptance until after Leopold and Maddock (1953) applied it to their study of the hydraulic geometry of stream channels. The geomorphic concept of dynamic equilibrium and its ideal, steady state, considers landscape features that result from the interaction of matter and energy passing through a system; it assumes a tendency toward a balance between process and form. The model is time independent in the sense that after dynamic equilibrium has been established, it cannot be ascertained how long that condition has prevailed. The concept of dynamic equilibrium is an open-system model because it assumes that during short periods of time, relative to a geomorphic time frame, equal amounts of material and energy enter and leave the system (Chorley, 1962).

Many of the studies that have used the open-system model were based on the assumption that a balance or dynamic equilibrium occurs between any two parts of a landscape. If the independent variables could be held reasonably constant, it followed that steady-state conditions would occur, and many data were collected without significant regard to process in an effort to demonstrate a steady state. In recent years, however, it has become increasingly apparent that either geomorphic steady states do not exist or else their existence cannot be demonstrated (Bull, 1975, p. 1490). Some landforms even can be shown not to conform to a dynamic-equilibrium model (e.g. Wahrhaftig, 1965; Crandell, 1971; Burkham, 1972; Wolman and Gerson, 1978). Dynamic equilibrium here is regarded as a system never truly in equilibrium but always tending towards a steady state and generally fluctuating about it.

To eliminate some of the weaknesses of the open-system model, Bull (1975, p. 1490) encourages geomorphologists to consider the "broader and more flexible" model of allometric analysis. Developed in the biological sciences, allometry is the study of relative rates of growth or change of two parts of an organism. As applied to geomorphology, allometric analysis is a quantitative expression of the manner in which one part or variable of a system changes with respect to another part or variable of the system. Here allometry is considered more a technique than a model. It is empiric in the sense that an allometric relation is assumed if a set of measurements for the two variables conform to the power function equation: $y = a x^b$. Conversely, if an allometric relation can reasonably be assigned to the correspondence between two variables of a geomorphic system, it is assumed that for the range of data a power function or similar equation can describe the correspondence. Although the basic allometric-analysis technique is fully empirical and does not consider the processes acting on a system, it has wide application to fluvial geomorphology and encourages a quantitative treatment of variables.

Simple power function equations, and generally by implication, allometry, have been used to describe a wide range of geomorphic and hydrologic relations (e.g. Lacey, 1930; Bull, 1964; Hedman, 1970; Carson and Kirkby, 1972; Graf, 1978; and Wolman and Gerson, 1978, figs. 1 and 5). A difficulty in using allometric analysis for earth-science studies, however, is the interference caused by unconsidered variables. Although the complexities of interrelated variables may be as pronounced for allometry in biological studies as it is for the earth sciences, the total effect or scatter caused by unconsidered variables often is much more significant in geomorphic studies than it is for biology. The relation between the volumes of the brains and bodies of deer, for example, may be affected by complicating variables, but the standard error of estimate for such a relation could be expected to be much lower than that of the relation between width and mean discharge for perennial streams. A relation for width-discharge data of alluvial streams is affected by both spatial and temporal variations in the flood histories, sediment loads and sizes, lithology, and riparian vegetation of the various streams. This relation could be misleading, therefore, if most of the smaller channels were sandy and wide relative to discharge and the larger streams were relatively narrow owing to high percentages of silt and clay in the channel material.

The problem of accounting for the effects of complicating variables has been reduced by the development for digital computer of easily used multiple-regression programs. Normally, however, multiple regression remains a purely empirical technique, discouraging consideration of the processes that occur in a system. In addition, least-squares regression, by minimizing the error sum of squares at each step of the computation, also serves to insure error in the power-function equation. In order to develop realistic equations, it is necessary to determine the relation (exponent) between two variables when the effects of other variables are held constant (Osterkamp, in press). Then the effects of other pertinent variables can be added to the known or invariable relation previously established. Thus, the fixed-exponent concept is the recognition that the relation between many geomorphic and hydrologic variables is an invariant power function for the range of data considered. The term invariant here refers only to exponents; the effects of other variables should be expressed in the coefficient, whether the relation has one or more than one independent variable.

The purpose of the present paper is to advocate the use of invariant exponents for empirical geomorphic studies in which the allometric-change technique appears applicable. The method, therefore, is considered both a special case and a sophistication of the allometric-change technique and retains the open-system attributes inherent in earth-science allometry (Bull, 1975). This suggestion is by no means new, but because of the development of computer capabilities, it is more practical than was the case as recently as a decade ago. Constant power-function exponents are advocated and used by White and Gould (1965) and Gould (1972) for paleontologic studies and by Bull (1975) in discussing application of allometry to the study of landforms. Numerous studies have implicitly advocated fixed exponents by their use without specific justification (e.g., Lane, 1957; Simons and Albertson, 1960, p. 50; and Lowham, 1976, p. 24-31).

The use of fixed exponents, either in simple or multiple power-function equations, can be justified in several manners. The most basic among these is that an allometric relation between two measurable variables must be considered invariant or else a quantitative correspondence cannot be defined. If a third variable is introduced, a complex relation among the three variables can be analyzed, but the third variable can have no actual effect on how the first two variables correspond to each other. That correspondence is shown solely by the rate at which one variable changes with respect to the other--the exponent of a power-function equation. The third variable, however, can have a separate allometric relation with each of the first two variables. It is this condition, or a more complex case of it, that properly can be described by a multiple power-function equation. An alternative, but more awkward, quantitative representation is a family of simple exponential equations in which each member describes the relation between the first two variables for a specific value or limited range of values for the third variable. In this case the effect of the third variable is indicated by different values of the coefficient; the exponent is fixed. Thus, the fixed-exponent concept serves to eliminate apparent conflict between regionalized data sets. The complicating influences in two different physiographic or political regions may differ, but a valid allometric relation, hence the exponent of that relation, between two variables must remain constant. Thereby quantification of the complicating variables becomes possible, as demonstrated by differences in the coefficients.

Recognition that other variables are expressed by the coefficients of simple power-function equations provides further justification for the invariant-exponent approach. Quantification of complicating influences tends to focus attention on the geomorphic and hydrologic processes operating in a fluvial system and consequently serves to reduce the empiricism of free bivariant analysis. The recognition emphasizes a fact minimized by free bivariant analysis: virtually never are two geomorphic or hydrologic variables totally interdependent, but their correspondence must be influenced by other variables. Thus, the approach of fixed exponents encourages a thoughtful, process-oriented use of multiple regression or similar techniques.

The use of invariant power functions also has practical benefits beyond the inferred increase in accuracy and sophistication that it provides to empirical studies. Preestablished exponents, based on numerous data, supply a measure of safety when relating and extrapolating very limited data. In addition, realistic comparisons become possible for results within a study or with other studies when the uniformity of fixed exponents is maintained. Bull (1975, p. 1492) notes, for example, that only when equations have the same units and exponents can coefficients, which are dependent on other variables as well as the units used, be compared validly.

INVARIANT POWER FUNCTIONS

The invariant power-function, or fixed-exponent, method is the use of the allometric-change technique when one (or more) exponent is treated as a constant. It includes the assumption that the relative rates of change between two geomorphic or hydrologic variables are unchanging for a specified range of data and that only external influences (complicating variables) can modify that relationship. Thus the method requires a determination of the value to be assigned to the exponent. This determination need not be made empirically, although in the earth sciences it generally will be.

Use of the Fixed-Exponent Technique

In its simplest form, the fixed-exponent technique is the establishment and subsequent use of a rate-of-change value between two measurable variables when all other known influences are held within narrow limits. If closely similar exponent values are obtained for two or more distinctly different groups of bivariant data, it can be assumed that an invariant power function describes the bivariant relation. The term "distinctly different" implies that each group of data is affected by a unique complex of other variables. The designated exponent value is then imposed on all subsequent uses of the bivariant relation, whether in simple or multiple power-function equations. The utility of an equation of this sort is generally for predictive or estimative purposes, in which case the predicted value is the dependent variable. For geomorphic-hydrologic systems, however, the designation tends to be arbitrary because often variables are interrelated rather than being dependent or independent.

Based on previous papers (i.e., Huxley, 1932; White and Gould, 1965; Woldenberg, 1966; and Gould, 1972), a summary of the theory on which the allometric-change model is founded is given by Bull (1975, p. 1490) and need not be repeated here. It is sufficient to note that data of an allometric relation plot in a straight line on logarithmic coordinates and that valid extrapolation of the line beyond the limits of the data range is sometimes possible. For low values of geomorphic and hydrologic data, however, an allometric relation may dissolve, and extrapolation from higher natural values should be made cautiously.

Once the exponent of a bivariant relation has been defined, its use permits the investigation of other relevant variables. Thus, the effect of other variables is quantified, and an investigator inevitably becomes concerned with explanations of the processes resulting in the observed relations. This concern encourages the elimination of redundant or statistically insignificant variables from multiple power-function equations. The use of invariant power functions, therefore, promotes both accuracy and simplicity.

The exponents of geomorphic and hydrologic relations in some cases can be defined theoretically (e.g., Leopold and Langbein, 1962; and Engelund and Hansen, 1967). More frequently, however, determinations are made empirically under natural, closely controlled, or laboratory conditions. Because laboratory and field data span a limited range, they might be unsuitable for defining an exponent. Laboratory and closely controlled field data, however, can be very beneficial by providing substantiation to simple exponential relations developed from a wider range of natural field data or by permitting extrapolation of a relation into low values (or high values for a negative exponent) of the data.

Relation to Other Models

The use of invariant power functions is advocated as a refinement of the allometric-change method. Assuming that any systematic approach to geomorphic study can be classed as either an open-system or closed-system model, the invariant power-function technique must be identified as an open-system approach. Like its parent, the allometric-change model (Bull, 1975), invariant allometry is time independent, assumes equal rates of inflow and outflow of matter and energy, and generally assumes a balance between process and form. It differs from the dynamic-equilibrium model and especially from its variation of steady state in the respect that invariant allometry assumes a condition of continual, hence instantaneous (rather than an average) equilibrium through time.

The aspect of time dependence can be clarified by considering an example of the manner in which alluvial-channel width of perennial streams might change with time. The example illustrates differences in the several open-system geomorphic models.

Application of the dynamic-equilibrium model, without regard to process, could suggest variation of a sort shown by the upper part of figure 1. For a specified mean discharge, some mean channel width or specified width range is assumed to correspond with a condition of dynamic equilibrium. Because short-term changes in both discharge and width are inevitable, the model acknowledges the necessity to consider averages or limited ranges. For the relatively brief periods during which width is constant (at, but not between, points A through H), a condition of steady state is assumed. Thus, steady state can occur at various channel widths, even if the channel is anomalously wide or braided (point E), and steady state does not necessarily require that a condition of dynamic equilibrium prevail.

The lower part of figure 1 represents how downstream width-discharge relations, as defined by Leopold and Maddock (1953), might change with time when invariant power-function allometry is used. For a specified mean discharge it is assumed that a minimum channel width can be obtained that fits the equation $\bar{Q} = aW^b$. Mean discharge ($\bar{Q}$) is treated here as the dependent

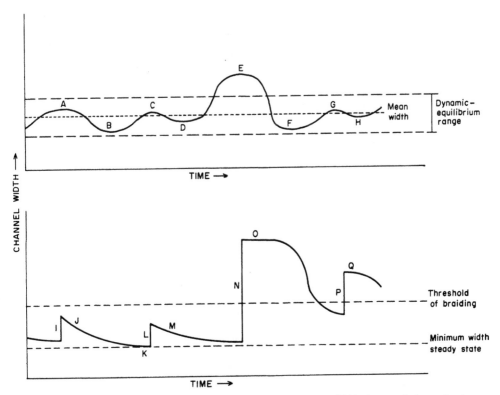

Figure 1. *Schematic comparison of use of dynamic-equilibrium and invariant-allometry techniques.*

variable to conform with previous studies that have used width measurements as a means of estimating flow characteristics in alluvial streams. Within the range of natural perennial streams, the exponent b is considered of constant value, about 2.0. Under natural conditions this minimum width cannot be reached, but if it were, a steady-state condition would be attained (point K). At points I and L a change in stream regimen has caused a nearly instantaneous channel widening. An expected cause would be erosive flood events (Schumm and Lichty, 1963; and Burkham, 1972), but other possible causes include changes in sediment supply or riparian vegetation (Graf, 1978). The change in channel geometry is expressed by a change in the coefficient a and ordinarily could be measured by changes in channel-material characteristics and channel gradient. After the widening events have occurred, narrowing again proceeds slowly (points J and M), accompanied by changes of channel material and gradient necessary to conform to $\overline{Q} = aW^{2.0}$. Thus, the coefficient a is the quantitative expression of all variables that influence the relation $\overline{Q} \sim W^{2.0}$.

If an unusually destructive flood occurs (point N), a previously stable channel can be widened beyond an assumed threshold of instability, and braiding prevails. When geomorphic thresholds are exceeded and landforms accordingly are highly unstable, it is doubtful whether invariant power functions reasonably can be applied. The temporal changes or processes (decreasing particle sizes and channel gradients) represented at points J and M, however, can be applied to other conditions to provide an understanding of the manner in which stability is restored. Point N represents an event when flooding caused extreme widening, winnowing of fines from channel material, and a large increase in median particle size of the channel sediment. At point 0 and beyond, floodplain reconstruction is occurring by a slow replacement of the fine channel material required for stable banks (Schumm and Lichty, 1963, p. 84-85). When this process has proceeded sufficiently to cause bank cohesiveness, relatively rapid channel narrowing is feasible. If another smaller flood occurs (point P), excessive width (point Q) again can occur because a mature growth of riparian vegetation is not available to limit bank erosion (Burkham, 1972). Replacement of fines must then start again.

Comparing the curves of figure 1, the perturbations of dynamic-equilibrium with time can be considered normal or expected fluctuations about some average value for the width-discharge relation. These fluctuations include no consideration of the processes causing deviation from the ideal. Analogous fluctuations (time equivalent) in the invariant-allometry technique are not viewed as deviations from equilibrium but as process-caused changes in the width-discharge relation that can be expressed quantitatively by considering other pertinent variables.

In many correlations of geomorphic-hydrologic variables (e.g., the relation between sediment yield and mean topographic slope for basins of varying climate and soil types, or between width-discharge relations of channels with differing sediment characteristics), doubt may develop as to whether constant exponents reasonably should be applied. The assumption that invariant exponents should be used for power functions of two or more sets of data taken from different populations is a specific case of the null hypothesis of statistics. If sufficient data define the power functions, much of the doubt can be eliminated by applying tests to accept or reject the null hypothesis.

For geomorphic purposes, a useful definition of null hypothesis is the assumption that significant difference does not occur between two (or more) sets of samples or measurements that are being compared statistically. In the general case of null hypothesis, observed differences are inferred to be random and not due to a systematic cause (Gary, McAfee, and Wolf, 1972, p. 486); as applied here, observed differences are assumed to be the result of complicating variables. These assumptions can be tested to determine if there are statistically significant differences between (a) variances about the power-function relations, (b) the exponents of the power-function relations, and (c) the coefficients of the relations. These three tests are described in various textbooks; the second and third, which are particularly pertinent to the present discussion, are summarized clearly in Snedecor and Cochran (1967, p. 432-438).

To test the possibility that two different power-function exponents should be the same, calculations give the contribution that the difference between the exponents makes to the sum of squares of deviations. The mean square of this difference is compared with the mean square within groups by an F-test (Snedecor and Cochran, 1967, p. 434-435). If it cannot be concluded that the exponents probably are different, the form of the null hypothesis represented by invariant power functions has been satisfied. Thus, a single exponent value for the two relations should be calculated, but the exponent for the ungrouped data must not be used. It is the error represented by that exponent that the use of invariant power functions avoids. Instead, the value of the fixed exponent should be a pooled estimate from the within-groups data (Dixon and Massey, 1957, p. 212-213; Snedecor and Cochran, 1967, p. 434-435), which is the same as a weighted average in which the weighting factor is the number of data pairs in each group. A coefficient (intercept) for each group of data then is calculated by imposing the same exponent value on each power-function equation.

The significance of the difference between two coefficients of power functions is tested in much the same manner as is the difference between exponents. The difference between the sum of squares of residuals for the power function of all data (with an imposed exponent) and the combined sum of squares of the residuals for the individual data sets are divided by the mean square of the pooled data, yielding an F value. A second two-tailed F-test indicates the significance of the observed difference between the two coefficients. If the coefficients are not significantly different, it can be concluded that the two data sets are from the same population or that complexities are exerting similar influences on the two bivariant relations. If, however, significantly different coefficients for the two relations with similar exponents are indicated, it can be assumed that the difference results from different effects of the variables influencing the power-function relations.

EXAMPLES

The use of invariant allometry and its suggested benefits can be demonstrated by considering several examples taken from studies of fluvial geomorphology. The examples are based on both real and artificial data and express relations between variables for which the fixed-exponent technique appears applicable. It is not suggested that allometry can describe correspondence between any two related geomorphic variables.

Width-Discharge Relations of Alluvial Stream Channels

Listed in table 1 are artificial width, mean-discharge, and channel-material data that were selected to illustrate an obvious but often overlooked hazard of relating two variables. Considering only the width and discharge data, simple least-squares regression yields the equation:

$$\bar{Q} = 0.010W^{1.88} , \qquad (1)$$

where $\bar{Q}$ is mean discharge and W is channel width. In many studies of the relative changes of geomorphic variables, the analysis does not continue beyond a simple power relation of this sort. If the percent silt-clay content (SC) of the channel material is considered by means of multiple regression or a similar type of analysis, the equation becomes:

$$\bar{Q} = 0.0061W^{2.00}SC^{0.10} . \qquad (2)$$

The data of table 1 and the resulting power-function equations are shown in figure 2. Consistent with real data (Osterkamp, 1977), it suggests that (a) different slopes of the power functions occur depending on whether channel-material data are considered and (b) width, relative to mean discharge, increases with decreasing channel siltiness (or increasing sandiness of the bed and bank material).

If the relations among mean discharge, width, and channel sediment were unaffected by other variables, the above equation would be accurate for natural channels. Many stream channels, however, show an increase in the

Table 1

Artificial mean-discharge ($\bar{Q}$) and channel-width (W)
data (undefined units) for various conditions of
channel sediment (silt-clay content, SC, in percent)

SC = 1%		SC = 10%		SC = 100%	
$\bar{Q}$	W	$\bar{Q}$	W	$\bar{Q}$	W
22.0	60.0	0.622	9.00	0.0139	1.20
29.9	70.0	0.768	10.0	0.0218	1.50
39.0	80.0	0.929	11.0	0.0313	1.80
49.4	90.0	1.11	12.0	0.0426	2.10
61.0	100	1.30	13.0	0.0557	2.40
73.8	110	1.51	14.0	0.0705	2.70
87.8	120	1.73	15.0	0.0870	3.00

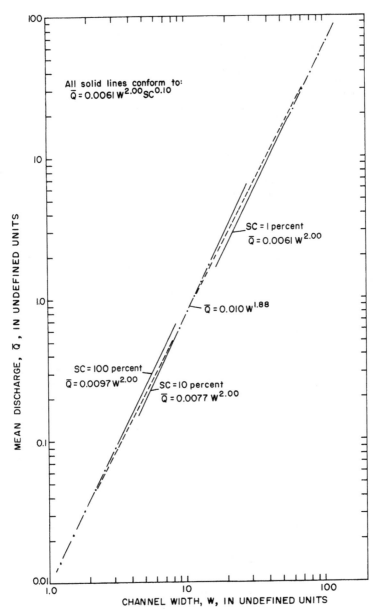

Figure 2. *Power-function relations for width (W) and mean discharge regard-*
less of silt-clay percent (SC) of channel material (dashed line),
and for different values of silt and clay (solid lines). Selected
data points, indicated by dots which break the solid line, are from
table 1.

43

amount of sand in the bed and bank material with increasing channel size and drainage area (Osterkamp, in press, p. 17). The sandiness and lack of co-hesiveness cause the channels to be more vulnerable to the widening effects of erosive discharges than are smaller channels formed of more cohesive alluvium. Thus, to simulate natural conditions, the data of table 1 were se-lected with a large amount of channel silt and clay for the low-discharge streams and a minor amount of silt and clay (large amounts of sand) in the channel material of the high-discharge streams. When channel-material data are disregarded, the resulting exponent is too low, which leads to a signi-ficant error for very high- and low-discharge streams.

To define the effect that channel-sediment characteristics have on a width-discharge relation, it is necessary to identify that relation. Three groups of perennial streams having generally stable channels were selected for study. A common feature of the groups was general channel stability – that is, a tendency to show minimal channel widening by high-discharge events. One group consisted of 32 channels of gradient exceeding 0.0080 and had low suspended-sediment discharge, high channel roughness, similar climatic con-ditions, and bank stability resulting from cobble and boulder armoring and mature woody riparian vegetation. The second group included 13 streams of eastern Kansas, all of which had at least 70% silt and clay in the bed mate-rial and had similar conditions of gradient, discharge variability, climate and vegetation, and suspended-sediment discharge. The third group, 18 spring-discharge channels of southern Missouri, had consistent conditions of climate and vegetation, a virtual absence of suspended sediment, and very stable discharge that results in an absence of erosive or channel-widening flow events.

Each group was selected to minimize the effects that all complicating var-iables have on the width-discharge relation and to provide a wide range of conditions represented. The range of conditions among groups resulted in power-function coefficients of 0.017, 0.042, and 0.011, respectively, for the three groups (when discharge is in cubic meters per second and width is in meters); whereas, the constancy of conditions within a group resulted in exponents of 1.98, 1.97, and 1.97 (Osterkamp, in press). It is assumed, therefore, that an exponent of 2.0, accurate to two significant figures, is an invariant expression of the relation between channel width and mean dis-charge. This exponent agrees with some of the early studies describing width-discharge relations.

A fixed exponent of 2.0 was applied to five groups of width and mean-discharge data collected from perennial streams of the Missouri River basin. The groups were defined according to differences in silt-clay content of the bed and bank material. Results (fig. 3) show increasing channel width, relative to mean discharge, with decreasing silt and clay in the bed and bank material. Widest streams, some of which are highly unstable and braided, occur when fine sizes are deficient in both the bed and banks (Osterkamp, 1977, p. 10-12). Analysis of all the data together, however, yields an ex-ponent of only 1.59 (fig. 3).

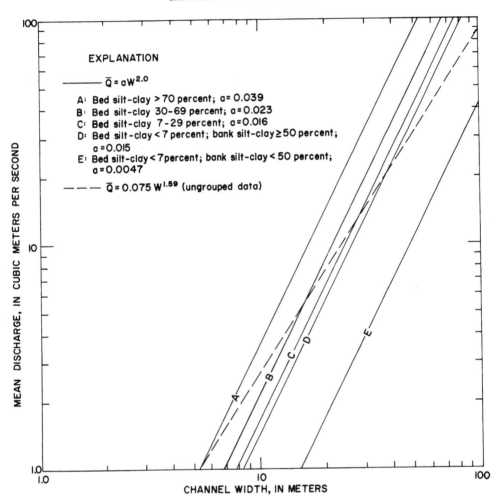

EXPLANATION

———— $\bar{Q} = aW^{2.0}$

A: Bed silt-clay > 70 percent; a = 0.039
B: Bed silt-clay 30-69 percent; a = 0.023
C: Bed silt-clay 7-29 percent; a = 0.016
D: Bed silt-clay < 7 percent; bank silt-clay ≥ 50 percent;
    a = 0.015
E: Bed silt-clay < 7 percent; bank silt-clay < 50 percent;
    a = 0.0047

— — — $\bar{Q} = 0.075\, W^{1.59}$ (ungrouped data)

Figure 3. Variation of width with discharge and sediment character.

Width-Length Relations of Channel Islands and Bars

In a study of the effects on channel morphology by tamarisk in the Colorado Plateau region, Graf (1978) presents power-function equations relating widths and lengths of channel islands and bars. The relations (fig. 4) are used to suggest that an adjustment between the frictional and turbulent energies of the streamflow occurs by formation of the channel features. It follows, therefore, that the adjustment can result in allometric geometries of the islands and bars. Graf (1978, p. 1498-1499) convincingly shows that power-function (allometric) relations can describe the variations of widths and lengths of the channel features and also suggests that the exponents of

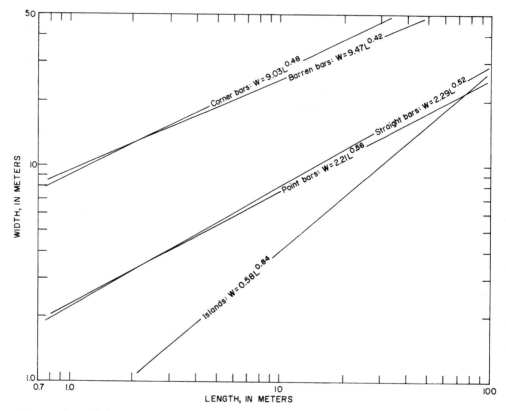

Figure 4. Width-length relations of channel islands and bars (modified from Graf, 1978).

the bar relations are lower than that for the islands owing to the differences in perimeter of a feature subject to accretion.

The slopes or exponents for the four groups of bar measurements fall within a relatively limited range (fig. 4). The slope of the relation for the channel islands is nearly twice that of the bars. The coefficients of the bar equations, however, show more than a four-fold range. Although Graf's (1978) results provide a good indication that the widths and lengths of in-channel geomorphic features are allometrically related, the question remains whether similar or different processes act to form those features. The width-length data of Graf (written commun., 1978) for corner bars, point bars, straight bars, and islands were analyzed for statistical significance to ascertain whether fixed exponents reasonably could be imposed on the equations for the bars. Owing to the data scatter, two-tailed F-tests of significance for the power-function equations of the corner and straight bars showed no significant difference between either the exponents (F = 3.92, d.f. = 1,65) or the coefficients (F = 4.60, d.f. = 1,65). Significant differences occur

46

between the relation for the channel islands and various bars. Although a linear relation for barren bars is shown on figure 4 (Graf, 1978, p. 1499), tests of significance were not applied to these features owing to the difficulty of identifying them from available data. Using techniques previously mentioned (Dixon and Massey, 1957; Snedecor and Cochran, 1967), a fixed exponent of 0.45 was calculated for the three groups of bar data. The resulting power-function equations (when width, W, and length, L, are expressed in meters) are:

$$\text{point bars} \qquad W = 3.19 \, L^{0.45} \qquad\qquad (3)$$

$$\text{straight bars} \qquad W = 4.04 \, L^{0.45} \qquad\qquad (4)$$

$$\text{corner bars} \qquad W = 5.56 \, L^{0.45} \qquad\qquad (5)$$

Although these results cannot be considered of more likely accuracy than those of Graf (1978), the tests of significance show that the equations are reasonable alternatives. Furthermore, the results suggest the possibility that the various types of bars defined by Graf (1978) are equally influenced by the variables complicating the width-length relation.

Reanalysis of Graf's data by digital computer gave an exponent of 0.90 for the width-length relation of the channel islands. Because of the possible errors inherent in equations 3 to 5, it must be considered coincidental that the island exponent is twice that of the bars, but this ratio supports the observation that the island widths change rapidly relative to those of bars because they have two sides for accretion as opposed to one for bars (Graf, 1978, p. 1499).

Gradient-discharge Relations of Alluvial Stream Channels

For the general power-function equation

$$G = fQ^e , \qquad\qquad (6)$$

relating gradient (G) to discharge (Q) of alluvial channels, values of the exponent, e, have been proposed by several researchers. Lane (1957) suggested different values of the coefficient f but the same exponent value of -0.25 for the gradient-mean discharge relations of meandering sand channels and braided streams. Using mean-discharge data from small unnatural channels, Ackers and Charlton (1971) proposed a value of -0.12 for the exponent of a relation separating low-gradient meandering and straight streams. Leopold and Wolman (1957) obtained a larger negative exponent, -0.44, for the gradient-bankfull discharge relation separating braided and meandering channels. By grouping streams of Kansas according to sinuosity and bed-material characteristics, Osterkamp (1978) proposed an exponent value of -0.25 for the gradient-mean discharge relations. Using free bivariant analysis, the exponents calculated from 76 data sets separated into five channel-pattern or sinuosity groups ranged from -0.24 to -0.36 (table 2, fig. 5). The same data, reorganized into four groups of similar bed-material sizes, yielded exponents of -0.23 to -0.32 (fig. 6). Selection of the fixed value of -0.25 was made by averaging techniques previously described and was influenced by the work of Lane (1957).

Table 2

Gradient-discharge relations for Kansas streams
(modified from Osterkamp, 1978)*

Sinuosity or sediment character**	Data sets	Equation, unimposed exponent	Equation, imposed exponent
Si $\geq$ 2.0	17	$G = 0.00068\overline{Q}^{-0.36}$	$G = 0.00070\overline{Q}^{-0.25}$
Si $= 1.7-1.9$	14	$G = 0.00074\overline{Q}^{-0.31}$	$G = 0.00071\overline{Q}^{-0.25}$
Si $= 1.4-1.6$	17	$G = 0.00074\overline{Q}^{-0.18}$	$G = 0.00079\overline{Q}^{-0.25}$
Si $= 1.1-1.3$	18	$G = 0.0011\overline{Q}^{-0.24}$	$G = 0.0011\overline{Q}^{-0.25}$
Braided	10	$G = 0.0019\overline{Q}^{-0.31}$	$G = 0.0017\overline{Q}^{-0.25}$
d_{50} <0.1mm	17	$G = 0.00051\overline{Q}^{-.24}$	$G = 0.00054\overline{Q}^{-0.25}$
S $\geq$ 10, SC < 50 percent	13	$G = 0.00085\overline{Q}^{-0.32}$	$G = 0.00083\overline{Q}^{-0.25}$
$d_{50} = 0.1-2mm$, S $\geq$ 3.0	23	$G = 0.00096\overline{Q}^{-0.26}$	$G = 0.00096\overline{Q}^{-0.25}$
$d_{50} = 0.1-2mm$, S < 3.0	23	$G = 0.0012\overline{Q}^{-0.23}$	$G = 0.0013\overline{Q}^{-0.25}$

* Data are expressed in meters per meter (gradient) and cubic meters per second (mean discharge).

** Si is sinuosity, channel length divided by valley length; d_{50} is median particle size; SC is percentage of silt and clay in bed material; S is sorting index:

$$S = \frac{1}{2} \left(\frac{d_{50}}{d_{16}} + \frac{d_{84}}{d_{50}} \right) .$$

Figures 5 and 6 show that grouping of the data by sinuosity or bed-material size produces a confusing indication of the effect that either exerts on the gradient-discharge relation. If fixed exponents are imposed on the power-functions of the sinuosity groups, an orderly increase of coefficients with decreasing sinuosity results (fig. 7). However, tests between the coefficients of the power functions for the three data groups of sinuosities exceeding 1.3 show questionable statistical significance. The lack of clear differences between the coefficients is illustrated by the close spacing of the linear representatives of the power functions (fig. 7). Sinuosity, therefore, appears to be a poor basis of grouping gradient-mean discharge data of alluvial stream channels for statistical analysis. Conversely, invariant power functions for the same data grouped by particle size have coefficients that, with one possible exception, differ significantly from each other (fig. 8). The exception per-

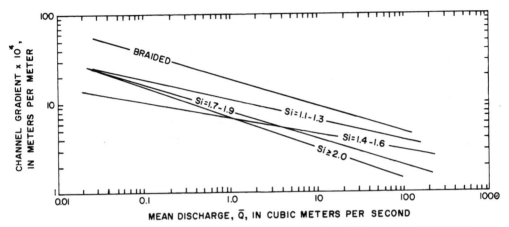

Figure 5. *Gradient-discharge relations for sinuosity (Si) groups of Kansas streams (modified from Osterkamp, 1978).*

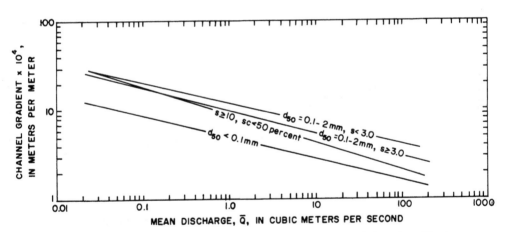

Figure 6. *Gradient-discharge relations using median particle size (d_{50}), sorting index (S), and silt-clay percent of bed material (SC) for sediment-character groups of Kansas streams (modified from Osterkamp, 1978).*

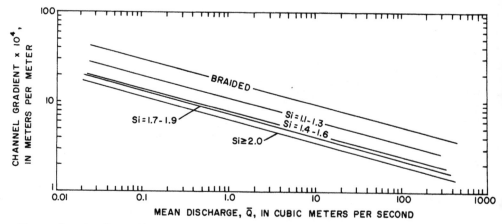

Figure 7. Gradient-discharge relations using imposed exponent for sinuosity
(Si) for groups of Kansas streams (modified from Osterkamp, 1978).

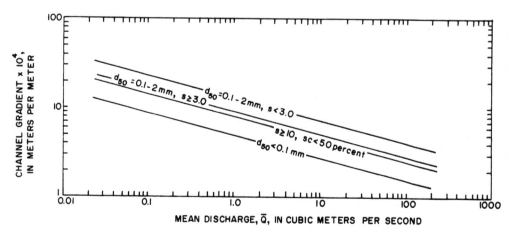

Figure 8. Gradient-discharge relations using median particle size (d_{50}),
sorting index (S), silt-clay percent of bed material (SC),
and imposed exponent for sediment-character groups of Kansas
streams (modified from Osterkamp, 1978).

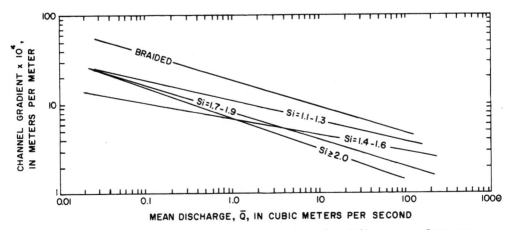

Figure 5. Gradient-discharge relations for sinuosity (Si) groups of Kansas
streams (modified from Osterkamp, 1978).

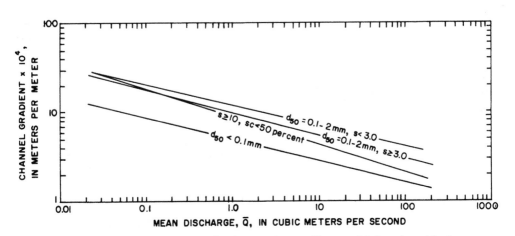

Figure 6. Gradient-discharge relations using median particle size (d_{50}),
sorting index (S), and silt-clay percent of bed material
(SC) for sediment-character groups of Kansas streams (modified
from Osterkamp, 1978).

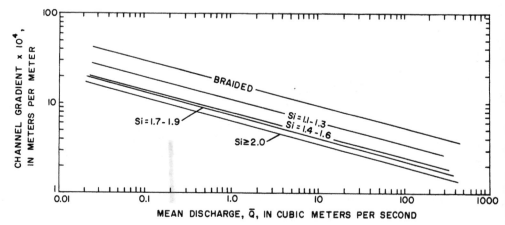

Figure 7. *Gradient-discharge relations using imposed exponent for sinuosity (Si) for groups of Kansas streams (modified from Osterkamp, 1978).*

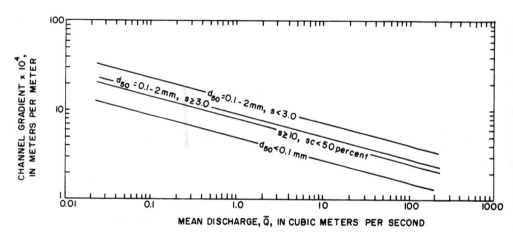

Figure 8. *Gradient-discharge relations using median particle size (d_{50}), sorting index (S), silt-clay percent of bed material (SC), and imposed exponent for sediment-character groups of Kansas streams (modified from Osterkamp, 1978).*

tains to two particle-size groups which are notably different, and therefore a comparison might not be justified. Relations such as those illustrated in figures 7 and 8 also provide information on the advisability of developing multiple power-function equations. For the gradient-discharge examples presented here, it appears unlikely that a statistically significant influence is provided by most sinuosities; whereas, the effect supplied by the bed-material characteristics may be too complex for treatment in multivariant equations.

DISCUSSION AND CONCLUSIONS

Invariant power functions, as advocated in this paper, are a special case of allometric analysis. Although the use of fixed exponents in power-function equations is feasible in a wide variety of empirical studies, it is particularly suited to relations of fluvial systems owing to their complexity. The technique described here is simple, a feature that has resulted in its tacit use by nume researchers, both in the earth sciences and in other disciplines. This simplicity, however, may be the reason that the technique has received little explicit attention.

Some of the justifications for the use of invariant power functions have been mentioned in the description and examples of the technique. These and other benefits include the following:

1. When the exponents of two or more power-function equations relating the same variables for different groups of geomorphic data are different, there is reasonable likelihood that one or more of the exponents are inaccurate. The use of fixed exponents, when they can be defined, reduces the error in the power-function equations and, therefore, increases the confidence that can be attached to the relations.

2. Invariant power functions have transfer power, thus eliminating the need for specific relations such as those defined for limited areas or conditions.

3. If, from other studies, a fixed exponent relating two variables has been identified, limited data that otherwise might be of questionable utility can be considered.

4. Fixed exponents provide a basis to compare results within a study or with other studies.

5. In studies of fluvial systems and other complex systems, attention is focused on processes, whereas process may be ignored in free bivariant analysis. For fluvial systems, process is often synonymous with the effect of complicating variables.

6. Because invariant power functions encourage a quantitative treatment of other variables that affect a bivariant relation, the technique helps identify possible geomorphic thresholds.

7. Invariant allometry maintains the advantages of an open-system approach to geomorphology but minimizes the generalization that often is inherent in a dynamic-equilibrium model. The technique clarifies the aptness of the dynamic-equilibrium model for real geomorphic data and suggests that for some relations, such as between width and discharge of alluvial channels, dynamic equilibrium can lead to a misleading simplification.

8. Invariant allometry encourages a thoughtful development and use of multiple power-function equations. The technique suggests the applicable range of secondary variables and helps prevent the inclusion of redundant variables in a multiple power-function equation.

Invariant allometry is advocated as a useful tool to describe correspondence between some, but not all, geomorphic and hydrologic variables. It should be used with caution because generally the appropriateness of a power function to describe a correspondence must be assumed. Furthermore, the conclusion that a fixed exponent is applicable to various groups of data with the same variables is generally a further assumption. The use of empirical and statistical tests, such as those previously described, to define an exponent merely indicates whether the assumptions are reasonable. Regardless of the curve-fitting technique that is applied to a set of geomorphic data, the equation that results must be regarded as an approximation relating the variables. For studies of fluvial geomorphology in particular, invariant allometry is proposed as one possible approximation of mathematical simplicity and accuracy.

REFERENCES

Ackers, P., and Charlton, F.C., 1971, The slope and resistance of small meandering channels: Institute of Civil Engineers Proceedings, Supp. XV, 1970, Paper 7362S, p. 349-370.

Bull, W.B., 1964, Geomorphology of segmented alluvial fans in western Fresno County, California: U. S. Geological Survey Professional Paper 352E, p. 89-129.

_____ , 1975, Allometric change of landforms: Geological Society of America Bulletin, v. 86, p. 1489-1498.

Burkham, D.E., 1972, Channel changes of the Gila River in Safford Valley, Arizona, 1846-1970: U. S. Geological Survey Professional Paper 655G, 24 p.

Carson, M.A., and Kirkby, M.J., 1972, Hillslope form and process: Cambridge, Cambridge University Press, 475 p.

Chorley, R.J., 1962, Geomorphology and general systems theory: U. S. Geological Survey Professional Paper 500B, 10 p.

Crandell, D.R., 1971, Postglacial lahars from Mount Rainier volcano, Washington: U. S. Geological Survey Professional Paper 677, 73 p.

Davis, W.M., 1889, The rivers and valleys of Pennsylvania: National Geographic Magazine, v. 1, p. 183-253.

Dixon, W.J., and Massey, F.J., Jr., 1957, Introduction to statistical analysis, 2nd edition: New York, McGraw-Hill, Inc., 483 p.

Engelund, F., and Hansen, E., 1967, A monograph on sediment transport in alluvial streams: Copenhagen, Teknisk Forlag, 62 p.

Gary, M., McAfee, R., Jr., and Wolf, C.L., eds., 1972, Glossary of geology: Washington, D.C., American Geological Institute, 805 p.

Gilbert, G.K., 1877, Report on the geology of the Henry Mountains: Washington, D.C., U. S. Department of the Interior, 160 p.

Gould, S.J., 1972, Allometric fallacies and the evolution of Gryphaea—a new interpretation based on White's criterion of geometric similarity; in Dobyhansky, T., Hecht, M.K., and Steere, W.C., eds., Evolutionary biology, v. 6: New York, Appleton-Century-Crofts, p. 91-119.

Graf, W.L., 1978, Fluvial adjustments to the spread of tamarisk in the Colorado Plateau region: Geological Society of America Bulletin, v. 89, p. 1491-1501.

Hedman, E.R., 1970, Mean annual runoff as related to channel geometry in selected streams in California: U. S. Geological Survey Water Supply Paper 1999E, 17 p.

Huxley, J.S., 1932, Problems of relative growth: London, Methuen, 276 p.

Kuhn, T.S., 1962, The structure of scientific revolutions: Chicago, University of Chicago Press, 172 p.

Lacey, G., 1930, Stable channels in alluvium: Institute of Civil Engineers Proceedings, v. 229, p. 259-384.

Lane, E.W., 1957, A study of the shape of channels formed by natural streams flowing in erodible material:

Leopold, L.B., and Langbein, W.B., 1962, The concept of entropy in landscape evolution: U. S. Geological Survey Professional Paper 500A, 20 p.

Leopold, L.B., and Maddock, T., Jr., 1953, The hydraulic geometry of stream channels and some physiographic implications: U. S. Geological Survey Professional Paper 252, 57 p.

Leopold, L.B., and Wolman, M.G., 1957, River channel patterns: braided, meandering, and straight: U. S. Geological Survey Professional Paper 282B, p. 39-85.

Lowham, H.W., 1976, Techniques for estimating flow characteristics of Wyoming streams: U.S. Geological Survey Water Resources Investigation 76-112, 83 p.

Osterkamp, W.R., 1977, Effect of channel sediment on width-discharge relations, with emphasis on streams in Kansas: Kansas Water Resources Board Bulletin 21, 25 p.

_____, 1978, Gradient, discharge, and particle-size relations of alluvial channels in Kansas, with observations on braiding: American Journal of Science, v. 278, p. 1253-1268.

_____, (in press), Variation of alluvial-channel width with discharge and character of sediment: Tenth International Congress on Sedimentology Proceedings, International Association of Sedimentologists.

Schumm, S.A., and Lichty, R.W., 1963, Channel widening and flood-plain construction along Cimarron River in southwestern Kansas: U. S. Geological Survey Professional Paper 352D, p. 71-88.

Simons, D.B., and Albertson, M.L., 1960, Uniform water conveyance channels in alluvial materials: American Society of Civil Engineers Proceedings, Journal of Hydraulics Division, v. 86, p. 33-71.

Snedecor, G.W., and Cochran, W.G., 1967, Statistical methods, 6th edition: Ames, Iowa State University Press, 593 p.

Wahrhaftig, C., 1965, Stepped topography of the southern Sierra Nevada: Geological Society of America Bulletin, v. 76, p. 1165-1190.

White, J.F., and Gould, S.J., 1965, Interpretation of the coefficient in the allometric equation: American Naturalist, v. 99, p. 5-18.

Woldenberg, M.J., 1966, Horton's laws justified in terms of allometric growth and steady-state in open systems: Geological Society of America Bulletin, v. 77, p. 431-434.

Wolman, M.G., and Gerson, R., 1978, Relative scales of time and effectiveness of climate in watershed geomorphology: Earth Surface Processes, v. 3, p. 189-208.

DYNAMIC ADJUSTMENTS OF ALLUVIAL CHANNELS

CHIH TED YANG

U.S. Bureau of Reclamation
Engineering and Research Center
Denver Federal Center
Denver, Colorado

CHARLES C. S. SONG

St. Anthony Falls Hydraulic Laboratory
University of Minnesota
Minneapolis, Minnesota

ABSTRACT

The dynamic adjustments of alluvial channels can be explained by the stream power concept which was used to derive the functional relationship between total sediment concentration and unit stream power. The change of velocity, slope, roughness, channel geometry, and patterns are self-adjustments a river can make to minimize its rate of energy dissipation. The theory of minimum rate of energy dissipation states that when a river is in an equilibrium condition, the rate of energy dissipation is at its minimum value compatible with the constraints applied to the river. If the rate of energy dissipation is not at its minimum value compatible with the constraints, a river will adjust itself in such a manner that the rate of energy dissipation can be minimized to regain the equilibrium condition. River data support the validity of the unit stream power concept and the theory of minimum rate of energy dissipation.

INTRODUCTION

A river is a dynamic system. It constantly adjusts its velocity, slope, roughness, geometry, and/or pattern to comply with the changing climatic, hydrologic, geologic, and man-made changes. Different theories and hypotheses are used to explain these adjustments. Most theories and hypotheses emphasize the local symptoms without sufficient explanation of the cause. Data used to support these theories emphasize the local response of a river within a short time span. This approach often leads to inconclusive arguments.

The theory of minimum rate of energy dissipation is used in this paper to explain the dynamic adjustments of alluvial channels. Field data are used to support the validity of the theory.

UNIT STREAM POWER AND SEDIMENT TRANSPORT

Lane (1955) studied the changes of rivers in response to the variation of water and sediment discharges. Lane's qualitative equation which describes

55

the general relationship between water discharge, Q, channel slope or energy gradient, S, sediment discharge, Q_s, and median particle size, d_{50}, can be expressed by

$$QS \sim Q_s d_{50} \tag{1}$$

Equation (1) indicates that the QS product should be proportional to the $Q_s d_{50}$ product. An increase of Q or S or both should have a corresponding increase of Q_s and d_{50}, or vice versa.

Yang (1972) applied the unit stream power concept to determine the rate of sediment transport. The basic relationship between total sediment concentration, C_t, and unit stream power, VS, where V = mean flow velocity, can be expressed by

$$\log C_t = I + J \log \left(\frac{VS}{\omega}\right) \tag{2}$$

in which C_t = total sediment concentration in ppm by weight; ω = terminal fall velocity of sediment particles; and I and J = coefficients. Yang (1973, 1979) further related I and J to some dimensionless sediment and fluid parameters to obtain two generalized unit stream power equations for total load. The accuracy of the generalized equations was verified by both laboratory and field data (Yang, 1973, 1979; Yang and Stall, 1976, 1978). Theoretical derivations were also presented by Yang et al. (in press) to show how and why total sediment concentration should be related to unit stream power. The unit stream power sediment transport equation in its basic form (2), and the theory of minimum stream power, are used in this paper to explain the dynamic adjustments of alluvial channels.

THEORY OF MINIMUM RATE OF ENERGY DISSIPATION

Equation (2) gives the basic relation between total sediment concentration and unit stream power under a steady equilibrium condition. It also provides a basic equation for predicting the general trend of adjustments of an alluvial channel once the balance between sediment concentration and unit stream power no longer exists. Questions remaining to be answered are:

(a) Under what condition will an alluvial channel be in equilibrium?

(b) If an alluvial channel deviates from its equilibrium condition, what general rule will the channel follow in the process of self-adjustment to regain the equilibrium?

The theory of minimum rate of energy dissipation and its simplified versions explain the dynamic adjustments of an alluvial channel. The theory states that an alluvial channel is in an equilibrium condition if its rate of energy dissipation is at its minimum under the given climatic, hydrologic, hydraulic, geologic, and man-made constraints. The minimum value depends on the constraints applied to the system. If an alluvial channel deviates from its minimum value, it will adjust its velocity, slope, roughness, channel geometry, and pattern in such a manner that the rate of energy dissipation can be minimized to regain the equilibrium. A special case of this theory can be derived from the equation of motion (Yang and Song, 1979) or from the theory of calculus of variation (Song and Yang, in press). In these latter papers the movement of sediment was not considered in the theoretical derivations. However, laboratory and field data indicate that the theory can be applied to gradually varied subcritical open channel flows with or without the movement of sediment. The general theory can be written as

$$E = \int \int \int_R \Phi \; dx \; dy \; dz = \text{a minimum} \tag{3}$$

in which E = total rate of energy dissipation; Φ = rate of energy dissipation per unit volume of fluid; x, y, z = coordinates in the longitudinal, vertical, and lateral directions, respectively; and R = boundary of the volume integral. When the velocity of the boundary is negligibly small, equation (3) can be simplified to

$$E = \int \int \int_R \gamma \; us \; dx \; dy \; dz = \text{a minimum} \tag{4}$$

in which γ = specific weight of water; u = time-averaged local velocity in the longitudinal direction; and s = local energy slope (Yang, 1978; Yang and Song, 1979, Song and Yang, in press). Equation (4) can be integrated to obtain

$$E = \int \gamma \; QS \; dx = \text{a minimum}. \tag{5}$$

If γ QS does not vary significantly along the channel, equation (5) becomes

$$QS = \text{a minimum} \tag{6}$$

in which the stream power QS (neglecting the constant γ) is the total rate of energy dissipation at a given cross-section. Equation (6) represents the theory of minimum stream power. This theory states that for gradually varied open-channel flow under an equilibrium condition, the stream power QS should be a minimum. If the flow is uniformly distributed over the cross-section, the total stream power γQS can be divided by the water-surface width and the weight of the depth-high water prism covering a unit bed area to give VS, called unit stream power. Equation (4) can then be simplified to

$$VS = \text{a minimum}. \tag{7}$$

Equation (7) represents the theory of minimum unit stream power which states that for uniform open channel flow under an equilibrium condition, the unit stream power should be a minimum. As before, the minimum value depends on the constraints applied to the system. Equation (6) was used by Chang and Hill (1977) to determine the variation of a delta stream width. Equation (7) was used by Yang (1976) to determine the channel depth with constant channel width. Equations (2), (6), and (7) will be used in this paper to explain the dynamic adjustment of some natural rivers.

DYNAMIC ADJUSTMENTS OF RIVERS

Ascalmore Creek

Ascalmore Creek in Mississippi was a meandering stream. It was straightened in the 1940's and became very unstable. During the 1950's and 1960's the U.S. Soil Conservation Service vegetated the stream banks and sand bars in the upstream reach and built several small dams to control the water and sediment inputs from 60 % to 70 % of the tributaries. The U.S. Soil Conservation Service's work was successful in stabilizing the stream banks. This work also reduced the sediment supply to the stream. According to the unit stream power concept stated by equation (2), a reduction of sediment concentration, C_t, should have a corresponding reduction in unit stream power VS. In order to regain the balance, sand bars were formed. The 1970 meander thalweg (fig. 1) had increased the actual length of flow, thereby decreasing the slope. This reduced the VS product, or the unit stream power.

South Fork Tillatoba Creek

The South Fork Tillatoba Creek in Mississippi is a tributary of the Tallahatchie River. The U.S. Army Corps of Engineers built several reservoirs in the general area and built improvement structures on the Tallahatchie River in the 1930's and 1940's. These actions shortened the Tallahatchie and lowered its bed elevation gradually. The river bed elevation at the confluence of the South Fork Tillatoba and the Tallahatchie eventually was lowered by 1.2 m. As of 1971 the South Fork Tallatoba Creek was still

Figure 1.   The meandering thalweg of Ascalmore Creek developed after channel straightening.

Figure 2.   The stable South Fork Tillatoba Creek in 1971.

stable (fig 2); however, shortly thereafter the bed degradation began progressing upstream along this creek and caused the banks to cave in. The South Fork Tillatoba Creek in 1973 (fig 3) was adjusting to the new constraints imposed by the change of the Tallahatchie River.

The two cases cited above taught us a lesson in river engineering. That is, river training works cannot be based on the free body diagram approach, in which an element of interest is isolated and examined from a static point of view. Instead, a river must be treated as a dynamic system. The upstream, downstream, and local conditions and their possible responses have to be evaluated in a systematic way before any change is made.

Lower Mississippi River

Extensive channel realignments were made on the Lower Mississippi River. The river's responses to these realignments were documented by the Vicksburg District of the U.S. Army Corps of Engineers (Winkley and Brooks, 1979). Four reaches shown in figure 4 were selected to study the river's response to man-made or natural events. The historical developments of the channel shape and the degree of navigation problems are summarized in table 1.

Figure 5 shows the variation of water surface slope with respect to water discharge. Compare the 1937 high discharge profile and the 1975 lower discharge profile. In the theory of minimum rate of energy dissipation, the minimum value of the rate of energy dissipation depends on the constraints applied to the river. Water discharge is one of the most important constraints.

Figure 3. The unstable South Fork Tillabota Creek in 1973.

Figure 5 shows that a higher discharge is associated with a higher rate of energy dissipation, or a higher slope in this case. According to figure 5, the Lower Mississippi River's response to changing water discharges agrees with the theory of minimum rate of energy dissipation.

Figure 6 compares the water surface profiles at approximately the same water discharge observed in four different years. Generally, water surface slope is a good approximation of energy slope. Because discharge is about the same for all the profiles, lower slopes mean a lower rate of energy dissipation (eq. (6)). According to the theory, a river is in its equilibrium or stable condition when the rate of energy dissipation, or the slope in this case, is at its minimum value under the given constraints. Records indicate that the Lower Mississippi River in 1975 was relatively stable while the 1937 river was not. The difference in slope between the 1975 and 1937 profiles (fig. 6) confirms the correctness of the theory.

LEGEND
▨▨ MAN MADE CUT-OFF

Figure 4. Location map of Lower Mississippi River reaches.

As explained before, the rate of energy dissipation or slope should increase with increasing discharge or stage. Figure 7 shows that on a meandering reach of the Mississippi River the response to the change of water stage agrees with the theory, while on a straight reach it does not. The channel maintenance records confirm that the meandering reach is relatively stable and maintenance-free, while the straight reach is highly unstable and require extensive maintenance work for navigation purposes. The data in figure 7 seem to support the statement made by Yang (1971a) that river meandering is a means for the river to minimize its rate of energy dissipation. These data suggest that the stability of a reach depends on whether the reach can adjust itself to respond to the changing constraints in accordance with the theory of minimum rate of energy dissipation.

Figure 8 compares the Greenville reach of the Mississippi River in 1933 before all the cut-offs were made and the same reach as surveyed in 1975. Cut-offs generally result in increased velocity and slope. If the water and sediment discharges from upstream remained unchanged after the cut-offs, a river should have adjusted itself in such a manner that the velocity-slope product, or the unit stream power, could be reduced in accordance with equation (2). After the cut-offs were made on the Greenville reach, the reach became highly unstable. Numerous dikes were built to restrict the lateral movement of the river. The river's response to the lateral restriction

Table 1

Summary of River Conditions of Different Reaches
of the Lower Mississippi River

Reach Number	Channel Pattern			Degree of Navigation Problems	
	Original (Before Cutoffs)	Intermediate (1950-1970)	Present (1977)	Past (1950-1970)	Present (1977)
1 (Ozark-Eutaw)	Very Sinuous	Sinuous	Sinuous	Minor	Nil
2 (Greenville)	Very Sinuous	Straight	Straight	Major	Major
3 (Kentucky Bend-Mayersville)	Sinuous	Straight	Sinuous	Major	Nil
4 (Baleshed-Ben Lomond)	Straight	Straight	Straight	Major	Major

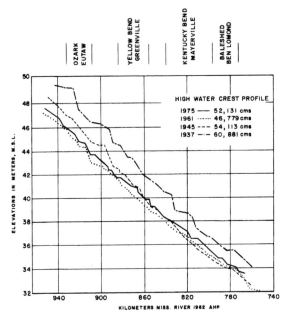

Figure 5. *Variation of water surface slope with respect to water discharge, Lower Mississippi River.*

61

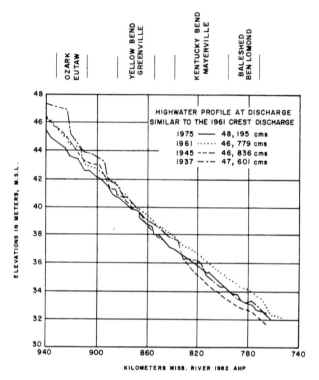

Figure 6. *Variation of water surface slope of the Lower Mississippi River at different times and equivalent water discharges.*

was the formation of a riffle-and-pool or undulating longitudinal bed profile, thereby reducing the rate of energy dissipation. This adjustment seems to support the conclusion of Yang (1971b) that the formation of riffles and pools is one way in which a river can adjust in the vertical direction to reduce its rate of energy dissipation. The riffles formed as a result of this process, as shown in the 1975 survey, have the same corresponding location as the crossings shown on the 1933 survey. Figure 8 strongly suggests that, regardless of what has been done to a river, the nature of a river cannot be changed. A river will adjust itself with whatever means available to comply with the theory of minimum rate of energy dissipation and the concept of unit stream power.

Figure 9 shows the relationship between slope and stage during the 1966-69 and the 1975-77 periods. The slight increase of slope with increasing stage observed in the 1975-77 period is in agreement with theory, while the 1966-69 relationship is not. Records show that the Kentucky Bend-Mayerville reach changed from an unstable reach in the 1960's to a more stable one in the 1970's. This fact suggests that if a river's response to varying constraints changes from disagreement to agreement with the theory and concept stated in this paper, it changes from unstable to stable conditions.

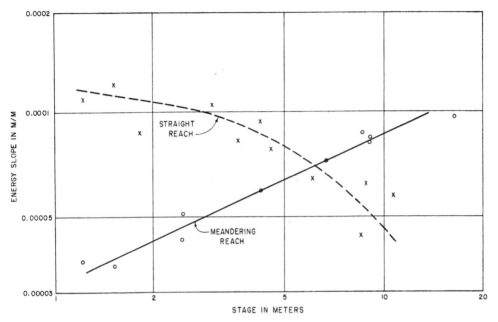

Figure 7. *Comparison of the stage-slope relationship between a meandering and a straight reach of the Lower Mississippi River.*

SUMMARY AND CONCLUSIONS

The concept of unit stream power and the theory of minimum rate of energy dissipation are briefly explained in this paper. Data collected from natural rivers are used to substantiate the concept and the theory. The conclusions drawn from this study are:

1. Natural rivers attempt to maintain a dynamic balance between total sediment concentration and unit stream power. If an adjustment is made in one of these, a corresponding adjustment will be made in the other to regain the balance.

2. A river can adjust its velocity, slope, roughness, geometry, and pattern by numerous means. The objective of all these adjustments is to minimize the rate of energy dissipation under the given climatic, hydrologic, hydraulic, geologic, and man-made constraints. The minimum value depends on the constraints applied to the river.

3. Field data support the validity of the unit stream power con-

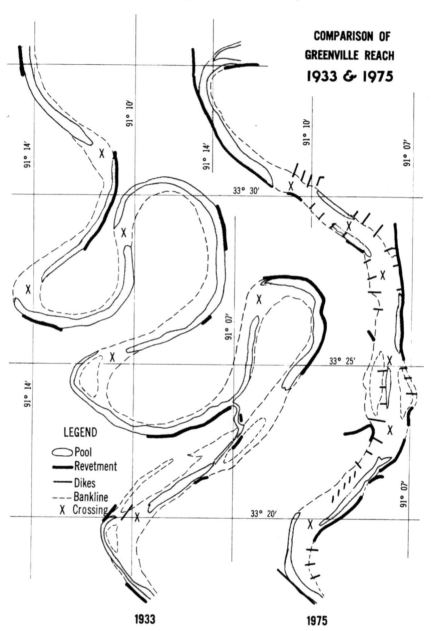

Figure 8. Comparison of the Greenville Reach of the Lower Mississippi River in 1933 before the cut-offs were made and in 1975 after the river had responded to the cut-offs and the construction of dikes by the formation of crossings.

cept in determining the total sediment concentration. The sediment transport equation, equation (2) establishes the minimum value of unit stream power under the sediment rate constraint. Further research is needed to establish similar equations relating the minimum values of unit stream power for other constraints.

4. The dynamic adjustments of the Lower Mississippi River in response to the variation of constraints agree with the theory of minimum rate of energy dissipation. The stability of a reach depends on whether the reach is in a state of minimum rate of energy dissipation under the given constraints.

5. A meandering reach is more stable than a straight reach under varying hydrologic and hydraulic constraints, because a meandering reach can easily adjust itself in accordance with the theory of minimum rate of energy dissipation.

6. When a reach of a river is changing from non-compliance to compliance with the theory of minimum rate of energy dissipation, it is in the process of changing from an unstable to a stable condition.

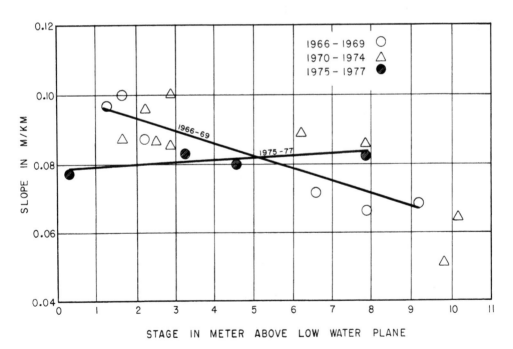

Figure 9. *Comparison of the slope-stage relationship of the Kentucky Bend-Mayersville Reach of the Lower Mississippi River between the periods 1966-69 and 1975-77.*

ACKNOWLEDGEMENTS

The photos and data used in this paper were provided by B.R. Winkley of the U.S. Army Corps of Engineers District, Vicksburg, Mississippi. His critical review and constructive suggestions for improvement are deeply appreciated.

REFERENCES

Chang, H.H., and Hill, J.C., 1977, Minimum stream power for rivers and deltas: American Society of Civil Engineers Proceedings, Journal of Hydraulics Division, v. 103, HY12, p. 1375-1389.

Lane, E.W., 1955, The importance of fluvial morphology in hydraulic engineering: American Society of Civil Engineers Proceedings, v. 21, paper J45, 17 p.

Song, C.C.S., and Yang, C.T., (in press), On the theory of minimum stream power: American Society of Civil Engineers Proceedings, Journal of Hydraulics Division.

Winkley, B.R. and Brooks, J.H., 1979 (in press), Navigation and flood control through geometric alignment: Vicksburg, Mississippi, U.S. Army Corps of Engineers Potomology Investigation Report 300-3.

Yang, C.T., 1971a, On river meanders: Journal of Hydrology, v. 13, p. 231-253.

_____, 1971b, Formation of riffles and pools: Water Resources Research, v. 7, p. 1567-1574.

_____, 1972, Unit stream power and sediment transport: American Society of Civil Engineers Proceedings, Journal of Hydraulics Division, v. 98, HY10, p. 1805-1826.

_____, 1973, Incipient motion and sediment transport: American Society of Civil Engineers Proceedings, Journal of Hydraulics Division, v. 99, HY10, p. 1679-1704.

_____, 1976, Minimum unit stream power and fluvial hydraulics: American Society of Civil Engineers Proceedings, Journal of Hydraulics Division, v. 102, HY7, p. 919-934.

_____, 1978, Minimum unit stream power and fluvial hydraulics [Closure]: American Society of Civil Engineers Proceedings, Journal of Hydraulics Division, v. 104, HY1, p. 122-125.

_____, 1979, Unit stream power equation for total load: Journal of Hydrology, v. 40, p. 123-138.

Yang, C.T., and Song, C.C.S., 1979, Theory of minimum rate of energy dissipation: American Society of Civil Engineers Proceedings. Journal of the Hydraulics Division, v. 105, HY7, p.

Yang, C.T., and Stall, J.B., 1976, Applicability of unit stream power equation: American Society of Civil Engineers Proceedings, Journal of Hydraulics Division, v. 102, HY5, p. 559-568.

Yang, C.T., and Stall, J.B., 1978, Applicability of unit stream power equation [Closure]: American Society of Civil Engineers Proceedings, Journal of Hydraulics Division, v. 104, HY7, p. 1095-1103.

Yang, C.T., Rooseboom, A., and Molinas, A., (in press). Total sediment load equations: American Society of Civil Engineers Proceedings, Journal of the Hydraulics Division.

HYDRAULIC ADJUSTMENT OF THE EAST FORK RIVER,

WYOMING TO THE SUPPLY OF SEDIMENT

EDMUND D. ANDREWS

U.S. Geological Survey
Water Resources Division
Denver, Colorado

ABSTRACT

A bedload trap located on the East Fork River, western Wyoming, has provided direct measurements of bedload transport and a unique opportunity to study the hydraulic adjustment of an alluvial stream to the particular combination of sediment and water it carries. The mountainous parts of the East Fork River basin have large water and small sediment yields. Conversely, the lowlands underlain by the erodible Wasatch Formation have small water and large sediment yields. Irrigation of the hay meadows along Muddy Creek, an East Fork tributary, has greatly accelerated erosion of the stream channel and banks, and, thus, has recently increased the sediment discharge of Muddy Creek into the East Fork River. Due primarily to the sediment contribution of Muddy Creek, the bedload-sediment discharge of the East Fork River in 1975 increased from approximately 200 t upstream from the confluence of Muddy Creek to more than 3,200 t at the bedload trap downstream.

The East Fork River slope decreases from 0.0011 immediately upstream from the confluence of Muddy Creek to 0.0084 in the vicinity of the bedload trap. There is no evidence to indicate that slope has adjusted to the increased sediment load. The bankfull-channel area of the East Fork River, however, increases downstream by approximately 30 percent, although the total available flood discharge is unchanged. As a result, the bankfull discharge is greater and a larger part of the available flood discharge is conveyed within the channel downstream of Muddy Creek. Consequently, at the respective bankfull discharges, the stream power per unit length and per unit area is approximately the same in both East Fork River reaches.

The downstream hydraulic geometry between the East Fork River study reaches differs considerably from the observed mean condition of other rivers in the region. Mean velocity and depth increase downstream more rapidly, while width increases less rapidly, than the corresponding mean values for other streams in the region. With the constraints that slope is independent and the concentration of sediment per unit width of active bed is constant, the measured hydraulic exponents for the East Fork River agree exactly with exponents computed by minimizing the variance of velocity, width, depth, and stream power per unit length and per unit area. Therefore, the hydraulic characteristics of the East Fork River, with the exception of slope, have adjusted mutually to transport the sediment contributed by Muddy Creek.

E. D. Andrews

INTRODUCTION

The concept of the graded stream (Mackin, 1948) suggests that the hydraulic and channel characteristics of the various tributaries within a basin are adjusted to the particular combination of sediment and water received. Although this idea is a well-accepted tenet of hydrology and geology, little is known about the way in which a stream channel adjusts to the quantity of water and sediment supplied to it. This situation is due largely to the great difficulties involved with measuring the bedload-sediment discharge of a river.

A bedload trap was constructed on the East Fork River near Boulder, Wyo., in the spring of 1973. A description of the bedload trap and data collected during 3 years has been published by Leopold and Emmet (1976, 1977). These data provide a unique opportunity to study the hydraulic adjustment of a natural alluvial channel.

The East Fork River basin also has the requisite areal variation of water and sediment yields. Mountainous parts of the basin have relatively large water yields but produce relatively little sediment. In contrast, lowland parts of the basin have relatively large sediment production but small water yields. As a result, the sediment load of the East Fork River increases significantly as the stream traverses the lowlands southwest of the mountain front, whereas the water discharge remains nearly constant. This paper is a study of the changes in channel and flow characteristics which accompany the increased sediment load without a proportionate increase of water discharge. The paper is arranged in two parts. The first part describes the supply of water and sediment to the channel network. The second part describes the hydraulic adjustment of the East Fork River channel to an approximately 16-fold increase in sediment load without an appreciable increase of water discharge.

WATER AND SEDIMENT DISCHARGE OF THE

EAST FORK RIVER AND MUDDY CREEK

Physiography and Geology

The bedload trap is located on the East Fork River in western Wyoming (fig. 1). The drainage area contributing to the project site is approximately 502 km². A major tributary, Muddy Creek, joins the East Fork River 4.0 km upstream from the bedload trap. At their confluence, the drainage area of the East Fork River is about 258 km² and that of Muddy Creek is about 238 km². Although the two subbasins have nearly the same drainage area, they are strikingly dissimilar in other aspects. As a consequence, the amounts of sediment and water each stream supplies to their confluence differ markedly, and the resulting channel downstream from their confluence is distinct from either of the upstream channels.

The main stem of the East Fork River drains glaciated granitic terrain within the Wind River Mountains (fig. 1). The majority of the basin northeast of the steep mountain front has a rather subdued topography of polished domes and alpine lakes at elevations between 3,050 and 3,500 m. Alpine

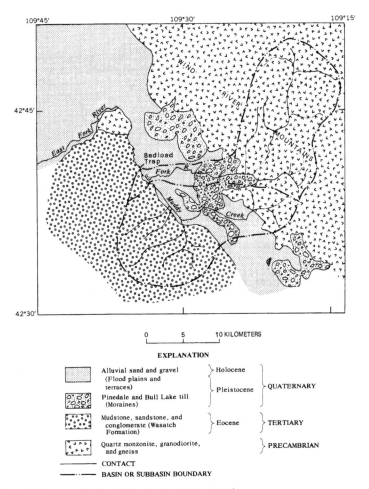

Figure 1. Geology of the East Fork River basin.

sources of sediment within the size range appropriate for fluvial transport are limited because Pleistocene glaciation of the Wind River Mountains thoroughly removed the mantle of weathered debris.

As the East Fork River flows from the mountain front, it traverses a series of terminal moraines of Bull Lake and Pinedale glacial stages. West (downstream) of the steep mountain front and moraines, outwash terraces of Tertiary and Pleistocene age dominate the topography (fig. 2). Although none of the Holocene terraces are extensive along the East Fork River, they are persistent and regular. The main Pinedale and Holocene terraces are cut into the Wasatch Formation of Eocene age, which underlies the basin southwest of the Wind River Mountains.

71

Figure 2. *Tertiary and Pleistocene terraces along the East Fork River. View is looking east from a Tertiary terrace across Pleistocene terraces towards the mountain front. (Photograph by R.H. Meade.)*

The Muddy Creek subbasin lies almost entirely southwest of the mountain front within the belt of Pleistocene moraines and gravel-outwash terraces (fig. 1). Relief is subdued. Bull Lake and Pinedale moraines are extensive in the central part of the basin, but are not as continuous as those along the East Fork River. Instead of traversing the moraines in one continuous cascade, Muddy Creek flows through numerous small, irrigated hay meadows which lie among the moraines.

The Wasatch Formation (Bradley, 1964) crops out throughout the southern and western parts of the Muddy Creek basin. This formation is predominantly a gritty mudstone, but also includes some extensive sandstone lenses. One of these sandy units crops out near the mouth of Muddy Creek and can be traced at least 4.0 km upstream. The sand is clean, friable, and easily eroded from cut banks along the outside of meander bends.

Size and Areal Distribution of Bed Material

The size and distribution of bed material in the East Fork River channel changes significantly at the mouth of Muddy Creek. Upstream from the mouth of Muddy Creek, the bed of the East Fork River is predominantly gravel with only occasional discontinuous streaks of sand. The median grain size of riffles decreases from about 70 mm immediately downstream from the moraine to 35 mm at the confluence with Muddy Creek.

The bed material of Muddy Creek is almost exclusively sand sized. Sieve analysis of bed material shows a median grain diameter of about 0.75 mm. Downstream from the confluence with Muddy Creek, the bed of the East Fork River becomes predominantly sand. The median grain size of bed material is 1.25 mm and less than 5 % is finer than 0.25 mm. Gravel-covered parts of the bed are limited to the outside of meander bends where the channel cuts into a gravel terrace, and immediately downstream of the bends where the gravel accumulates in bars. In addition, small patches of gravel commonly appear along the base of banks in straight reaches. A study of scour and fill in the East Fork River (Andrews, 1979) indicates that the gravel forms an extensive and possibly continuous base below the sand.

Runoff Characteristics

Upstream from the bedload trap, the mean elevation of the East Fork River subbasin is in excess of 3,200 m, compared to approximately 2,200 m for the Muddy Creek subbasin. Precipitation occurs mainly in the form of winter snowfall and the amount increases significantly with elevation. Consequently, the snowpack within the Muddy Creek subbasin prior to the spring runoff is less than 20 % of the snowpack within the mountains. Due to high infiltration and low relief, a smaller percentage of the Muddy Creek snowpack appears as runoff during the spring flood. As a result, despite nearly equal basin areas at their confluence, Muddy Creek has a bankfull discharge of 1.1 m^3/sec compared to 16.1 m^3/sec for the East Fork River upstream from the confluence with Muddy Creek. At bankfull stage, the East Fork River channel is 12 to 24 m wide and 0.9 to 1.4 m deep (fig. 3). By contrast, Muddy Creek is only 4.6 to 5.5 m wide and 0.45 to 0.75 m deep (fig. 4).

Due to the elevation difference between the two subbasins, the snowmelt floods of the two streams do not coincide. The snowmelt flood on Muddy Creek may begin as early as the last week of March, and the subbasin is usually clear of snow by early May. The East Fork River flood normally follows a month or so later--between the last week of May and the end of June.

The discharge of Muddy Creek is greatly augmented by irrigation return flow. Between mid-June and early August, approximately 2.8 m^3/sec are diverted from the East Fork River and used to irrigate the hay meadows along Muddy Creek. As a result, Muddy Creek discharges 1.1 to 1.4 m^3/sec during the irrigation season, in contrast to a base flow of about 0.42 m /sec before and after irrigation.

Bedload-Sediment Discharge of Muddy Creek and the East Fork River

To measure the downstream hydraulic adjustment of the East Fork River to the quantity of bedload sediment contributed by Muddy Creek, three study reaches were established: (1) on the East Fork River upstream from the confluence of Muddy Creek (called the "Clothesline reach"); (2) on Muddy Creek just upstream from its mouth; and (3) on the East Fork River in the vicinity of the bedload trap, approximately 4.0 km downstream from the confluence of Muddy Creek (called the "Bedload Trap reach") (fig. 5).

*Figure 3. East Fork River looking towards cross section 1 in the Clothesline
reach.*

Each study reach consisted of: (a) several cross sections which were re-
surveyed frequently during the runoff period, (b) a continuous-stage recorder, and (c) one cross section with the necessary facilities for making water-
and sediment-discharge measurements. The cross sections in each study reach
were selected to be representative of the range of channel and hydraulic characteristics. These sections were monumented and surveyed with rod and
level before the spring runoff began. At high flow, it was impossible to wade
the East Fork River; therefore, beaded cableways were installed at each section. The sections were sounded at stations spaced every 0.61 m across the
channel by measuring the depth of flow with a wading rod from a light aluminum boat attached to the cableway. Each cross-section survey was plotted
and the channel area, width, mean depth, and mean velocity were determined.
The results of the cross-section surveys were summarized by Andrews (1977).
The eight cross sections located in the vicinity of the bedload trap were surveyed during the spring of 1974. All other parts of the investigation were
conducted in the spring of 1975.

In the Muddy Creek study reach, water- and sediment-discharge measurements were made from a foot bridge. Because of the greater channel width of
the Clothesline reach, a cableway was installed. The cableway carried a bedload sampler (described below) to any chosen distance from the river bank.
The bedload sampler was lowered and raised by a winch controlled from the

Figure 4. Muddy Creek upstream from the gaging station. The channel is approximately 4.5 m wide and 0.5 m deep.

shore. Water-discharge measurements at the Clothesline reach were made from a light aluminum boat attached to the cableway. Water- and sediment-discharge measurements for the Bedload Trap reach were provided by Leopold and Emmett (1976, 1977).

Measurements of the bedload-transport rate through the Clothesline and Muddy Creek study reaches were made with a Helley-Smith bedload sampler. Emmett (1979) calibrated the sampler (7.6 cm orifice, 27.2 kg) against the bedload trap (Leopold and Emmett, 1976) and concluded that the Helley-Smith bedload sampler measured transport rates comparable to those obtained with the East Fork River bedload trap. Hence, the Helley-Smith bedload sampler could be used with confidence under the flow conditions experienced on the East Fork River--cold and clear water, mean velocities from 0.6 to 2 m/sec, and medium to coarse sand in transport.

Measurements at the bedload trap have shown that even with constant hydraulic conditions the rate of bedload transport is quite variable, both in location across the channel and with time at a given cross-channel station. When sampled at several stations across the channel, the maximum transport rate is frequently 3 to 5 times the whole channel average. The zone of maximum transport is not fixed and shifts back and forth across the channel (Leopold and Emmett, 1976). Much of the variation at a given station appears

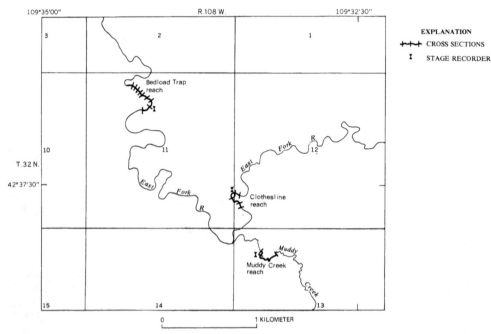

Figure 5. *Study reaches showing location of cross sections and stage recorder.*

to be due to the passage of dunes, because the transport rate varies in a cyclic manner with a period of 10 to 20 min. In order to average such variations, the bedload-transport rate must be sampled at several stations across the channel, as well as during a reasonable period of time. Emmett (1979) discussed the sampling procedure which was adopted for this study. Bedload-discharge measurements were made at the Foot Bridge section on Muddy Creek by sampling 13 stations, spaced 0.30 m apart across the 4.6-m-wide channel. Normally, two traverses of the cross section were combined into one bagged sample, hence, 2 times 13 stations. Three such bagged samples were usually collected at a given discharge to determine the bedload-transport rate.

The duration of sampling time at each station was varied so that a minimum of 1 kg was collected per sample. As the sediment discharge increased, the sampling time at each cross-section station was reduced to keep the sample to a manageable size. However, the minimum sampling time per station was never less than 30 seconds.

Bedload-transport-rate measurements in the Clothesline reach were made in a similar manner. Because of the wider channel, 16 stations spaced at 0.61-m intervals constituted the standard traverse. Commonly, each traverse was treated as a separate sample, and 4 to 6 samples were collected each day.

Adjustment to Sediment Supply

Several water-discharge measurements were made in each reach, normally in conjunction with the bedload-transport measurements. When combined with the stage records, these water-discharge measurements give a continuous record of the mean depth, velocity, width, and discharge. For each of the three reaches, the measured bedload-transport rates per foot of channel width, i_b, were plotted against the corresponding mean velocities, $\bar{u}$ (fig. 6). The i_b versus $\bar{u}$ relations used to compute the total bedload-sediment discharges at the three gaging stations for the period of record were fitted by eye (fig. 6).

The total transport of bedload sediment past each gaging station was calculated by simple integration over the respective hydrographs, using the appropriate i_b versus $\bar{u}$ relation. Water-stage recorders were operated on Muddy Creek and the East Fork River study reaches from April 10 to September 2, 1975. Although this period of record is less than 5 months, the rate of bedload transport through each study reach during the remainder of the year was quite small; i_b was less than 0.0045 kg/sec/m. Thus, the calculated bedload-sediment discharges for the 5-month period are reasonable estimates of the annual quantities in 1975.

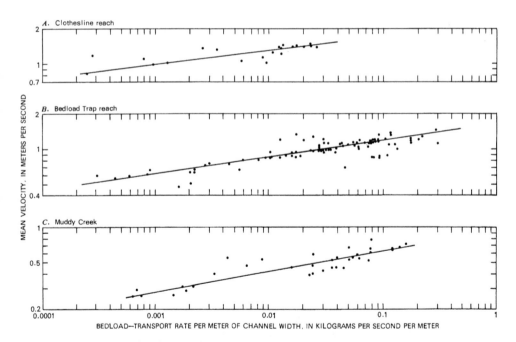

Figure 6. Graphs showing bedload-transport rate versus mean velocity relations for the (A) Clothesline and (B) Bedload Trap reaches on the East Fork River and (C) Muddy Creek. Sediment loads are expressed as dry weight. Data from: A, Andrews (1977); B, Leopold and Emmett (1976, 1977); and C, Andrews (1977).

The total quantities of bedload sediment transported past the three gaging stations in 1975 are compared in figure 7. During the period of record, April 10 to September 2, 1975, approximately 3,000 t of bedload sediment were transported through the Muddy Creek reach. By comparison, only 200 t of bedload sediment passed the Clothesline reach, although it had a much greater water discharge than Muddy Creek. The sum of these quantities, 3,200 t, is believed to be a good estimate of the 1975 contribution of bedload sediment to the confluence of Muddy Creek and the East Fork River.

Direct contributions of bedload-size material to the East Fork River between the confluence of Muddy Creek and the bedload trap are primarily from cutbanks in the higher terraces and ephemeral gullies. Bluffs composed of the Wasatch Formation (fig. 8) occur along the south side of the East Fork River downstream from the mouth of Muddy Creek. Erosion of these cutbanks appears to take place primarily during the winter. By late November, the East Fork River freezes over, and any material which falls from the cutbanks accumulates on the river ice. When the East Fork River was inspected in late March 1975, the river was still frozen, and a considerable deposit of debris was lying on the ice along the base of the cutbanks.

Several dozen freeze-and-thaw cycles undoubtedly occur during the winter on the east-facing cutbanks in the Wasatch Formation. This circumstance, combined with the majority of the annual precipitation, probably accounts for the large quantity of debris shed from the cutbanks during the winter months.

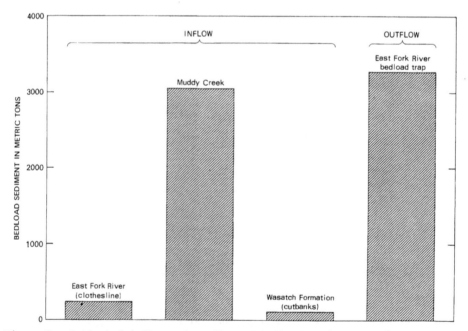

Figure 7. *Estimated inflow and outflow of bedload sediment to the East Fork River between the confluence of Muddy Creek and the bedload trap during 1975.*

Figure 8. A cutbank in the mudstone member of the Wasatch Formation along the East Fork River. The mudstone member contains less than 20 % by weight of material larger than 0.25 mm. (Photograph by R.H. Meade.)

The total volume of accumulated debris between the confluence and the bedload trap during 1975 was estimated to be 190 m^3. This value was determined by measuring several representative cross sections of the deposit and pacing the length of the cutbanks. The average density of four samples of the Wasatch Formation was 2,050 kg/m^3. Thus, approximately 390 t of debris from the Wasatch cutbanks were supplied to the East Fork River when the ice melted. However, not all of this material would be transported as bedload. The bluffs are composed of the gritty mudstone member of the Wasatch Formation which disaggregates into particles of clay, silt, and sand after a few days in water. A sample of the mudstone was soaked for 1 week and wet sieved. Less than 20 % of the sample was retained by the 0.25-mm sieve. Therefore, most of the debris would be transported out of the sediment-budget reach as suspended load. Thus, 20 % of 390 t, or 78 t, appears to be a reasonable estimate of the quantity of bedload material supplied to the East Fork River from the Wasatch bluffs in 1975.

Ephemeral gullies draining both sides of the East Fork River are another source of sediment to the channel. Because most of the area drained by the gullies is underlain by the mudstone member of the Wasatch Formation, sediment contribution to the East Fork River by the gullies is primarily in the form of suspended load.

Slightly more than 3,200 t of bedload sediment were transported past the bedload trap in 1975 (fig. 7). This sum closely approximates the 3,300 t of bedload sediment supplied to the East Fork River channel between the Clothesline reach and the bedload trap. Therefore, Muddy Creek is by far the major source of the bedload sediment transported through the Bedload Trap reach.

The quantity of bedload sediment transported past the bedload trap will not always equal the quantity of bedload sediment supplied to the East Fork River channel upstream of the bedload trap in a particular year. The inflow of bedload sediment to the East Fork River is a result of flow conditions in Muddy Creek, while the outflow of sediment is controlled by flow conditions in the East Fork River. Because the discharge of Muddy Creek is so small compared to that of the East Fork River, even relatively large flows in Muddy Creek have no appreciable effect upon the rate of sediment transport in the East Fork River. For reasons which will be discussed later, the discharge of Muddy Creek is only indirectly related to the discharge of the East Fork River. Assuming that the Muddy Creek and East Fork River channels and flood plains are stable at this time, and there is no evidence to the contrary, the outflow of bedload sediment at the bedload trap should equal the inflow from the upper East Fork River and Muddy Creek plus the supply of sediment to the intervening reach of channel, averaged over a period of years. A computation shows that approximately 10 times the annual outflow of bedload sediment is stored in and moving through the East Fork River channel upstream from the bedload trap. Therefore, yearly fluctuations in the inflow of sediment should have only a minor effect on the supply of material available for transport past the bedload trap.

Sources of Bedload Sediment to Muddy Creek and the East Fork River

Approximately 60 % of the East Fork River basin (most of which lies within the mountains) supplies approximately 200 t or 6 % of the bedload sediment passing the bedload trap. Pleistocene glaciation thoroughly stripped the Wind River Mountains of soil, regolith, and other weathered debris, thereby removing sources of sediment of an appropriate size for fluvial transport. Since the end of the Pleistocene, weathering of the granitic bedrock has not produced significant amounts of sediment. The limited quantity of sediment which is produced of an appropriate size for fluvial transport, and which enters a channel, is rarely transported very far before it is trapped in a lake.

Southwest of the mountain front, active cutbanks in the Pleistocene and Holocene terraces are uncommon along the East Fork River. Although the channel meanders, the stream rarely flows along the Pleistocene terraces because of the wide flood plain. Even then, retreat of the cutbanks supplies only a minor amount of bedload sediment, as the material eroded is mainly the mudstone member of the Wasatch Formation.

These observations apply equally to Muddy Creek throughout most of its length, except for the lower 4 km. Cutbanks within the moraine are not active today.. Elsewhere, Muddy Creek is primarily a deep, sod-lined channel meandering through hay meadows, which contribute little, if any, sediment to the channel.

Muddy Creek appears to receive the majority of its large sediment load in the 4.0 km of channel between the last hay meadow and its confluence with the East Fork River. The Muddy Creek flood plain is relatively narrow, about 6 channel widths at most, and nearly every meander cuts into a terrace (fig. 9). A friable sandy member of the Wasatch Formation crops out in the down-stream part of the Muddy Creek basin. The material supplied to the Muddy Creek channel by erosion of cutbanks along this 4.0 km reach is medium to coarse sand, $D_{50} = 0.6-2.0$ mm, and thus is transported as bedload rather than suspended load. Erosion of these cutbanks can be rapid. Bank-retreat pins placed in six cutbanks in August 1974 indicated an average annual erosion of 0.15 m. The downstream 4.0 km of Muddy Creek had an estimated 2,420 m^2 of active cutbanks in 1974-75. Therefore, approximately 800 t of sand entered Muddy Creek in that period from bank retreat alone. Numerous gullies drain-ing the low hills south of Muddy Creek are also significant contributors of sand.

Historic Changes in the Hydrology of Muddy Creek

A narrow flood plain and rapid cutbank retreat indicate a historic change in the hydrology of Muddy Creek. Shortly after 1900, the natural grass meadows along Muddy Creek were homesteaded and three irrigation ditches were constructed to convey water from the East Fork River. Presently, nearly 2.80 m^3/sec are diverted out of the East Fork River from mid-June to late July or early August and spread across the meadows along Muddy Creek.

Figure 9. A typical cutbank about 4 m high, in the friable sand member of the Wasatch Formation along the lower 4.0 km of Muddy Creek.

Runoff from the irrigated fields returns to the East Fork River by way of Muddy Creek, and increases the discharge of Muddy Creek by approximately 300 %. During the irrigation season, Muddy Creek discharges between 1.1-1.4 m^3/sec, in contrast to a base flow of about 0.42 m^3/sec before and after irrigation.

The significance of irrigation return flow on the annual sediment discharge of Muddy Creek is illustrated by figure 10. The top graph shows the weekly sediment discharge (histogram) and cumulative percentage of the total sediment discharge (solid curve) of Muddy Creek between April 13 and August 31, 1975. Two distinct periods of significant weekly sediment discharge are shown. The snowmelt flood occurred from April 17 to May 18, and approximately 510 t or 17 % of the annual sediment load was transported during this period. Irrigation began on June 8 and lasted until about August 3. Approximately 2,180 t or 73 % of the annual sediment load was transported during the period of irrigation return flow. Thus, the large sediment load supplied by Muddy Creek to the East Fork River channel was transported primarily by irrigation return flow.

The middle graph in figure 10 shows the weekly sediment discharge (histogram) and cumulative percentage of the total 1975 sediment discharge of the East Fork River at the bedload trap (dashed line). Because of the higher mean elevation of the East Fork River basin, its snowmelt flood normally follows the Muddy Creek flood by 4 to 5 weeks. In 1975, nearly the entire sediment load, 99 percent, was transported between May 25 and July 14. The two distinct peaks in the histogram of weekly sediment discharge reflect several days of wet, cool weather during the spring.

The lower graph in figure 10 compares the cumulative percentage of the total 1975 sediment loads of Muddy Creek (solid line) and the East Fork River at the bedload trap (dashed line). The snowmelt flood on Muddy Creek contributed approximately 510 t of sediment to the East Fork River between April 20 and May 25. Because sediment transport at the bedload trap was negligible during this period, the sediment from Muddy Creek was stored in the East Fork River channel. Subsequently, during the snowmelt flood of the East Fork River, the quantity of bedload sediment transported past the bedload trap greatly exceeded the quantity being supplied by Muddy Creek, and the East Fork River channel was gradually depleted of bed material. Relatively large weekly sediment discharges from Muddy Creek, however, continued until the middle of August. By the end of August, enough sediment had been supplied by Muddy Creek to replenish the East Fork River bed material.

HYDRAULIC ADJUSTMENT OF THE EAST FORK RIVER

The changes in size and distribution of bed material in the East Fork River channel that were associated with the historic increase of sediment supplied by Muddy Creek have already been described. These, however, were not the only channel changes. In order to maintain a quasi-equilibrium channel, adjustments of the East Fork River morphology and flow characteristics have accompanied the increased sediment load. Channel roughness, depth, width, and slope tend to be mutually adjusted within the imposed constraints in order to provide just the velocity necessary to transport the sediment supplied to

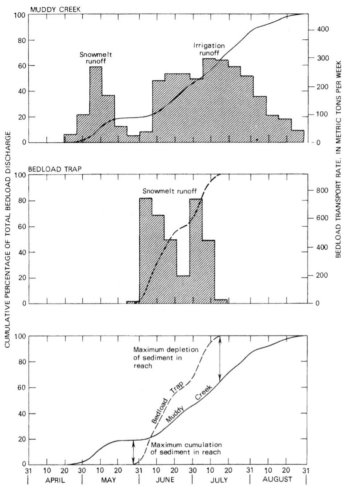

Figure 10. *Weekly bedload-sediment discharges (histograms) and cumulative percentage of the total sediment discharges (curves) of Muddy Creek and the East Fork River at the bedload trap, April-August 1975.*

the channel with the available discharge.

Thus far this paper has dealt primarily with the independent variables--sediment and water discharge--of the East Fork River and Muddy Creek. Water and sediment are not derived uniformly in time and space from the drainage basin, nor are the source areas for sediment and water the same. Yet, in spite of large changes in the water-and-sediment mixture it carries, the East Fork River remains in a quasi-equilibrium state. The second part of this paper will be devoted to the hydraulic adjustment of the East Fork River to the approximately 16-fold increase in annual sediment load downstream from the confluence of Muddy Creek. The goal will be to understand the hydraulic adjustment, as well as to describe it. This is not a simple task, due to the indeterminate nature of the river system. Langbein and Leopold (1964), however, suggested that there is a tendency toward a minimum-variance condition; that is, within the physical constraints or limitations placed upon the river system, the effects of a change in the independent variables are distributed among the dependent variables so as to minimize the required adjustment.

Independent Variables

The major independent variables of the river system are the sediment and water discharge, bedload-sediment size, suspended-sediment concentration, and water temperature. The downstream variations of the sediment and water discharge of the East Fork River were discussed in detail previously. Bedload-sediment size, suspended-sediment concentration, and water temperature are important, but they had nearly identical values in both East Fork River reaches. Therefore, these factors do not appear to be responsible for the observed channel changes.

River slope is probably the most difficult hydraulic variable to evaluate. Mackin's (1948) definition of the graded stream identifies slope as the primary dependent variable, whose adjustment maintains a balance between the supply and transport of sediment. Since then, however, several authors, notably Wolman (1955), and Miller (1958), have concluded that changes in river slope appear to be a long-term hydraulic adjustment. Though river slope appears to adjust to new hydrologic conditions, the rate of adjustment is too slow to be of primary significance in maintaining a quasi-equilibrium alluvial channel when the change in hydrologic conditions is relatively recent.

The longitudinal profile of the East Fork River (fig. 11) does not indicate that slope has adjusted to the large increase in sediment load downstream from the confluence of Muddy Creek. Instead of steepening, as the graded-stream concept would lead one to expect, the average East Fork River slope flattens from 0.0011 through the Clothesline reach to 0.00084 through the Bedload Trap reach. In addition, there is no field evidence which indicates that the slope has changed appreciably in response to the increased sediment load contributed by Muddy Creek. Therefore, it appears that the East Fork River slope is independent of the sediment supplied by Muddy Creek.

Downstream Change in Channel Characteristics

The bankfull discharge for the Clothesline and Bedload Trap reaches was calculated from each station's stage-discharge relation after measuring bankfull stage at the field sites. The computed bankfull discharge for each East Fork River cross section is shown in column 2 of table 1. The mean bankfull discharge of the Bedload Trap reach, 23.2 m³/sec, was much larger than the mean bankfull discharge of the Clothesline reach, 16.1 m³/sec.

The minor water contribution of Muddy Creek could not account for the increase of about 7.1 m³/sec in bankfull discharge of the East Fork River

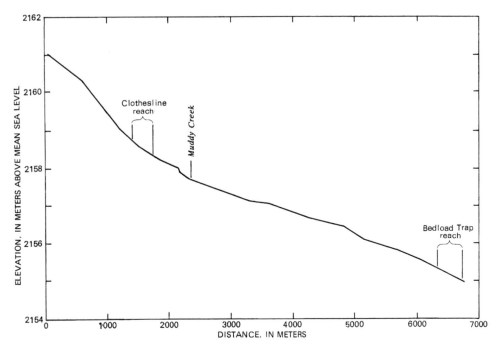

Figure 11. Longitudinal profile of the East Fork River from the Clothesline reach to the Bedload Trap reach.

downstream from Muddy Creek. Because the total water discharge, whether flowing within the channel or over the flood plain, was virtually the same for both the Clothesline and Bedload Trap reaches, other factors must be responsible for the larger channel size and bankfull discharge downstream from the mouth of Muddy Creek. Sediment load and slope are the two independent variables which do change between the Clothesline and Bedload Trap reaches. Therefore, the bankfull discharge of the East Fork River has adjusted to the downstream increased sediment load and decreased slope in order to maintain the stream in quasi-equilibrium.

An important consequence of the larger channel downstream from the confluence of Muddy Creek is revealed by comparing the stream power in the two reaches. Stream power per unit length of channel, θ, is

$$\theta = \gamma\, QS,$$

where:

γ = unit weight of water, kg/m^3;

Q = discharge, m^3/sec; and

S = slope (dimensionless).

Table 1

Comparison of Bankfull Hydraulic Characteristics
of the Clothesline and Bedload Trap
Reaches, East Fork River

Cross section number	Discharge	Stream power per unit area[1]	Stream power per unit length[1]	Velocity	Width	Depth	Darcy-Weisbach friction factor[2]
	Q	$\gamma QS/W$	γQS	$\bar{u}$	W	$\bar{d}$	ff
	(m^3/sec)	$[(kg/sec)/m]$	(kg/sec)	(m/sec)	(m)	(m)	
Clothesline Reach							
1	17.5	1.43	19.3	1.13	13.4	1.16	0.064
2	15.6	1.25	17.2	1.22	13.7	.93	.049
3	15.3	.82	16.9	.90		.83	.081
4	16.1	1.15	17.8	1.03	15.5	.99	.070
Mean	16.1	1.16	17.8	1.07	15.8	0.98	0.066
Bedload Trap Reach							
1	22.7	0.73	18.2	0.96	25.9	.91	0.058
5	24.1	1.01	20.0	1.21	19.5	1.02	.045
6	23.3	.94	19.8	1.13	20.7	.99	.047
7	23.8	1.07	20.4	1.12	18.9	1.13	.050
8	24.1	1.27	20.0	1.22	15.9	1.25	.046
9	22.1	1.21	18.4	1.11	15.2	1.31	.059
10	22.1	1.54	18.4	1.30	12.2	1.42	.045
11	23.8	1.27	19.8	1.21	15.5	1.26	.050
Mean	23.2	1.13	19.4	1.16	18	1.16	0.050

[1] γ = unit weight of water = 1,000 kg/m^3; S = slope.

[2] $ff = 8gDS/\bar{u}^2$.

Because the East Fork River channel is larger downstream from Muddy Creek, a greater discharge flows within the channel of the Bedload Trap reach than within the channel of the Clothesline reach, whenever the available flood discharge exceeds 16.1 m3/sec. The slope of the Bedload Trap reach, however, is approximately 25 % less than the slope through the Clothesline reach. The histogram of figure 12 compares the stream power per meter of channel length expended daily within the channels of the two study reaches during the 1975 spring flood. The total stream power expended within the two channels during the entire flood season is approximately equal, because the greater bankfull discharge of the Bedload Trap reach compensates for the decrease in slope.

Bankfull Hydraulic Characteristics

The effects of an increased sediment load in the East Fork River were studied by comparing the hydraulic characteristics of the two study reaches at their respective bankfull discharges (table 1). In this analysis the bankfull discharge is assumed to be the effective channel-forming discharge (Wolman and Miller, 1960). The greater bankfull discharge of the Bedload Trap reach is distributed proportionally among its component parts--velocity, depth, and width. Thus, on the average, flow through the Bedload Trap reach is slightly faster, deeper, and wider than flow through the Clothesline reach. Flow resistance given by the Darcy-Weisbach friction factor also changes slightly. The higher flow velocity in the Bedload Trap reach combined with a decreased slope and only slightly deeper flow necessitates a smoother channel.

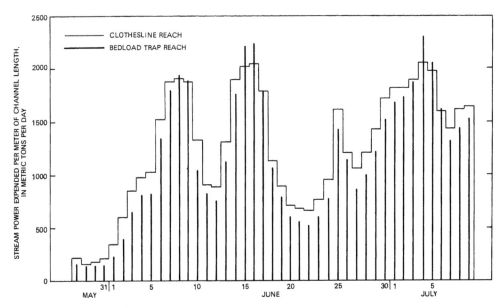

Figure 12. *Comparison of stream power expended per foot of channel length in the Clothesline and Bedload Trap reaches, East Fork River, May-July 1975.*

The stream power expended per unit length of channel (γQS) and unit area of channel ($\gamma QS/w$) also vary only slightly between the two reaches (table 1). The tendency to minimize the downstream change of these two variables has been noted before. Langbein and Leopold (1964, p. 748) observed that, "The most probable condition [of a river] would be a compromise or intermediate state between two opposing tendencies toward, (a) uniformly-distributed rate of energy expenditure, and (b) minimum total work expended in the system." The distribution of energy expenditure is uniform when the stream power per unit area is constant, and the total work is at a minimum when stream power per unit length is constant. Whereas it may not be possible to prove that the observed variances of the two stream-power variables are, in fact, mutually minimized, comparison of the mean values listed in table 1 shows that data from the East Fork River are in agreement with the above-stated conditions. Furthermore, any change in the approximate bankfull discharge (slope is independent) to improve the agreement with the tendency to minimize the variance of stream power per unit area results in a greater disagreement with the tendency to minimize the variance of stream power per unit length.

Downstream Hydraulic Geometry

The downstream adjustment of the bankfull hydraulic characteristics of the East Fork River to the increased sediment load contributed by Muddy Creek may be described concisely by the scheme of the hydraulic geometry. Leopold and Maddock (1953) stated that the variation of the hydraulic factors--mean velocity ($\overline{u}$), mean depth ($\overline{d}$), width (w), roughness (ff), and slope (S)-- varied as power functions of discharge (Q).

Downstream hydraulic exponents between the two East Fork River reaches were computed from the mean reach values of the hydraulic factors at the respective bankfull discharges (see table 1). The downstream hydraulic exponents of the East Fork River are compared in table 2 with the mean regional values, computed from measurements at approximately 100 gaging-stations in Wyoming, Montana, and Kansas (Leopold et al., 1964). The significantly larger velocity exponent of the East Fork River is the most striking difference between the values for the river and the mean regional values. Mean velocity increases downstream between the two East Fork River study reaches at a rate over twice as rapidly as the regional condition. Because the hydraulic exponents are functionally related, an exponent cannot vary without complimentary changes in one or more of the other exponents. The larger depth and smaller friction factor exponents show that the East Fork River becomes deeper and smoother between the two study reaches more rapidly than the mean regional values would indicate. Width, however, does not increase downstream as rapidly as expected.

Minimum-Variance Hydraulic Geometry

Attempts to derive the hydraulic exponent values theoretically fail because the known hydraulic relations describing flow in an alluvial stream are insufficient (Langbein, 1964). Langbein noted, however, that there was a tendency in the river system to distribute a change in the independent variables among the dependent variables so as to minimize the change required by any one variable within the physical limitations. To derive the most probable

Table 2

Downstream Exponents in the Hydraulic Geometry at Bankfull
Discharge Between the Clothesline (Q_{bkf}=16.1 m^3/sec)
and Bedload Trap (Q_{bkf}=23.2 m^3/sec) Reaches, as
Compared with the Mean Regional Values

Hydraulic exponent[1]	East Fork River	Mean regional values[2]
m	0.23	0.10
f	.46	.40
b	.31	.50
y	-.77	-.55
z	-.77	-.75

[1] $\frac{1}{u} \propto Q^m$, $\bar{d} \propto Q^f$, $w \propto Q^b$, $ff \propto Q^y$, $s \propto Q^z$.

[2] Leopold, Wolman, and Miller (1964, p. 244).

downstream hydraulic exponents which agreed with selected mean empirical exponents, Langbein minimized the variance of velocity, depth, width, and stream power per unit length and unit area. It must be stressed that there is only a tendency for the stream to adjust toward the minimum-variance exponents. The hydraulic exponents of a particular stream may differ due to constraints imposed upon this tendency by lithology, structure, and, most significantly, time. The major difficulty with applying the minimum variance concept to a specific example, such as the increased sediment load of the East Fork River, is identifying the physical constraints.

The dependent hydraulic variables appear to adjust at different rates; that is, the adjustment is time-dependent. For example, channel roughness adjusts to changes in sediment and water discharge rapidly; whereas, the vast quantity of sediment which must be moved to effect a change in the longitudinal profile requires a much longer time. To maintain a balance between the supply and transport of sediment, the readily adjustable hydraulic factors must initially absorb the entire effect of a change. The length of time required to make an adjustment is a constraint when a change in external conditions has occurred relatively recently. In the instance of the East Fork River, 40 yrs have elapsed since the increased sediment contribution of Muddy Creek without a detectable change in slope. Thus, the downstream variations in slope may be regarded as independent of the present sediment and water discharge. The measured decrease in slope from the Clothesline reach to the Bedload Trap reach corresponds to z = -0.77, which is very close to the mean empirical value of $z = -0.75$.

A second constraint, suggested by Langbein (1965), is that the bedload-transport rate per meter of width of the active movable bed was approximately equal in streams with the same discharge and bed-material size distribution. The difference in total bedload sediment transported through the Clothesline and Bedload Trap reaches was due largely to the part of the channel bed that was active. In the Clothesline reach, bedload moved only in a narrow strip, covering less than 10 % of the bed width; whereas, moving bedload, covered nearly the entire bed width in the Bedload Trap reach. As a result, concentration of bedload sediment per meter of active bed width was approximately equal in both study reaches at a given discharge. Hence:

$$I_b/Q^* \approx \text{constant},$$

where:

I_b = whole channel transport, in kg/sec; and

Q^* = water discharge over the active part of the bed, in m^3/sec.

Empirically, the bedload transport per meter of width is proportional to the cube of mean velocity $\bar{u}$, when the size of material in transport is constant. Thus:

$$I_b/W^* \propto (\bar{u})^3,$$

where:

W^* = width of active bed, in meters.

Hence:

$$I_b \propto Q^*,$$
$$I_b/W^* \propto Q^*/W^* \propto (\bar{u})^3,$$

and

$$\bar{d} \propto (\bar{u})^2.$$

Expressed as hydraulic exponents, this relationship is

$$f = 2m.$$

The hydraulic relations and constraints that were used to derive the minimum-variance downstream hydraulic geometry of the East Fork River between the Clothesline and Bedload Trap reaches are summarized in table 3. The hydraulic relations shown satisfy the requirements of continuity of discharge, the Darcy–Weisbach friction factor, and the constraint of constant concentration. Relation 4 of table 3 minimizes the variance of velocity, depth, width, stream power per unit length and stream power per unit area. (The variance of a hydraulic factor is proportional to the square of its exponent.) Thus,

Table 3

Hydraulic Relations Used to Derive the Minimum-Variance Downstream Hydraulic
Geometry of the East Fork River. [Dependent variables--velocity,
width, depth, and friction factor; constraints--uniform
sediment concentration and slope $\propto Q^{-0.77}$; minimum
variance--velocity, width, depth, stream power
per unit area and stream power per unit length]

Hydraulic relations	Equations[1]	Relations among exponents[2]
1. Continuity	$Q = \overline{u}dw$	$m + f + b = 1$.
2. Friction factor	$ff = 8gDS/\overline{u}^2$	$y = f + z - 2m$.
3. Uniform concentration	$I_b/Q^* \simeq$ constant	$m = 1/2\ f$.
4. Minimum variance		$m^2 + f^2 + b^2 + (m + f + z)^2 + (z + 1)^2 \rightarrow$ minimum.

[1]Symbols: Q, discharge; $\overline{u}$, mean velocity; $\overline{d}$, mean depth; w, width; ff, Darcy-Weisbach friction
factor; I_b, whole channel bedload-transport rate; S, slope.

[2]$\overline{u} \propto Q^m$, $\overline{d} \propto Q^f$, $w \propto Q^b$, $ff \propto Q^y$, $S \propto Q^z$.

$$m^2 + f^2 + b^2 + (1 - 0.77)^2 + (m + f - 0.77)^2 \rightarrow minimum.$$

Substituting the known equations b = 1 - m - f and f = 2m into relation 4 and setting the first derivative equal to zero yields

$$\frac{d}{dm} (23m^2 - 10.62m) = 0.$$

Solving for m, and then for the other hydraulic exponents, gives m = 0.23, f = 0.46, b = 0.31, and y = -0.77.

The computed minimum-variance exponents agree exactly with the downstream hydraulic exponents for the East Fork River shown in table 2. The hydraulic characteristics of the East Fork River have adjusted to the large sediment contribution of Muddy Creek in accordance with the minimum-variance principle. The adjustment is not complete, however, as slope has not changed. The minimum-variance concept suggests that the slope of the East Fork River will eventually adjust, over an as-yet-unknown length of time, until the effects of the increased sediment load are distributed mutually among roughness, depth, width, velocity, and slope.

CONCLUSIONS

The sediment load of the East Fork River in 1975 increased from approximately 200 t transported past the Clothesline reach upstream from the mouth of Muddy Creek to slightly more than 3,200 t at the bedload trap, 4.8 km downstream. This large increase in load was primarily the contribution of Muddy Creek. Small ephemeral gullies and cutbanks each supplied a minor quantity of sediment. Muddy Creek, however, added an insignificant water discharge to the East Fork River--less than 3 % of the latter's flood discharge. Thus, the increased sediment load of the East Fork River downstream from the confluence of Muddy Creek was transported without an appreciable increase in water discharge. The hydraulic characteristics of the East Fork River downstream from the mouth of Muddy Creek have adjusted to the increased sediment load in order to maintain a quasi-equilibrium channel.

In spite of the considerable length of time required for complete adjustment of the dependent variables, a quasi-equilibrium channel is commonly maintained throughout the period of adjustment. Initially, only roughness and depth adjust to accommodate a change in the independent variables. Then, after a number of years, channel width gradually changes until the hydraulic adjustment is distributed mutually among roughness, depth, and width. This is the condition in which the East Fork River is observed today. Slope will adjust eventually if the independent variables remain approximately constant for several centuries or more, so that eventually roughness, depth, width, and slope are mutually adjusted. Thus, over a considerable length of time, perhaps thousands of years, the effects of a change in the independent variables are distributed among the variables so as to minimize the adjustment made by any one variable; that is, the hydraulic geometry tends toward the minimum-variance condition.

Because the observed hydraulic characteristics of a stream are time-dependent, they are not uniquely determined for a given water and sediment discharge. This fact, no doubt, accounts for a great deal of the scatter common-ly found in hydraulic data, especially river slope. Because slope requires the longest period of adjustment, it is least likely to be adjusted to the existing supply of water and sediment. Therefore, the adjustment of river slope often is incomplete, as much a relic of the past as it is a product of the present.

ACKNOWLEDGEMENTS

This research was accomplished with the assistance and encouragement of many individuals. I am indebted to Luna B. Leopold for many thoughtful discussions and the unique measurements of bedload transport that made this investigation possible. His counsel and guidance throughout have been in-valuable. W.W. Emmett contributed most of the field equipment, as well as continual encouragement. I was ably assisted in the field work by Tom Lisle, Paul Butler, William Haible, and Robert Myrick.

REFERENCES

Andrews, E.D., 1977, Hydraulic adjustment of an alluvial stream channel to the supply of sediment, western Wyoming [Ph.D. dissertation]: Berkeley, University of California, 152 p.

_____, 1979 (in press), Scour and fill in an alluvial sediment chan-nel, East Fork River, Western Wyoming: U.S. Geological Survey Pro-fessional Paper 1117.

Bradley, W.H., 1964, Geology of Green River Formation and associated Eocene rocks in southwestern Wyoming and adjacent parts of Colorado and Utah: U.S. Geological Survey Professional Paper 496A, 86 p.

Emmett, W.W., 1979, A field calibration of the sediment-trapping characteris-tics of the Helley-Smith Bedload Sampler: U.S. Geological Survey Open File Report 79-411, 96 p.

Langbein, W.B., 1964, Geometry of river channels: American Society of Civil Engineers Proceedings, Journal of Hydraulics Division, v. 90, HY2, p. 301-312.

_____, 1965, Geometry of river channels: American Society of Civil Engineers Proceedings, Journal of Hydraulics Division, v. 91, HY3, p. 297-313.

Langbein, W.B., and Leopold, L.B., 1964, Quasi-equilibrium states in channel morphology: American Journal of Science, v. 262, p. 782-794.

Leopold, L.B., and Emmett, W.W., 1976, Bedload measurements, East Fork River, Wyoming: Proceedings of the National Academy of Science, v. 73, p. 1000-1004.

_____, 1977, 1976 bedload measurements, East Fork River, Wyoming: Proceedings of the National Academy of Science, v. 74, p. 2644-2648.

Leopold, L.B., and Maddock, T., 1953, The hydraulic geometry of stream channels and some physiographic implications: U.S. Geological Survey Professional Paper 252, 57 p.

Leopold, L.B., Wolman, M.G., and Miller, J.P., 1964, Fluvial processes in geomorphology: San Francisco, Freeman, 522 p.

Mackin, J.H., 1948, Concept of the graded river: Geological Society of America Bulletin, v. 59, p. 463-512.

Miller, J.P., 1958, High mountain streams -- Effects of geology on channel characteristics and bed material: New Mexico Bureau of Mines and Mineral Resources Memoir 4, 51 p.

Wolman, M.G., 1955, The natural channel of Brandywine Creek, Pennsylvania: U.S. Geological Survey Professional Paper 271, 56 p.

Wolman, M.G., and Miller, J.P., 1960, Magnitude and frequency of forces in geomorphic processes: Journal of Geology, v. 68, p. 54-74.

DISTRIBUTION OF BOUNDARY SHEAR

STRESS IN RIVERS

JAMES C. BATHURST

Formerly of
Engineering Research Center
Colorado State University
Fort Collins, Colorado

ABSTRACT

Field measurements over cross sections in cobble-bed rivers have allowed patterns of boundary shear stress distribution to be delineated at various discharges. In straight pool reaches the cross-sectional distribution of shear stress is characterized by peaks and troughs. These may result from the action of multicell, stress-induced, secondary circulation systems which create alternate regions of upwelling and downwelling flow. Upwelling and downwelling flow produces alternate regions of low and high shear stress respectively. At sections just upstream of riffles the shear stresses have a major peak in the center of the channel. This peak may be caused by the accentuating effect of flow acceleration on the shear stress peak associated naturally with the core of maximum velocity. At sections across bends, shear stress peaks are associated with the core of maximum velocity and with downwelling of flow caused by skew-induced secondary circulation.

Little change in shear stress with discharge was observed over cross sections on straight reaches, but at bends the positions and relative magnitudes of the peaks change with discharge.

Uniformity of shear stress distribution over the channel cross-section generally increases as Reynolds number increases. This is because at high Reynolds number the capacity of secondary circulation to decrease the uniformity of distribution is relatively reduced.

Scouring can occur where there is a longstream increase in shear stress. At bends the positions of such regions change with discharge, so the position of maximum scour along a channel bend is likely to depend on the dominant range of discharges.

INTRODUCTION

Changes in river channel shape and pattern are brought about by erosion and deposition. These processes are linked to the magnitude and distribution of the boundary shear stress in the channel and the variation of those parameters with discharge. It is therefore necessary to understand the patterns of shear stress distribution if channel adjustments are to be successfully predicted.

Various laboratory studies of shear stress distribution have been made but, until recently, there have been few field studies. Also, there is a lack of knowledge concerning variations in the distribution with discharge. The object of this study, therefore, is the employment of field data, collected at sections across rivers with cobble beds, to delineate patterns of shear stress distribution and the variations which occur with changes of discharge.

THEORY

The magnitude of the boundary shear stress which acts on a fluid flowing past a boundary is directly related to the near-wall velocity gradient in the fluid. Higher shear stresses appear where the near-wall isovels of the primary flow are relatively compressed. This happens in regions of relatively high velocity and where there is downwelling caused by secondary circulation. Isovel compression may also occur where there is acceleration of flow. The near-wall fluid responds faster to acceleration than does fluid further from the wall (Clauser, 1956), so the overall velocity profile becomes more uniform and the near-wall velocity gradient becomes steeper.

Distribution of shear stress at a section is therefore related to the pattern of flow. In particular the distribution varies according to the relative influences of any secondary flows (dependent on channel cross-sectional shape, pattern and roughness) and of the primary flow (dependent on discharge or Reynolds number). The primary flow affects the shear stress distribution through the creation of regions of high and low velocity and of regions of acceleration and deceleration. The secondary flow affects the distribution through the creation of regions of upwelling and downwelling and by moving the positions of shear stress peaks across the channel.

Secondary circulation is stress-induced at straight reaches and skew-induced at bends (Prandtl, 1952; Perkins, 1970; Bradshaw, 1971). Stress-induced circulation develops where there is anisotropic turbulence and a non-uniform distribution of boundary shear stress. Skew-induced circulation develops where the velocity profile is skewed by, for example, centrifugal forces at bends. In both cases secondary flow effects are stronger over rough boundaries than over smooth boundaries (Ippen et al., 1962; Launder and Ying, 1972). However, the influence of secondary flow relative to that of the primary flow changes with discharge or Reynolds number. In straight reaches the stress-induced circulation becomes steadily less effective as Reynolds number increases (Leutheusser, 1963). At bends skew-induced circulation seems to be relatively weak at low and high discharges and strongest at medium discharges (Bhowmik and Stall, 1978; Bathurst, Thorne and Hey, in press). In all cases then, primary flow effects are likely to dominate shear stress distributions at high discharges.

FIELD MEASUREMENTS

Most of the data analyzed in this paper were gathered from British rivers by the author. However, a few data collected at bends of concrete irrigation channels in Colorado by Al-Shaikh Ali (1964) are also analyzed. The latter data were initially gathered for reasons other than delineation of shear stress distributions.

River Sites

Measurements were made on the upper River Severn in Wales and the upper
River Swale in England. The bed material in each river is coarse gravel and
cobbles. The banks of the Severn are composed of fine alluvium overlying
coarse gravel and those of the Swale consist of sand and gravel. Railway
Straight sections 1 and 2 and the Morfodian section on the Severn and section 200 on the Swale are on generally straight reaches of pools. Llanidloes
section and Maes Mawr section 3 on the Severn and section 100 on the Swale
are on generally straight reaches just upstream of riffles. (Maes Mawr section
3 is on the inflexion point between two meander bends.) The bend sites are
Rickety Bridge section (apex), Llandinam sections 2 (apex) and 3 (between
apex and exit) and Maes Mawr sections 1 (apex) and 2 (exit) on the Severn.
Details of the sites are given in table 1 and elsewhere (Bathurst, 1977).

Boundary Shear Stress Calculations

Boundary layer theory shows that for the near-wall region of a flow (the
bottom 10 % to 15 % of the flow), point mean velocity, u, at a distance y from
the wall is given by the equation:

$$\frac{u}{u_*} = \frac{2.305}{K} \log(y) + \text{constant} \tag{1}$$

where

$$u_* = \left(\frac{\tau_o}{\rho}\right)^{0.5} \tag{2}$$

and τ_o = point boundary shear stress; u_* = point shear velocity; ρ = the
density of the fluid; and K = the von Kármán constant, which has a value of
0.41 (Bradshaw, 1971).

Strictly, equation (1) applies only to two-dimensional flows. However, it
appears that the near-wall characteristics of three-dimensional flows can be
approximately described by two-dimensional near-wall laws (Pierce and Zimmerman, 1973). Thus, equation (1) applies reasonably well even in the presence
of secondary circulation (Eskinazi and Yeh, 1956; Aly, Trupp and Gerrard,
1978; Bridge and Jarvis, 1977).

Equations (1) and (2) show that point values of shear stress can be calculated from velocity measurements. Velocity profiles were therefore obtained
at several verticals at each river section, the sections being perpendicular to
both banks in the straight reaches and to the outer bank at bends. Each
velocity measurement was made with an Ott C-31 current meter during a period
of one minute. The results are accurate to within 2-10 %, depending on the
velocity and depth of flow (Bathurst, 1977). The vertical intervals between
points were 0.01 to 0.05 m near the bed and 0.1 to 0.2 m near the water surface. The verticals were at horizontal intervals of 0.1 to 0.5 m near the
outer banks of bends and at intervals of 1 to 3 m elsewhere. Where direct
measurements of velocity in the bottom 10 % of the flow could not be made,
values were obtained by extrapolation of the other measured points on a linear

Table 1

Parameters of Planform and Bed Sediment at River Sites

Site	Radius of curvature of outer bank at bends (m)	Arc angle at bends (deg)	Median axis (mm) of sediment bigger than or equal to n% of median axes by count	
			n = 84	n = 50
Llanidloes (straight riffle)	-	-	45.4	19.25
Morfodian (straight pool)	-	-	103.0	57.0
Railway Straight 1 (straight pool)	-	-	39.0	19.5
Railway Straight 2 (straight pool)	-	-	45.0	24.5
Rickety Bridge (bend apex)	44.0	50	93.0	63.0
Llandinam 2 (bend apex)	70.5	62	60.0	40.0
Llandinam 3 (bend, apex to exit)	70.5	62	57.5	36.3
Maes Mawr 1 (bend apex)	95.0	38	50.0	31.5
Maes Mawr 2 (bend exit)	95.0	38	57.0	32.0
Maes Mawr 3 (straight, riffle)	-	-	47.0	27.5
Swale 200 (straight pool)	-	-	139.0	78.0
Swale 100 (straight, riffle)	-	-	142.0	85.0

graph of velocity against depth. This method was also used to augment the data which were gathered in that region. A comparison of shear stresses derived from velocities measured in the bottom 10 % of the flow at eight profiles with those derived from extrapolations of the outer regions of those profiles, showed the resulting difference to be less than 10 %.

For each vertical, between five and ten values of mean point velocity, u, were plotted against the logarithm of y over the bottom 10-15 % of the flow. (The origin for y was assumed to be at the top of the boundary roughness.) Shear velocity and shear stress were obtained from the gradient of the result-ing straight line using equations (1) and (2). Mean values of those para-meters were then obtained by graphical integration of the point values. The quantities so obtained are not related to each other by equation (2) since the distributions of shear stress are not uniform.

Irrigation-Channel Sites

Velocity profiles were obtained by Al-Shaikh Ali (1964) at sections at bends of concrete-lined, trapezoidal irrigation channels in Colorado. For this study, data from sections at the apices and exits of Bends 1, 2, 6, and 7 are analyzed. At Bends 1 and 2 the bed slope is 0.0002 and the bed width is 2.13 m. At Bends 6 and 7 the bed slope is 0.0013 and the bed width is 3.66 m. In all cases the sides lie at an angle of $\tan^{-1}(0.8)$, or 38.66°, to the horizontal.

The profiles are not as detailed as those measured in the rivers, so were analyzed less rigorously. Taking the two velocity measurements, u_1 and u_2, nearest the wall (at perpendicular distances y_1 and y_2 respectively from the wall) manipulation of equations (1) and (2) allowed point values of shear stress to be calculated as:

$$\tau_o = \rho \left[\frac{(u_2 - u_1)}{5.62 \log(y_2/y_1)} \right]^2 \qquad (3)$$

The analysis then proceeded as with the river data, except that for all the integrations required in calculating mean parameters, a computer method based on the trapezoidal rule was employed. The results are likely to be more in error than are those for the river sections, partly because of the method of analysis and partly because the two velocity measurements nearest the wall did not always lie in the bottom 10 % of the flow. Individual results therefore may not be very accurate, but the trends are believed to be correct.

RESULTS

Some of the isovel patterns and their associated shear stress distributions (based on the ratio of point shear stress to mean shear stress) are presented in figures 1-8. Space considerations do not allow all the data to be presented in this fashion, but the remaining data are given elsewhere (Bathurst, 1977; Bathurst, Thorne and Hey, in press). Distributions obtained at given sections at similar discharges but on separate occasions are in good agreement (figs. 7 and 8), indicating that the method used in their derivation is reliable. Details of the river flows are given in table 2, the river shear stress distri-butions appear in table 3, and table 4 lists the data for the Colorado irriga-tion channels.

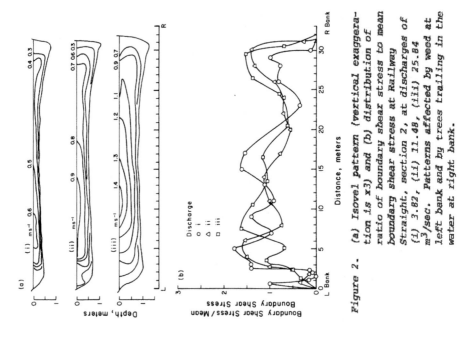

Figure 2. (a) Isovel pattern (vertical exaggeration is x3) and (b) distribution of ratio of boundary shear stress to mean boundary shear stress at Railway Straight, section 2, at discharges of (i) 3.82, (ii) 11.48, (iii) 25.84 m³/sec. Patterns affected by weed at left bank and by trees trailing in the water at right bank.

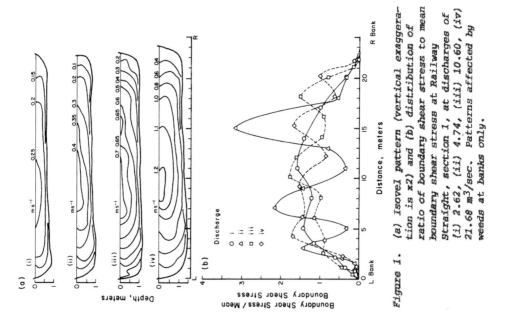

Figure 1. (a) Isovel pattern (vertical exaggeration is x2) and (b) distribution of ratio of boundary shear stress to mean boundary shear stress at Railway Straight, section 1, at discharges of (i) 2.62, (ii) 4.74, (iii) 10.60, (iv) 21.68 m³/sec. Patterns affected by weeds at banks only.

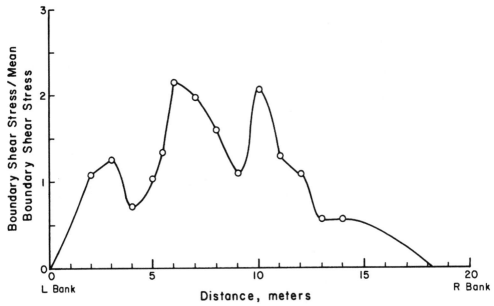

Figure 3. Distribution of ratio of boundary shear stress to mean boundary
shear stress at Swale section 200 at discharge of 0.71 m^3/sec.
Isovel patterns are not readily available for publication.

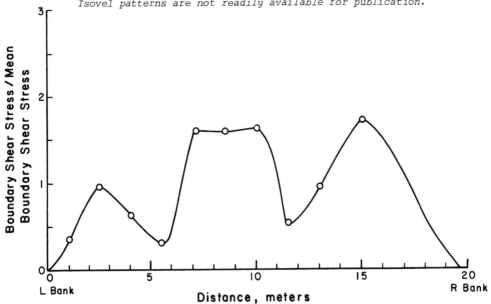

Figure 4. Distribution of ratio of boundary shear stress to mean boundary
shear stress at Swale section 100 at discharge of 7.84 m^3/sec.
Isovel patterns are not readily available for publication.

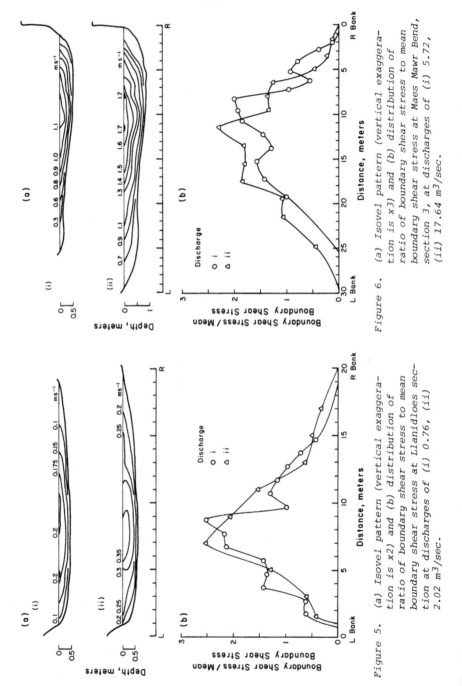

Figure 6. (a) Isovel pattern (vertical exaggeration is x3) and (b) distribution of ratio of boundary shear stress to mean boundary shear stress at Maes Mawr Bend, section 3, at discharges of (i) 5.72, (ii) 17.64 m³/sec.

Figure 5. (a) Isovel pattern (vertical exaggeration is x2) and (b) distribution of ratio of boundary shear stress to mean boundary shear stress at Llanidloes section at discharges of (i) 0.76, (ii) 2.02 m³/sec.

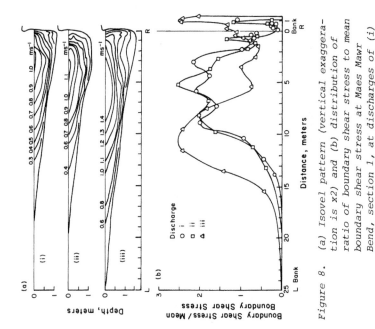

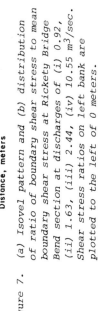

Figure 8. *(a) Isovel pattern (vertical exaggeration is x2) and (b) distribution of ratio of boundary shear stress to mean boundary shear stress at Maes Mawr Bend, section 1, at discharges of (i) 5.22, (ii) 6.84, (iii) 15.48 m3/sec. Shear stress ratios on right bank are plotted to the right of 0 meters.*

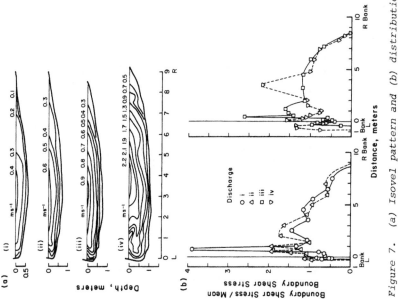

Figure 7. *(a) Isovel pattern and (b) distribution of ratio of boundary shear stress to mean boundary shear stress at Rickety Bridge Bend section at discharges of (i) 0.92, (ii) 1.63, (iii) 2.44, (iv) 10.55 m3/sec. Shear stress ratios on left bank are plotted to the left of 0 meters.*

Table 2

Parameters of Flow and Channel Shape Determined
from Velocity Measurements at River Sites

Site	Discharge (m³/sec)	Mean velocity (m/sec)	Water surface slope (× 10³)	Hydraulic radius (m)	Cross-sectional area (m²)	Surface width (m)	Reynolds number $\overline{U}R/\nu$ (× 10⁻⁴)
Llanidloes	0.76	0.144	0.010	0.293	5.30	18.0	3.20
	2.02	0.251	0.018	0.415	8.05	19.0	7.34
Morfodian	0.80	0.280	0.940	0.166	2.86	17.0	3.55
Railway Straight 1	2.62	0.188	0.160	0.601	13.94	22.4	8.19
	4.74	0.276	0.215	0.735	17.20	22.3	16.56
	10.60	0.482	0.290	0.927	22.00	22.55	35.18
	21.68	0.783	0.380	1.125	27.70	23.2	65.74
Railway Straight 2	3.82	0.311	0.440	0.388	12.31	31.2	9.85
	11.48	0.608	0.622	0.584	18.87	31.3	28.63
	25.84	0.943	0.735	0.826	27.40	31.9	56.44
Rickety Bridge	0.92	0.276	0.470	0.371	3.32	8.75	7.82
	1.63	0.394	0.656	0.440	4.14	8.8	11.52
	2.44	0.545	0.790	0.498	4.32	8.4	22.43
	10.55	1.351	1.300	0.760	7.90	9.1	81.77
Llandinam 2	1.12	0.489	2.900	0.242	2.30	9.4	9.03
	2.36	0.563	2.900	0.342	4.20	12.0	13.73
Maes Mawr 1	5.22	0.563	1.120	0.504	9.27	18.0	20.27
	6.84	0.633	1.220	0.532	10.80	19.4	24.42
	15.48	0.935	1.544	0.620	16.56	25.0	37.64
Maes Mawr 2	1.20	0.288	0.555	0.287	4.16	14.4	6.51
	14.30	1.041	1.510	0.515	13.74	26.2	37.76
Maes Mawr 3	5.72	0.741	1.159	0.310	7.72	24.5	16.41
	17.64	0.964	1.610	0.618	18.36	29.3	41.95
Swale 200	0.71	0.123	0.235	0.314	5.76	18.2	3.11
Swale 100	7.84	0.769	1.800	0.505	10.20	19.6	27.73

Table 3

Parameters of Boundary Shear Stress Distribution at River Sites*

Site	Mean shear stress (N/m^2)	Mean shear velocity (m/sec)	Mean velocity over mean shear velocity $\overline{U}/\overline{u_*}$	Mean value of u/u_* $\overline{(u/u_*)}$	$\left\| (\overline{U}/\overline{u_*}) - \overline{(u/u_*)} \right\|$ over $(\overline{U}/\overline{u_*})$
Llanidloes	0.236	0.0139	10.37	7.64	0.263
	0.618	0.0225	11.18	9.50	0.150
Morfodian	0.758	0.0251	11.16	8.54	0.235
Railway	0.369	0.0183	10.27	9.63	0.062
Straight 1	0.588	0.0222	12.43	13.27	0.068
	2.144	0.0440	10.96	10.77	0.017
	5.416	0.0687	11.41	11.63	0.019
Railway	0.538	0.0214	14.53	16.49	0.135
Straight 2	2.102	0.0430	14.16	14.74	0.041
	4.666	0.0657	14.36	14.46	0.007
Rickety Bridge	0.446	0.0196	14.05	11.49	0.182
	0.814	0.0273	14.44	11.88	0.177
	1.136	0.0327	16.63	13.88	0.165
	13.823	0.1072	12.60	12.75	0.012
Llandinam 2	2.203	0.0449	10.88	9.11	0.163
	3.081	0.0537	10.48	8.25	0.213
Maes Mawr 1	2.234	0.0416	13.51	8.70	0.356
	1.853	0.0370	17.10	11.08	0.352
	5.500	0.0606	15.43	11.10	0.281
Maes Mawr 2	0.403	0.0188	15.24	11.43	0.250
	6.327	0.0725	14.36	11.39	0.207
Maes Mawr 3	6.009	0.0687	10.79	7.56	0.299
	10.165	0.0904	10.66	9.56	0.103
Swale 200	0.127	0.0104	11.85	9.47	0.201
Swale 100	2.663	0.0463	16.58	14.76	0.110

*Note: Sections are ranked as in Table 2

105

Table 4

Parameters of Channel Planform, Flow and Shear Stress
Uniformity at Colorado Sites

Bend	Radius of curvature of centerline (m)	Arc angle (deg)	Discharge (m^3/sec)	Reynolds number $\overline{UR/\nu} \times 10^{-4}$ at		$\left\| (\overline{U/u_*}) - (\overline{u/u_*}) \right\|$ over $(\overline{U/u_*})$ at	
				apex	exit	apex	exit
1	116.4	28.67	6.09	82.7	83.0	0.107	0.858
			10.17	116.0	116.6	0.210	0.275
			15.04	140.2	139.0	0.170	0.268
2	124.7	28.63	6.15	91.7	91.9	0.086	0.210
			10.20	122.4	124.9	0.221	0.150
			15.04	148.2	146.8	0.138	0.253
6	43.7	73.15	8.58	111.7	112.7	0.181	0.032
			13.57	148.7	148.7	0.254	0.210
			19.20	191.8	194.2	0.153	0.133
			22.88	221.6	228.0	0.246	0.229
			32.46	279.2	284.7	0.102	0.140
7	87.3	48.63	5.52	82.5	83.9	0.403	0.389
			10.68	133.3	135.2	0.258	0.285
			18.97	199.5	201.9	0.231	0.230
			18.80	203.7	–	0.304	–
			22.20	224.3	224.3	0.253	0.222
			25.55	246.7	–	0.178	–
			25.94	249.2	248.5	0.225	0.232
			31.32	282.3	337.2	0.127	0.307
			36.76	314.2	317.2	0.210	0.128

Straight Reaches on Pools

The shear stress distributions at Railway Straight sections 1 and 2, Swale section 200 (figs. 1-3) and Morfodian section (not shown) have several peaks but in most cases no obvious maximum. There is little variation with discharge in the magnitude of the ratios of the peak shear stresses to the mean shear stresses at individual sections and, with one or two exceptions, the ratios have values of 1.4 to 2.1. The positions of the peaks vary with discharge but not in a decipherable fashion.

A possible cause of the pattern could be a multicell system of stress-induced secondary circulation triggered by deformations of flow (Perkins, 1970). In such a system, each cell rotates in the opposite sense to that of its neighbor, so there are alternate regions of upwelling and downwelling with associated troughs and peaks respectively in the shear stress distribution. At the river sections the measured interpeak distances (which would correspond to two cell widths) range from four to thirteen times the mean depth at each section. This agrees well with cell widths of two to four thicknesses of flow measured at corners of ducts by Brederode and Bradshaw (1978).

Straight Reaches Upstream of Riffles

At Llanidloes section, Maes Mawr section 3 and to some extent, Swale section 100, the shear stress distribution has a central peak with smaller shoulder peaks (figs. 4-6). (The relatively pronounced shoulder peaks at Swale section 100 may be due to the fact that section 100 lies just downstream of section 200.) The central peak seems to be generally associated with the core of maximum velocity. The pattern is maintained over the observed range of discharges, although the position of the peak shifts slightly. The ratio of the peak shear stress to the mean shear stress has values in the range of 1.6 to 2.5.

The cause of the distribution may be related to the changes in the flow pattern which occur as the bed rises at the approach to a riffle. The flow must either accelerate or, if the width increases, undergo lateral diversions. Acceleration of the flow would tend to accentuate the compression of the isovels in the region of the core of maximum velocity, thereby causing a pronounced shear stress peak. Alternatively, lateral diversion of the flow might trigger a secondary circulation system which could produce the observed shear stress distribution.

Bends

Analysis of the secondary circulation patterns at the bends on the River Severn has been carried out elsewhere and has given some indication of their effect on the shear stress distribution (Bathurst et al., 1977; Bathurst et al., in press). A deeper analysis of the effects is presented here.

Shear stress peaks over the channel cross-section are associated with the core of maximum velocity and with downwelling near the outer bank (figs. 7 and 8). The downwelling occurs at the junction of the main skew-induced secondary circulation cell typical of bends and the small circulation cell (rotating in the opposite sense to that of the main cell) which occurs next to steep outer banks (Bathurst et al., 1977; Bathurst et al., in press). The ratios of the peak shear stresses associated with the core of maximum velocity to the mean shear stresses lie in the range 1.5 to 2.5. The ratios for the shear stress peaks associated with downwelling vary from about 0.75 to 4, depending on the strength of the outer bank cell.

Considerable variations in the pattern occur as discharge changes. To explain these it is necessary to first describe the characteristics of the flow through a bend. At the entrance to a bend there is a tendency for free vortex flow to appear, keeping the core of maximum velocity near the inner bank. In the bend secondary circulation develops and breaks down the free vortex. At a point which depends on the strength of the circulation and the bend arc angle, the circulation carries the core of maximum velocity across the channel towards the outer bank. Thus the region of high shear stress lies near the inner bank at the bend entrance and then crosses towards the outer bank somewhere in the bend or just downstream of the bend (Ippen and Drinker, 1962; Al-Shaikh Ali, 1964; Hooke, 1975; Bridge and Jarvis, 1976; Naas, 1977).

The region of high shear stress usually lags behind the core of maximum velocity in crossing the channel (fig. 8). This is partly because the secondary circulation encourages the core of high-velocity surface flow to cross the channel but inhibits the similar movement of the high-velocity near-wall flow (which is related to the region of high boundary shear stress). (This is not to imply that the near-wall fluid moves towards the outer bank.) Also inertial effects, related to the movement of the flow from the shallow inner region to the deeper outer region of the bend, tend to decrease the near-wall velocity gradient. This slows the crossover of the region of high boundary shear stress (Dietrich et al., in press; J.D. Smith, University of Washington, personal communication, 1979).

Field measurements have shown that skew-induced secondary circulation is strongest at medium discharges (Bhowmik and Stall, 1978; Bathurst et al., in press). Therefore the crossover region of the high shear stress should lie further upstream at medium discharges than at low or high discharges. Hooke's (1975) laboratory measurements, which show that the crossover region moves downstream as discharges become high, appear to confirm this. Thus at any given section the shear stress peak associated with the core of maximum velocity should lie nearer to the outer bank at medium discharges than at low or high discharges. This appears to be borne out by the field measurements which show the position of that shear stress peak to move further from the outer bank as discharges become high (fig. 8). However, this should not be construed to mean that the shear stress at the outer bank is higher at medium than at high discharges. Figures 7 and 8 show that the overall increase in shear stress with discharge means that the bank shear stress is higher at high discharges than at medium discharges, despite the removal of the shear stress peak from the vicinity of the bank.

Shear stress distributions in the Colorado irrigation channels show a similar trend. Figure 9 shows that as discharge increases, the shear stress peaks generally move further from the outer bank. The trend is most obvious at Bend 6 which is the bend where the flow is least influenced by upstream disturbances. The flows at the other bends are affected by the secondary circulation patterns of upstream bends with opposite senses of curvature. Ippen et al. (1962) have shown that, in such cases, the shear stress peak in the downstream bend is pushed close to the inner bank throughout the bend. Where the upstream bend is of the same sense of curvature, the shear stress peak lies close to the outer bank in the downstream bend. The effect of upstream bends is most noticeable in channels of uniform cross sections. In cobble-bed rivers there is often a riffle between bends which largely destroys the secondary circulation of the upstream bend.

According to the model outlined there should be a general longstream increase in shear stress along the outer half of a bend. This is certainly shown by the laboratory results of Ippen and Drinker (1962) and Hooke (1975). Comparison of shear stress distributions obtained at similar discharges (at approximately the threshold of sediment movement) at Maes Mawr section 1 (the bend apex) and section 2 (the bend exit) also suggests that

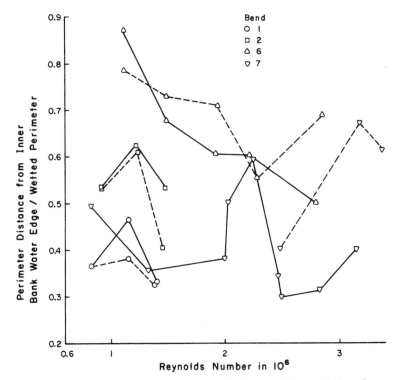

Figure 9.   *Variation with Reynolds number of the position of the shear stress peak associated with the core of maximum velocity at the sections across the Colorado irrigation channels. Solid line indicates conditions at a bend apex, dashed line indicates conditions at a bend exit. Conditions at Bend 7 exit and apex are very similar at the lower Reynolds numbers.*

such an increase exists (fig. 10). Similarly figure 9 shows that at Bend 6 of the Colorado channels, at least at high discharges when presumably the crossover region of high shear stress is in the downstream part of the bend, the shear stress peak lies nearer to the outer bank at the bend exit than at the apex. The result is a longstream increase in shear stress near the outer bank.

The relative strengths of the shear stress peaks associated with the core of maximum velocity and with downwelling vary with discharge. At low discharges, because primary and secondary flow effects are both weak, either peak can be the maximum for the section. Thus at Rickety Bridge Bend, where the outer bank cell is strong, the peak associated with downwelling is the maximum (fig. 7). At Maes Mawr Bend, where the outer bank cell is weaker, the peak associated with the core of maximum velocity is the maximum (fig. 8). The same state may occur at medium discharges since primary and secondary flow effects are then both stronger. However, at high discharges primary flow effects are stronger than secondary flow effects, so the shear

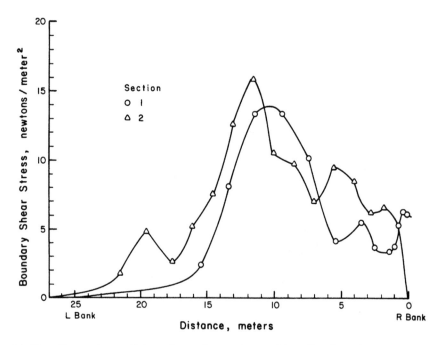

Figure 10. *Comparison of boundary shear stress distribution at Maes Mawr Bend, sections 1 and 2, at discharges of 15.48 and 14.30 m^3/sec respectively. Right bank is the outer bank.*

stress peak associated with the core of maximum velocity is the maximum at all the sections studied.

Uniformity of Boundary Shear Stress

In order to investigate the cross-channel uniformity of the shear stress distributions with discharge, a parameter of uniformity was defined as the difference between $\overline{U}/\overline{u}_*$ and the mean value of $\overline{u}/u_*$ at a section. $\overline{U}$ = the mean velocity of flow at the section; $\overline{u}_*$ = the mean shear velocity at the section; $\overline{u}$ = the mean depth-integrated velocity at a vertical; and u_* = the point shear velocity at the base of that vertical. Where the distributions of velocity and boundary shear stress are uniform, the difference is zero and the uniformity parameter is zero. Values of the absolute difference as a percentage of $\overline{U}/\overline{u}_*$ [i.e., $|(\overline{U}/\overline{u}_*) - \overline{(\overline{u}/u_*)}| / (\overline{U}/\overline{u}_*)$, where $\overline{(\overline{u}/u_*)}$ is the mean value of $\overline{u}/u_*$] were calculated (Table 3).

Figure 11·shows the variation of the uniformity parameter with Reynolds number, $\overline{U}R/\nu$, where R = hydraulic radius and ν = the kinematic viscosity of water. A decrease in the uniformity parameter represents an increase in the uniformity of the shear stress distribution across the section. Because of the uncertainty attached to the Colorado channel results, the trends for those channels are illustrated by bands rather than lines.

In all cases there is an increase in uniformity as Reynolds number increases. The points for the straight reaches (apart from Maes Mawr section 3) lie approximately about the eye-fitted line:

$$\frac{\left|\left(\dfrac{\bar{U}}{u_*}\right) - \left(\overline{\dfrac{\bar{u}}{u_*}}\right)\right|}{\left(\dfrac{\bar{U}}{u_*}\right)} = \frac{10^4}{\left(\dfrac{\bar{U}R}{\nu}\right)} \tag{4}$$

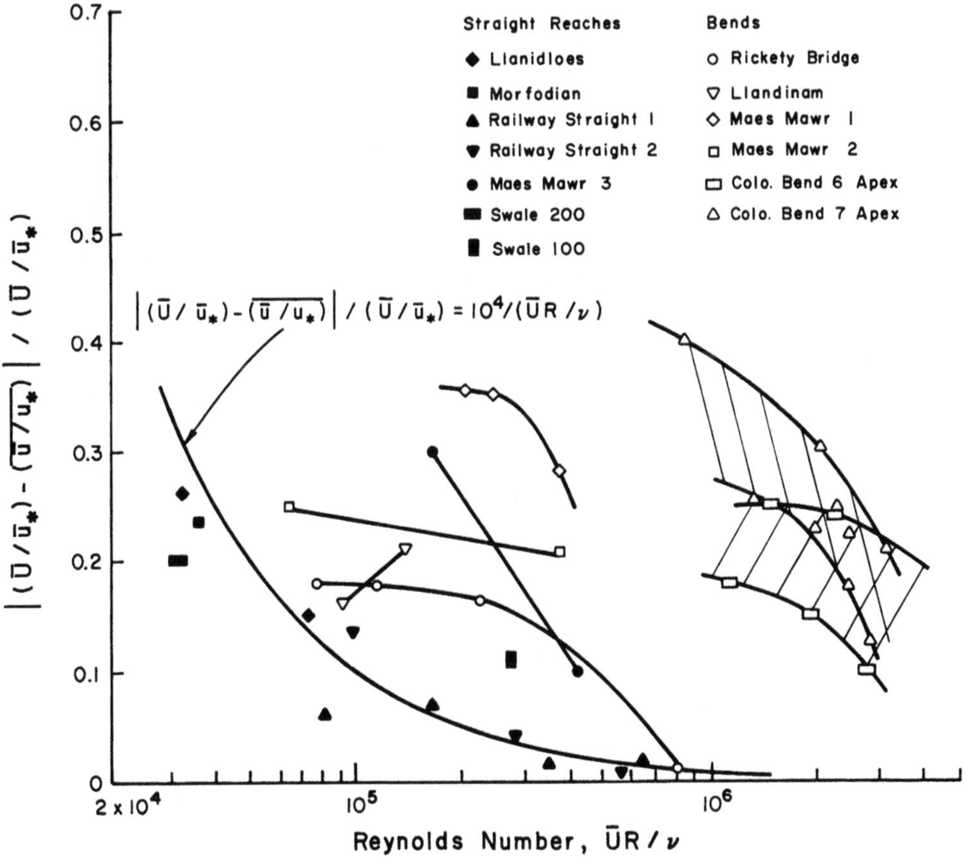

Figure 11. Variation with Reynolds number of the parameter of uniformity of boundary shear stress distribution. Bands in right-hand part of graph represent trends for the apices of the Colorado irrigation channels. In order to avoid confusion, data points for the bend exits of the Colorado channels are not presented.

111

The exception of the Maes Mawr results may be related to the fact that section 3 lies just downstream of a bend. Skew-induced secondary circulation from that bend may still have some effect on the uniformity of shear stress distribution at the section.

Equation (4) is very tentative and is chosen for convenience rather than mathematical accuracy. However, since the points for the straight reaches appear to lie about a common line, the factors influencing the shear stress distribution in straight reaches may depend purely on Reynolds number. This in turn suggests that such factors are flow effects, and that other influences such as channel cross-sectional shape (at least for large ratios of width to depth) are negligible. Support for this concept is provided by the observations of Rajaratnam and Muraldihar (1969) which show that in channels such as those of this study (approximately rectangular with width-to-depth ratios greater than about 15), shape has little effect on the shear stress distribution. Further, if the nonuniformity in the shear stress distribution is connected with the effects of stress-induced secondary circulation, then equation (4) mirrors the decrease in influence of such circulation which is expected to accompany an increase in Reynolds number (Leutheusser, 1963).

At bends the uniformity of the shear stress distribution is affected by skew-induced secondary circulation as well as by Reynolds number. Different bends support different degrees of skewing, so the uniformity of distribution varies between bends even at the same Reynolds number. More data are needed to show how the various parameters of bend cross section and pattern affect the distribution. In general, though, it seems likely that the stronger the secondary circulation in any given flow, then the higher the Reynolds number at which primary flow effects become most important and therefore the slower the rate of increase in uniformity with Reynolds number.

The greatest difference in uniformity between distributions in straight reaches and bends occurs at medium discharges. Skew-induced secondary circulation then has its strongest effect on shear stress distributions at bends. At very low and high discharges, when skew-induced circulation is relatively weak, shear stress distributions in straight reaches and bends have comparable degrees of uniformity.

APPLICATIONS

Shear stress distributions can indicate areas of potential scour and deposition and thence the likely course of channel adjustments brought about by natural or artificial means.

For scour to occur at a point, the flow away from that point must be able to carry more sediment than the flow approaching the point. This is possible only where there is a longstream increase in shear stress. Similarly, a longstream decrease in shear stress should result in deposition.

At river bends the longstream increase of shear stress along the outside of the bend implies that scour should occur there. The longstream decrease of shear stress along the inside of the bend implies that that region should be one of deposition. Such a pattern is illustrated at Maes Mawr Bend.

Tracer measurements at that bend reported by Thorne and Lewin (1979) show that, at the bend entrance, scour of sediment is most pronounced at the channel center or nearer the inner bank. At the bend apex, scour occurs mainly near the outer bank. Downstream of the bend there is no scouring but there is a uniform longstream movement of sediment. The longstream increase in shear stress near the outer bank. evident in figure 10, presumably accounts for the scour hole near the bend apex. In the center of the channel shear stresses are higher, but there is a less significant longstream increase and consequently little scour, even though the rate of sediment transport should be greater there.

The position of the region of maximum scour is likely to depend on the dominant position of the crossover region of high shear stress. In cobble-bed rivers at least, more sediment is moved in a given period of years by high rather than by medium discharges, so presumably it is the position of the crossover region at high discharges which is important. Since skew-induced secondary circulation is relatively weak at high discharges, the crossover region should lie relatively far downstream. Thus, the longstream increase in shear stress is likely to be most pronounced in the downstream half of the bend or even downstream of the bend if the bend arc angle is small. Consequently the position of maximum scour is likely to occur downstream of the bend apex. Assuming such factors as bank composition are constant, the maximum bank erosion should occur in the same region, resulting in the downstream movement of bends, as observed by Leopold and Wolman (1960).

If artificial conditions are imposed on the channel by, for example, river regulation or interbasin transfers, the position of maximum erosion will change. An increase in medium flows at the expense of high flows would mean that secondary circulation would be at its strongest more often. The crossover region of high shear stress would therefore lie near its upstream limit more often. Depending on the percentage reduction in high flows and assuming that sediment transport occurs at medium discharges, the position of maximum scour and bank erosion could shift upstream, thereby altering the pattern of bend development.

An increase in medium discharges at the expense of low discharges might strengthen the shear stress peak associated with downwelling near steep outer banks. Local scour along the base of the bank could then increase. This would lead to a more rapid removal of the products of bank collapse and therefore a faster rate of bank erosion (Thorne and Lewin, 1979).

CONCLUSIONS

Measurements of velocity in rivers and irrigation channels have allowed patterns of boundary shear stress to be delineated and their changes with discharge to be studied. Because of the small number of data the results are not definite, but the following conclusions may be drawn.

1) The distribution of shear stress at a section depends mainly on primary and secondary flow effects. The stress-induced secondary circulation of straight reaches generally decreases in strength relative to the primary flow as Reynolds number increases. At bends the skew-induced circulation

is strongest relative to the primary flow at medium discharges. Thus, primary flow effects are generally stronger than secondary flow effects at high discharges.

2) At the straight reaches observed, the shear stress distribution appears to be affected by stress-induced secondary circulation and by acceleration (and deceleration) of flow. Little change in the overall pattern with discharge was observed.

3) At bends the magnitudes and positions of shear stress peaks are determined by the relative influence of skew-induced circulation. Consequently the shear stress peak associated with the core of maximum velocity is carried towards the outer bank earliest in the bend at medium discharges. At high discharges that peak (caused by a primary flow effect) is greater than the peak associated with the downwelling caused by secondary circulation.

4) The cross-sectional distribution of shear stress becomes more uniform as Reynolds number increases. At the straight reaches observed, the degree of uniformity varied solely with Reynolds number. At bends the degree of uniformity also depends (inversely) on the relative strength of skew-induced secondary circulation.

5) Longstream changes in shear stress cause erosion and deposition. Artificial alterations to the natural distribution of discharges may change the dominant longstream shear stress distribution and, consequently, the position of the dominant area of erosion relative to its natural position. Prediction of the resulting channel adjustments requires a more complete understanding of the subject. Further field studies should be carried out.

ACKNOWLEDGEMENTS

The research described in this paper was carried out by the author as part of his work for the degree of Doctor of Philosophy at the School of Environmental Sciences, University of East Anglia, Norwich, U.K. (Bathurst, 1977). Financial support was provided by the Natural Environment Research Council, U.K., and the University of East Anglia. The author would like to thank Richard D. Hey for supervising the research, Ruh-Ming Li and Hsieh W. Shen (Colorado State University) for kindly reviewing this paper and J. Dungan Smith (University of Washington) for his helpful communications on shear stress distributions.

REFERENCES

Al-Shaikh Ali, K.S., 1964, Flow dynamics in trapezoidal open channel bends [Ph.D. thesis] : Fort Collins, Colorado State University, 133 p.

Aly, A.M.M., Trupp, A.C., and Gerrard, A.D., 1978, Measurements and prediction of fully developed turbulent flow in an equilateral triangular duct: Journal of Fluid Mechanics, v. 85, p. 57-83.

Bathurst, J.C., 1977, Resistance to flow in rivers with stony beds [Ph.D. thesis] : Norwich, University of East Anglia, 402 p.

Bathurst, J.C., Thorne, C.R., and Hey, R.D., 1977, Direct measurements of secondary currents in river bends: Nature, v. 269 (October, p. 504-506.

——————————————————————, (in press), Secondary flow and shear stresses at river bends: American Society of Civil Engineers Proceedings, Journal of Hydraulics Division.

Bhowmik, N.G., and Stall, J.B., 1978, Hydraulics of flow in the Kaskaskia River; in American Society of Civil Engineers Proceedings, Special Conference on Verification of Mathematical and Physical Models in Hydraulic Engineering: University of Maryland, August 9-11, p. 79-86.

Bradshaw, P., 1971, An introduction to turbulence and its measurement, 1st edition: New York, Pergamon Press, 218 p.

Brederode, V. de, and Bradshaw, P., 1978, Influence of the side walls on the turbulent center-plane boundary-layer in a square duct: American Society of Mechanical Engineers Transactions, Journal of Fluids Engineering, v. 100, p. 91-96.

Bridge, J.S., and Jarvis, J., 1976, Flow and sedimentary processes in the meandering River South Esk, Glen Cova, Scotland: Earth Surface Processes, v. 1, p. 303-336.

——————————————————————, 1977, Velocity profiles and bed shear stresses over various bed configurations in a river bend: Earth Surface Processes, v. 2, p. 281-294.

Clauser, F.H., 1956, The turbulent boundary layer: Advances in Applied Mechanics, v. 4, p. 1-51.

Dietrich, W.E., Smith, J.D., and Dunne, T., (in press), Flow and sediment transport in a sand bedded meander: Journal of Geology.

Eskinazi, S., and Yeh, H., 1956, An investigation on fully developed turbulent flows in a curved channel: Journal of Aeronautical Science, v. 23, p. 23-34, and 75.

Hooke, R. le B., 1975, Distribution of sediment transport and shear stress in a meander bend: Journal of Geology, v. 83, p. 543-565.

Ippen, A.T., and Drinker, P.A., 1962, Boundary shear stresses in curved trapezoidal channels: American Society of Civil Engineers Proceedings, Journal of Hydraulics Division, v. 88, HY5, p. 143-179.

Ippen, A.T., Drinker, P.A., Jobin, W.R., and Shendin, O.H., 1962, Stream dynamics and boundary shear distributions for curved trapezoidal channels: Massachusetts Institute of Technology, Hydrodynamics Laboratory Report 46, 81 p.

Launder, B.E., and Ying, W.M., 1972, Secondary flows in ducts of square cross section: Journal of Fluid Mechanics, v. 54, p. 289-295.

Leopold, L.B., and Wolman, M.G., 1960, River meanders: Geological Society of America Bulletin, v. 71, p. 769-794.

Leutheusser, H.J., 1963, Turbulent flow in rectangular ducts: American Society of Civil Engineers Proceedings, Journal of Hydraulics Division, v. 89, HY3, p. 1-19.

Naas, S.L., 1977, Flow behavior in alluvial channel bends [Ph.D. thesis]: Fort Collins, Colorado State University, 171 p.

Perkins, H.J., 1970, The formation of streamwise vorticity in turbulent flow: Journal of Fluid Mechanics, v. 44, p. 721-740.

Pierce, F.J., and Zimmerman, B.B., 1973, Wall shear stress inference from two and three-dimensional turbulent boundary layer velocity profiles: American Society of Mechanical Engineers Transactions, Journal of Fluids Engineering, v. 95, p. 61-67.

Prandtl, L., 1952, Essentials of fluid dynamics, 1st edition: London, Blackie, 452 p.

Rajaratnam, N., and Muraldihar, D., 1969, Boundary shear stress distribution in rectangular open channels: La Houille Blanche, v. 24, p. 603-609.

Thorne, C.R., and Lewin, J., 1979, Bank processes, bed material movement and planform development in a meandering river; in Rhodes, D.D. and Williams, G.P., eds. Adjustments of the fluvial system: Dubuque, Iowa, Kendall/Hunt Publishing Co., p. 117-137.

BANK PROCESSES, BED MATERIAL MOVEMENT

AND PLANFORM DEVELOPMENT IN

A MEANDERING RIVER

COLIN R. THORNE

School of Environmental Sciences
University of East Anglia
Norwich, United Kingdom

JOHN LEWIN

Department of Geography
University College of Wales
Aberystwyth, United Kingdom

ABSTRACT

Field observations of bank processes and bed material movement at a meander bend on the River Severn, U.K., are presented and discussed in the light of historical evidence for channel change in the last 150 years. Mechanisms of bank failure, in general dependent on bank structure and composition, are here dominated by fluvial undercutting and mechanical failure of cantilevers in the upper bank. Failed material accumulates at the bank foot from where it is removed by fluvial entrainment. Tracer experiments show that bank retreat rates are fluvially controlled though failure mechanisms are not. Measured retreat rates are around 0.5 m/yr comparing with up to 0.7 m/yr historically on the same reach.

Contrasting forms of planform development are apparent, including complex loop formation, neck and chute cut-offs, and the rapid abandonment of lengths of cut bank and deep channel. Upstream changes and large, but infrequent, discharge events have profound effects which are difficult to predict.

Channel change can be usefully regarded as a sediment transfer process involving bank failure, sediment entrainment, transport, and deposition. However, both field studies and documentary analysis are needed for a good understanding of the full range of contemporary channel change characteristics in space and time.

INTRODUCTION

The development of natural river channels in alluvial materials proceeds by complex processes of bed and bank material erosion, transport, and deposition. Each process responds with varying magnitude and frequency to the flow regime of the river and its associated spatial distributions of velocity and shear stress. It has been concluded that significant alluvial landforms are formed by frequently occurring events of moderate intensity and not by rare floods of unusually high magnitude (Wolman and Miller, 1960). In this context the bankfull discharge has been nominated as the flow doing most work and as such, the dominant discharge in forming the channel and its associated alluvial features. There is, however, some disagreement as to whether a single flow may be used to represent the effects of variable flows (see for example Ackers and Charlton, 1970; Pickup and Warner, 1976). Close observation of geomorphic processes appears to be required before the relative importance of different processes and events of different magnitude and frequency in the formation of specific channel features can be evaluated.

There have been a number of field-based studies of such channel features. For example Wolman (1959) and Twidale (1964) studied bank erosion, Leopold and Emmett (1976) bed material transport, and Bridge and Jarvis (1976) and Jackson (1975) the relationship between channel flows and sedimentation. There have also been numerous studies of rates and patterns of channel planform change (Lewin, 1977). However, it has seldom proved possible to link such changes to contemporary processes and the transfer of material from one location to another, in the context of the magnitude and frequency of flow events responsible for the processes. This is the approach adopted in this paper. Channel change is seen as the cumulative result of sediment transfer. Emphasis is placed upon the importance of linking field measurements of channel process to the documentary study of planform change. This may allow the effectiveness of event magnitude and frequency in determining channel forms to be evaluated over both short and longer terms.

Study Area

The reach studied lies on the River Severn below Caersŵs in Wales (National Grid Reference SO 033 917). At Caersŵs the River Severn has a catchment area of 375 km². Bankfull discharge is approximately 70 m³/sec. At bankfull, the average width is 30 m, the water surface slope is 0.175%, and the width-to-depth ratio is 30. The river is actively meandering across the valley floor (fig. 1) with a sinuosity (Schumm, 1963), measured over 1500 m of valley, of 1.83 in 1975. The bed is formed in coarse alluvial deposits (gravel and cobbles) having a mean grain size of 40 mm. The standard deviation of the grain size distribution for the bed material is 0.22. Channel bars consist of sandy gravel and cobbles with D_{50} = 28 mm and std. dev. = 0.25. The upper strata of cut banks are formed from cohesive sandy silt with some clay.

Plan of Study

To relate current processes of bank erosion and bed material movement to the short term development of channel planform, field measurements of channel processes were undertaken between September 1976 and April 1977 at a meander bend 1500 m downstream from Caersŵs Bridge (bend D in fig. 1). Observations consisted of (a) the identification and monitoring of erosion processes and mechanisms of bank failure in the study reach and (b) an experimental investigation of bed material movement, using tracers.

Rates and patterns of change at this and adjacent bends were considered in relation to large scale vertical aerial photographs from surveys flown in 1948, 1969, 1972, and 1975. The longer-term development of the channel was examined using historical maps available for the last century and a half. Bank pro-

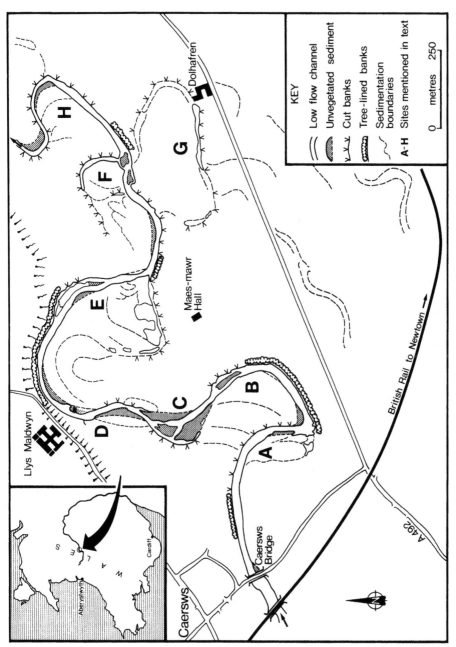

Figure 1. Study area. Bends discussed in text are lettered. General location of channel reach shown in inset.

cesses, bed material movement and longer-term channel changes are considered in turn.

BANK PROCESSES

Nature of Banks

Mechanisms of bank failure depend on the structure of the bank and the engineering properties of its constituent materials (Thorne, 1978). Banks may be classified accordingly as being non-cohesive, cohesive, or 'composite'. The banks in the study reach on the River Severn have a composite structure. This is common for alluvial banks formed where a meandering river is flowing across its floodplain (cf. Turnbull and others, 1966). The banks consist of non-cohesive sand, gravel, and cobbles formed from relic channel bars and cohesive sandy silt sediments, deposited by overbank flow and on top of emergent bars. The characteristic composite bankform shows cohesionless deposits overlain by a capping of cohesive material (fig. 2). The gravel and cobbles are imbricated according to the flow under which they were deposited. Prior to the operation of sub-aerial processes, the spaces between the coarse particles are closely packed with sand. The cohesive material of the upper bank has a clear ped fabric due to the lateral swelling and shrinkage associated with wetting and drying cycles. Peds are separated by vertical fissures which extend down nearly to the summer water table. The actual depth depends on the intensity of drying and the tensile strength of the soil (Terzaghi and Peck, 1948). The fissures have a polygonal pattern in plan view. Even when fully closed such fissures represent planes of weakness in the soil, due to preferred orientation of clay particles on the faces of the peds (van Olphen, 1963). This weakness is especially evident in the tensile strength of the soil (Lo, 1970).

At bend D on the River Severn the interface between the upper (cohesive) and lower (non-cohesive) parts of the bank is well defined (fig. 2). Observations on many other British and American rivers suggest that this is usually the case. The position of the interface relative to the present river channel depends on the degree to which the river is incised into its floodplain   the local height of the relic gravel bar, and the depth of scour close to the bank. At bend D the Severn is not incised but the height of the relic bar and the depth of scour vary considerably over short distances. The height of the interface above the bed ranges from over 2 m to less than 0.3 m as a result.

Previous qualitative observations on composite banks have shown erosion processes to operate by a combination of failure mechanisms. In their study on the Lower Mississippi, Turnbull and others (1966) recognized that failures of the upper bank were precipitated by the cumulative effect of a number of lower bank failures. It was clear that the failures of the upper bank were derived internally, and that consequently, they were not directly associated with the application of fluid stresses or fluvial processes per se. Preliminary field observations on the River Severn suggested that this was also the case there.

Study Procedure

Bank erosion pins of the type described by Wolman (1959) were installed in the upper and lower portions of banks at fourteen sections spaced through-

Figure 2. Composite bank structure. Non-cohesive sand and gravel overlain by cohesive sandy silt. River Severn at Maes Mawr.

out the study reach. Some sections of cut bank in this area have been abandoned by the river and are now quite remote from the present channel. These were known to be completely unaffected by fluvial processes during the experimental period. Four sections were established in such abandoned areas of bank, to act as a control, allowing the rates of erosion associated with non-fluvial processes to be evaluated. Consideration of rates of retreat of the upper and lower portions of the bank under the operation of fluvial and sub-aerial erosion allowed the relative contributions of these processes to the total rate to be assessed. The processes dominant in bringing about significant and systematic development of the channel planform could then be identified.

Sub-aerial Bank Erosion

Data from sections of bank which were not subject to fluvial processes indicated that sub-aerial erosion was quite effective in eroding the steep, unvegetated face of composite banks. These processes have been studied in detail elsewhere (Carson and Kirkby, 1971) and will not be dealt with here. Rates of retreat on the steep portion of the banks were in the range 15-25 mm/yr. There was a clear winter peak in the rate of retreat, the period November to April averaging 30 mm/yr, whilst that for the summer was only 12 mm/yr. During two periods of very cold weather frost action was especially effective in loosening crumbs or aggregates of soil from the upper bank. This process has also been observed in field studies elsewhere (Wolman, 1959).

Weathering of the lower bank proceeded by the removal of interstitial sand and loosening of the coarser fraction. This led to the failure of individual grains in the gravel and cobble size range, reducing the packing density and hence the angle of internal friction of the non-cohesive deposits (Carson, 1971). Internally-derived failures occurred as a result (cf. Carson, 1971). Eroded material collected as a talus slope at the foot of the bank owing to the absence of fluvial activity. This accumulation of loosened material as a basal concavity in the bank profile tended to stabilize the lower part of banks. Colonization of the lower bank by vegetation also occurred, reducing the rate of weathering by several orders of magnitude and, in the short term at least, stabilizing the bank (fig. 3). These are effects which have also been noted elsewhere (Carson and Kirkby, 1971; Thorne, 1978).

Figure 3. Abandoned cut-bank on River Severn. Basal accumulation of weathered material and colonization by vegetation tend to stabilize bank.

Fluvial Erosion of Banks

At ten experimental sections where the banks were subject to fluvial attack, the erosion pins indicated more rapid rates of retreat. The stability of a particle at the surface of a bank which is being attacked fluvially depends on an internally derived stress which is equal and opposite to that applied by the flowing water (White, 1940). Hence the erosion of a bank by fluvial entrainment depends on the engineering properties of the bank material and the hydraulic parameters of the flow. The interaction of flowing water and sediment particles is, however, a highly complex phenomenon. Together with the stochastic nature of the magnitude of instantaneous stress peaks associated with fully turbulent river flow, this complexity has so far precluded the development of a satisfactory stability analysis based on a static equilibrium approach (Task Committee, 1966; Graf, 1971; Yalin, 1977; Thorne, 1978). It is however clear from some of the field observations made on the River Severn that the sand, gravel, and cobbles of the lower part of a composite bank are highly susceptible to fluvial entrainment. At the outer bank on bends, rates of retreat of the lower bank often exceeded 350 mm/yr and were exceptionally as high as 600 mm/yr.

This was not the case for the cohesive material of the upper bank. Not only was this highly resistant to fluvial erosion, but also its higher position in the bank meant that only rarely was it attacked by the flow. As a result, only 28 mm/yr of retreat occurred by this process at even the most active bends. Nearly all of this retreat occurred during the winter months, corroborating the conclusion drawn by Wolman (1959) that cohesive banks are most vulnerable in the winter. Close inspection of the surface of actively eroding cohesive sections of bank revealed that fluvial erosion proceeded mainly by the entrainment of crumbs or aggregates of soil in the sand or gravel size range. This was also noted in the independent studies carried out by Sundborg (1956), Karasev (1964), and Krone (1976), and it may help to explain why erodibility indices based on aggregate stability seem to explain rates of fluvial erosion reasonably well, and in many cases more satisfactorily than indices based on shear strength (Bryan, 1977).

The large disparity in the rates of retreat of the upper and lower portions of the bank led directly to the generation of overhanging bank sections or cantilevers in the upper bank (fig. 4). The dimensions of overhanging blocks and their stability depend on the thickness of the cohesive stratum and the engineering properties of the soil. Fluvial forces are insignificant compared with those of weight, friction, and cohesion in the case of overhanging peds (Thorne, 1978). The stability of a cantilever may be analysed using the basic principles of soil mechanics and the elementary theory for the bending of beams. The ped fabric present in the soil is very important in controlling the mode and time of failure of an overhanging block. When a ped block is revealed at the bank face the lateral earth pressure is removed from its outermost side (Terzaghi and Peck, 1948). This allows the interped fissure on its innermost side to spring open. The development of this crack, also promoted by drying and desiccation of the ped surface, is often the critical factor in bringing about failure (Thorne, 1978). The roots and rhizomes of meadow and reed grasses on the flood plain reinforce the soil and the interped fissures. By inhibiting the development of cracks, vegetation can enhance the stability of a bank considerably. The effects of cracks, fissures, and vegetation must be taken into account in any stability analysis if serious errors are to be avoided (Terzaghi and Peck, 1948; Turnbull and others, 1966; Bradford and Piest, 1977; Imeson and Jungerius, 1977).

In this study three modes of cantilever failure were identified (fig. 5). Shear failures occur in sandy soils which have a low degree of cohesion. Beam failures occur in highly cohesive soils where the thickness of the cohesive layer is small. Tensile failures may occur in the lower portions of thick cantilevers.

Cantilever failures generate blocks of soil, typically with dimensions which are of the order of 0.5 m. These blocks are quite cohesive and may be strongly bonded by roots and rhizomes. Following failure the blocks come to rest, together with the products of sub-aerial and fluid shear failures, at the foot of the bank. Their removal from this area depends entirely on fluvial entrainment. Once entrained as bed load, soil ped blocks eroded rapidly by attrition and broke up under their own weight and momentum, disintegrating to wash load.

Rates of upper bank retreat by cantilever failure were controlled by rates of lower bank erosion by fluvial undercutting. Maximum rates were of the

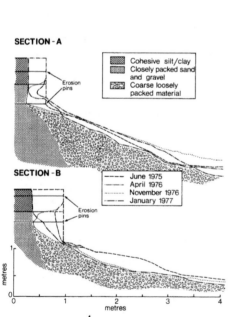

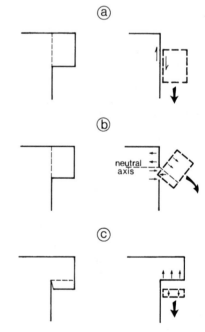

Figure 4. Generation of a cantilever overhang by fluvial undercutting of composite bank. Location of sections shown on figure 1.

Figure 5. Modes of failure of cantilever overhangs in a composite bank. (a) Shear failure, (b) Beam failure, (c) Tensile failure.

order of 600 mm/yr and minimum rates were around 300 mm/yr. These rates, with the additional effects of fluvial and sub-aerial erosion of the upper bank, necessarily match the rate of undercutting of the bank. Cantilevers were not clearly generated where the bank experienced sub-aerial processes alone. Rates of bank erosion associated with undercutting and cantilever failure were an order of magnitude greater than those associated with other bank erosion processes. Clearly the processes of fluvial undercutting and cantilever failure were responsible for planform development in the short-term.

The rapidity with which undercutting proceeds and cantilevers are generated depends on the removal of failed blocks and bank-erosion debris from the basal area. Hence fluvial processes control the bank profile and retreat rate even though the mechanism of failure is determined by the engineering properties of the bank. Consequently the plan shape of the bank depends on the distribution of fluvial scour in the channel. Therefore it was necessary to investigate the distribution of basal scour at a bend in order to explain the distribution of rapid bank retreat by undercutting and cantilever failure and hence meander development.

BED MATERIAL MOVEMENT

Method of Study

The distribution of scour and deposition at a bend was investigated in a tracer experiment carried out at bend D in the experimental reach (fig. 1). The tracers used were pebbles and cobbles collected from the channel and painted a distinctive yellow color using road-lining paint. This paint is well suited to the task, being both conspicuous and durable. Tracers were prepared in sets of 100. The size distribution of each set was matched to that of a Wolman sample of the bed, carried out over the relevant area (Wolman, 1954). This procedure ensured that the tracer set truly represented the range of bed material sizes and shapes present.

Five tracer sets were placed in the channel along five cross-sections around bend D, from the point of inflexion upstream to that immediately downstream. The tracers were evenly spaced across sections which extended between the trim lines on opposite banks. Each stone could be identified by a number painted on it (1 to 100) and each set by the color in which the numbers were painted. The ends of the sections were monumented and their locations marked on a compass and tape map of the reach. Hence the initial location of every tracer stone was accurately known (fig. 6a). The tracers were put in place on 22 September 1976). Their positions were determined on two subsequent occasions, 19 November 1976 and 16 April 1977. These data are presented in figure 6b through 6f.

Transporting Flows

A small flood occurred only 4 days after installation of the tracers. Some movement of the stones was observed after this event and this, together with observations made in an earlier pilot experiment, defined the threshold discharge for sediment transport as approximately 17 m^3/sec. On this basis five floods above the threshold value occurred in the period between 22 September

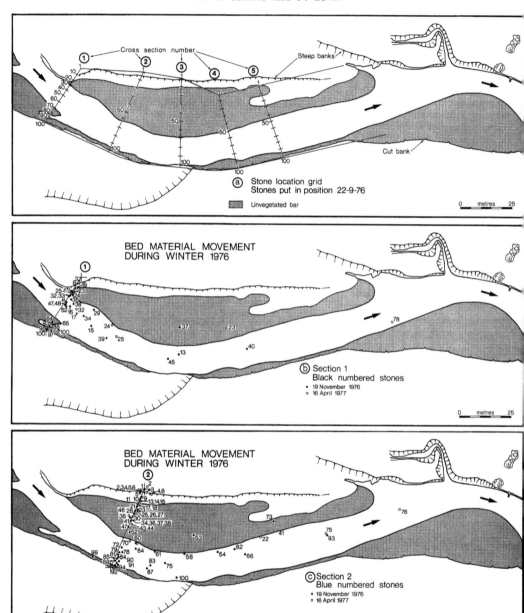

Figure 6. *Movement of bed material tracers at bend D, River Severn near Caersŵs, 1976-1977. (a) Initial location of tracers 22 September 1976; (b) section 1: upstream inflexion point (black painted stones); (c) section 2: bend entrance (blue painted stones);*

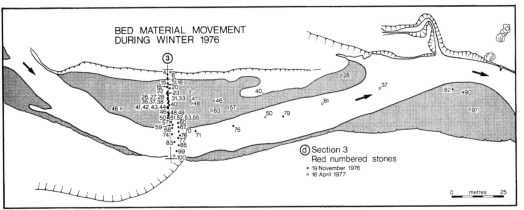

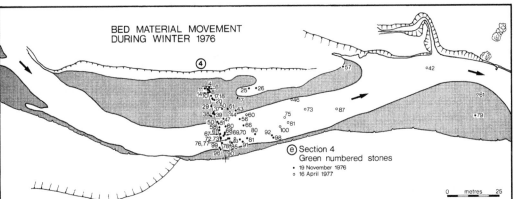

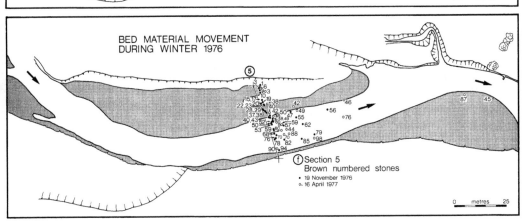

Figure 6 (continued). (d) section 3: bend apex (red painted stones); (e) section 4: bend exit (green painted stones); (f) section 5: downstream inflexion point (brown painted stones).

and 19 November 1976. These had peak discharges of 17, 21, 31, 33, and 35 m³/sec. Forty-seven percent of the tracers were located at the end of this period. More than twenty floods above the threshold value occurred in the second period, up to 16 April 1977. The largest event had a peak discharge of approximately 80 m³/sec. The large amount of sediment movement resulted in only 15% of the tracers being relocated at the end of the second period.

Results

During the first period, at the entrance to the bend (sections 1 (fig. 6b) and 2 (fig. 6c)) scour was confined to the area around the center line of the channel. There was little erosive activity at the base of either bank. Indeed there was a tendency for tracers at the right bank to be buried in situ by material derived from the bank above them. This suggests that the banks in the entrance to the bend do not experience scour at their base under floods of up to 35 m³/sec. This was also the case for sections in the exit of the bend (sections 4 (fig. 6e) and 5 (fig. 6f)). Here there was little entrainment of material, and basal accumulation was dominant. Although few of the tracers were removed from sections 4 and 5, tracers entrained further upstream in the bend were able to pass through these sections, to be deposited further downstream on the point bar in the next bend. Clearly the inflexion point was a section of channel which was just capable of transmitting the sediment load arriving from upstream. Some tracers from upstream were deposited around center channel on the bar at section 5. This probably reflected temporary storage on the bar, prior to 'permanent' deposition on the point bar in bend E (fig. 1).

At the bend apex, section 3 (fig. 6d), scour was concentrated near the base of the outer bank. There was little sediment entrainment on the point bar at the left bank. One or two tracers on the point bar actually moved upstream against the flow. This could be taken to be the result of reverse flow in a zone of separation at the inner bank (Leeder and Bridges, 1975), but a more probable explanation here is that the tracers were disturbed by livestock crossing the river.

In the bend, material entrained as bed load tended to move laterally up the point bar whilst being carried along the channel. This lateral movement is in accord with direct measurements of secondary currents made here (Bathurst, Thorne, and Hey, 1977) and general models for bend sedimentation (Allen, 1970; Bridge, 1976). Material eroded from the outer bank at a bend was deposited on the point bar in the next bend downstream, on the same side of the channel. This confirms Friedkin's (1945) proposal. Eroded material did not cross the channel to the point bar opposite, as was proposed by Leliavsky (1966).

Under the higher flows which occurred during the second period (November 1976 to April 1977), scour is evident all along the right bank at the bend. Eroded material tended to be deposited around the center of the channel at the inflexion point or on the point bar in the next bend downstream (bend E).

The bank profiles at sections around bend D reflect the varying frequency of fluvial scour at the base. At the inflexion points the profiles have distinct basal concavities (fig. 7a, b, d, and e). These are wedges of bank-derived

128

material which have accumulated because of the excess input by bank failures over longstream removal. By contrast the profile at the bend apex shows no basal concavity (fig. 7c). Here, over the year, the flow removed all the products of bank failures and so maintained a steep and very active profile over the full height of the bank. Seasonal fluctuations, representing the temporary accumulation of material which can occur between floods (cf. Brunsden and Kesel, 1973), are evident in the profiles. The bed load tracers showed that quite small floods of about only half the bankfull capacity generate scour at the bend apex whilst rather large floods are required to scour the inflexion points. The frequency of the smaller floods is of course much greater than that of larger events. The bank at the bend apex therefore is maintained in an active state whilst for long periods erosion of the bank at the inflexion points is inhibited. The observed rates of retreat illustrate this. At the bend apex the rate of retreat in the period September 1976 to September 1977 was approximately 500 mm/yr whilst at the inflexion points it was only 200 mm/yr (fig. 7).

The short term planform development at bend D thus consisted of meander growth and increasing sinuosity. This development occurred in response to a number of flow events up to and including bankfull discharge, but excluding any larger floods. This development in response to events of moderate size generally accords with conclusions drawn by Wolman and Miller (1960) concerning magnitude, frequency, and landform development.

CHANNEL PLANFORM DEVELOPMENTS

In the previous sections bank processes and bed material movement in the short term at a single bend have been considered. At the outset this bend seemed not untypical of conditions found widely in this part of the River Severn, but it is useful to set these observations in a wider context by considering two points. Firstly, in what manner and at what rate have the processes observed over one winter season combined to produce channel change over a longer period of years? Secondly, do the observed processes and changes adequately represent the whole range of erosion and sedimentation processes likely to occur at such a site?

Some answers to these questions may be obtained through examination of the range of historical maps and aerial photographs available for the area. These are listed in table 1. Map sources have been replotted where necessary to a common scale of 1 : 10,560 for comparison, whilst channel detail from the air photographs has also been transferred to base maps of the same scale, using a Rank-Watts Stereo-sketch plotter. Taken together the maps and air photographs cover over a century and a half, the photographs in particular providing detailed information for the last decade. Thus channel development on the River Severn for about 2 km down-valley of Caersŵs may be examined over an extended period.

The earliest sources, 1832/4, and 1846/7, show the site of the experimental bend (D in fig. 1) to be inside the inner (right, convex) bank of a single major meander loop centered slightly up-valley of Maes Mawr Hall. By the end of the century this loop had become 'P' shaped with an almost straight, south-north upstream limb. This upstream limb had a small right bank accumulation

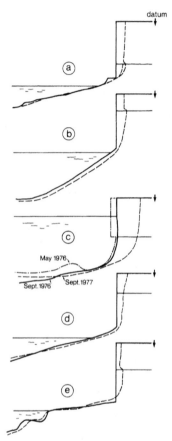

Figure 7. Profiles at outer (right) bank of bend D, River Severn near Caersŵs, 1976-1977. (a) Section 1: upstream inflexion point. (b) Section 2: bend entrance. (c) Section 3: bend apex. (d) Section 4: bend exit. (e) Section 5: downstream inflexion point.

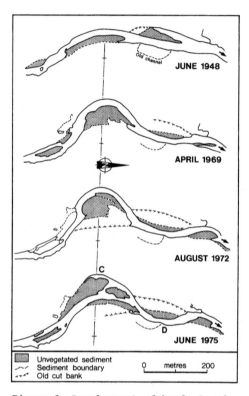

Figure 8: Development of bends C and D, 1948 to 1975.

of unvegetated sediment at C (fig. 1) in 1884, and at both C and D in 1901. However, it was certainly not until after about 1948 that the upstream loop of the major Maes Mawr bend became genuinely dismembered into a rapidly evolving secondary loop at C and the less striking downstream bend at D. Possible earlier (mid 19th-century?) multiple loops of this type on the same

Table 1

Available Channel Surveys

Source	Survey date
Maps	
Ordnance Survey, first edition one-inch	1832-4
Llandinam and Llanwrog Tithe maps	1846-7
Ordnance Survey, first edition six-inch	1884
second edition six-inch	1901
provisional edition six-inch	(?) 1948
Air photographs	
S41/RAF/59	June 1948
BKS	April 1969
Ordnance survey	August 1972
Meridian Airmaps	June 1975

reach are indicated by the right bank loop (now filled with fine sediment) which intersects the present cut bank at section 3 in figure 6.

Developments at C and D resulting in the formation of alternate point bars and cut bank arcs thus appear to have occurred essentially within the last 30 years (fig. 8). The complex point bar at C (which air photographs of various dates show to consist of one or more bar head lobes, a marginal and downstream bar tail, and an enlarging, grass-covered, inner surface at a slightly lower elevation than the floodplain) is followed downstream by smaller and simpler bars on the left and right banks in turn. Point bar sandy gravels have thus been deposited either as continuous sheets accreting both laterally and downstream without strong sedimentary discontinuities, or by processes involving flow separation and avalanche fronts and lobes.

Rates of bank recession between the dates of air photography for both bends C and D are given in table 2. These show that the development of bend D tends to have post-dated that of bend C and it seems likely that the changes upstream are themselves in part responsible for those downstream. This is significant and illustrates the point that changes on any one meander loop are not independent of upstream changes. It has been shown earlier that rates of bank erosion depend on the pattern of basal scour, which itself responds to the distribution of boundary shear stress and to primary and secondary velocities. These may be altered and realigned following upstream channel changes with alterations in the rates of bank retreat or even the abandonment of sections of cut bank.

The 1948-1975 average rate of bank recession for that point on bend C which suffered most erosion was 0.7 m/yr; this compares closely with the

131

Table 2

Rates of Cut Bank Recession

| | | Rate of recession (m^2/yr) | |
Period	Years	Bend C	Bend D
1948–1969	21	26.7	–
1969–1972	3	70.0	16.7
1972–1975	3	20.0	30.0

field-measured rate of around 0.5 m/yr. Comparisons between air photo plots show that the locus of maximum erosion shifts, in the short term, as bend geometry changes. Thus the 0.7 m/yr rate is not constant over time. Rates over shorter time spans cannot be obtained with sufficient precision from the photography available, but the general magnitude of change between photo and field data is clearly comparable.

Notable destruction of the evolving loop at C took place between 1972 and 1975 (fig. 8). It seems very probable that a flood on 6 August 1973 was largely responsible for this destruction. The peak discharge was 180 m^3/sec, no other flood between June 1972 and 1975 having exceeded 65 m^3/sec. A point bar at the left bank appears to have been moved bodily downstream in a manner reminiscent of one on the River Rheidol in 1951-2 (Lewin, 1978). This sediment blocked off the mainstream channel, diverting water into a chute across the former point bar complex at C.

Thus both in its initial rate of development and in its dismemberment, loop C illustrates planform changes which could not be anticipated from a short-term examination of loop D alone. Furthermore, and in addition to the chute cut-off of bend C, two other forms of channel abandonment are characteristic of this reach of the Severn.

A neck cut-off is illustrated by the major compound loop of abandoned channel downstream of Maes Mawr Hall (G in fig. 1). This loop was connected at both ends to the river with but a short breach channel in 1832/4, but disconnected thereafter. Opposite this loop another (F in fig. 1) was created within about the last 50 years and cut off between 1969 and 1972.

Major cut-offs are common on the Severn downstream of Caersŵs, with 9 curving cut-off channel segments of relatively fresh appearance within approximately 6 km of valley. In addition, crop markings and soil patterns show other, less distinct, abandoned channels well away from the present river (fig. 1), though these are of uncertain age.

These true cut-off forms may be contrasted with non-linear and irregularly shaped floodplain depressions (often with fine sediment infilling and with standing water) representing channel pools abandoned without the deposition of coarse sediment during channel movement. As on the Rheidol (Lewin, 1978) these may give a barbed appearance to meander planforms. Such depressions occur at A in figure 1, where contemporary downvalley migration of the channel loop leaves the progressively abandoned apex scour pool unfilled with sediment, and at E (fig. 1) where the narrow loop orientated upvalley was short-cut at some time in the first half of this century, in the manner illustrated in figure 9a. It is possible, though not certain, that the cut-off occurred in response to severe snow melt floods in 1947. Figure 9 also shows that the rapid development of loop F is probably associated with channel developments immediately upstream.

The pattern and dating of sedimentation downstream from Caersŵs within the period 1847-1975 is shown by figure 9b. As has been suggested, the depositional processes within the time periods may be complex. Channel sinuosity increased from 1.41 in 1847 to a maximum of 2.0 in 1969, decreasing slightly thereafter as cut-offs occurred. Loop development here cannot be inferred from point bar ridges and swales in the manner proposed by Hickin (1974), because such patterns are not prominent on the floodplain surface. However, the pattern of multiple loops seen in the historical evidence is similar to that identified by Brice (1974) and Hickin himself. Loop asymmetry develops in a number of ways. Loop A (fig. 9b), growing in a downvalley direction has

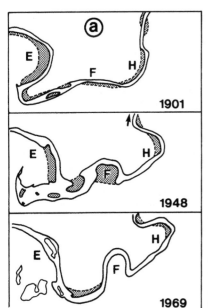
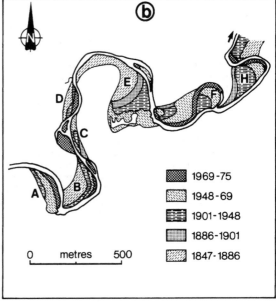

Figure 9. *Channel planform developments. (a) Bends E - H, 1901 to 1969. Showing abandonment, without sedimentation, of upvalley loop at E and rapid development of loop F. (b) Sedimentation in study area by age zones, 1847-1975.*

developed at the expense of B. The apex of loop H has moved rapidly down-valley with little truncation of age zones on the upstream limb. Development of smaller loops on larger ones has occurred at C, D, E, and F. The later development of the "P" form at E interestingly shows what may happen after hooked forms of Hickin's type B (Hickin, 1974) are abandoned.

In summary, the longer-term perspectives provided by the historical evidence show that the processes involved in the piecemeal development of the cut bank and point bar as observed at D have to be seen in the context of additional channel form changes. In particular these may be triggered by events of high magnitude but low frequency. Changes include chute and neck cut-offs and the formation of non-sedimented abandoned channel reaches which are not strictly cut-offs. The whole reach is characterized by loop developments which occur at a variety of spatial and temporal scales and which may individually be highly dependent on upstream change. Many of these developments are difficult to anticipate and incorporate into an observational program because they occur in response to rare events. The historical evidence does however confirm, on a broader scale, the significance of analyzing routes of sediment transport from bank source to depositional sink.

DISCUSSION AND SUMMARY

Channel changes on the River Severn are accomplished by the fluvial entrainment of basal sandy gravel and the mechanical failure of cantilever overhangs in the cohesive top sediments of composite cut banks. Bed scour of such sediments and the redeposition of coarse particles on channel diagonal and point bars have been documented in a tracer experiment, to the extent that the mechanics of the sediment transfer system at a bend under flows up to bankfull are reasonably well understood. However, historical evidence indicates that a number of additional processes and associated form changes are important to the development of alluvial landforms. These include situations where cut-banks are not developed and channel scour depressions are not re-sedimented by simple bars.

The distribution of bed scour through the channel varies spatially and temporally. Quite low discharges may exceed transport threshold at a bend apex, whilst intermediate flows are required at inflexion points. On a magnitude-frequency basis it is floods of around bankfull discharge which are responsible for an orderly development of meandering planforms. However, extreme events have associated with them magnitudes and distributions of velocity and shear stress which appear out of balance with previous channel geometry. Form-change thresholds are crossed and phenomena such as rapid bank recession, the abandonment of lengths of cut bank, cut-offs, areas of non-sedimentation marginal to the evolving channel, and flood chutes may all occur. Bankfull discharges alone therefore do not produce the full range of fluvial channel forms found in the field.

At quite moderate discharges, modifications to channel alignment are likely to condition future changes on adjacent downstream reaches. Indeed such changes may be coupled with modifications arising from alignment shifts that are in any case occurring differentially on adjacent meander loops. Both the effects of infrequent events and of upstream changes on downstream loops

appear to be important on the Severn. However, it is difficult to record the mechanics involved with a precision equivalent to that of the frequent bank, bed, and channel sediment movement discussed previously. For the present, both field process studies and documentary analysis appear necessary for a good understanding of the full range of contemporary planform changes in a reach in space and time.

ACKNOWLEDGEMENTS

The field measurements reported here were undertaken by Colin R. Thorne as part of his work for the degree of Doctor of Philosophy under the supervision of Richard D. Hey. The authors would like to thank Professors K.M. Clayton and K.J. Gregory for their helpful comments on the manuscript. Financial support was provided by the Natural Environment Research Council, U.K., whilst discharge data were supplied by the Severn-Trent Water Authority, Directorate of Operations, Malvern, England.

REFERENCES

Ackers, P., and Charlton, F.G., 1970, Meander geometry arising from varying flows: Journal of Hydrology, v. 11, p. 230-252.

Allen, J.R.L., 1970, Physical processes of sedimentation: London, Allen & Unwin, 248 p.

Bathurst, J.C., Thorne, C.R., and Hey, R.D., 1977, Direct measurements of secondary currents in river bends: Nature, v. 269 (October), p. 504-506.

Bradford, J.M., and Piest, R.F., 1977, Gulley wall stability in loess derived alluvium: Soil Science of America Journal, v. 41, p. 115-122.

Brice, J.C., 1974, Evolution of meander loops: Geological Society of America Bulletin, v. 85, p. 581-586.

Bridge, J.S., 1976, Mathematical model and Fortran IV program to predict flow, bed topography and grain size in open channel bends: Computers and Geosciences, v. 2, p. 407-416.

Bridge, J.S., and Jarvis, J., 1976, Flow and sedimentary processes in the meandering river South Esk, Glen Clova, Scotland: Earth Surface Processes, v. 1, p. 303-336.

Brunsden, D., and Kesel, R.H., 1973, Slope development on a Mississippi river bluff in historic time: Journal of Geology, v. 81, p. 576 - 598.

Bryan, R.B., 1977, Assessment of soil erodibility. New approaches and directions; in Toy, T.J., ed., Erosion research techniques, erodibility and sediment delivery: Norwich, Geobooks, p. 57-72.

Carson, M.A., 1971, The mechanics of erosion: London, Pion Press, 174 p.

Carson, M.A., and Kirkby, M.J., 1971, Hillslope form and process: Cambridge, Cambridge University Press, 475 p.

Friedkin, J., 1945, A laboratory study of the meandering of alluvial rivers: Vicksburg, Mississippi, U. S. Army Corps of Engineers Waterways Experiment Station, 40 p.

Graf, W.H., 1971, Hydraulics of sediment transport: London, McGraw-Hill, 513 p.

Hickin, E.J., 1974, The development of meanders in natural river channels: American Journal of Science, v. 274, p. 414-442.

Imeson, A.C., and Jungerius, P.D., 1977, The widening of valley incisions by soil fall in a forested Keuper area, Luxembourg: Earth Surface Processes, v. 2, p. 141-152.

Jackson, R.G. II, 1975, Velocity-bedform-texture patterns of meander bends in the lower Wabash River of Illinois and Indiana: Geological Society of America Bulletin, v. 86, p. 1511-1522.

Karasev, I.F., 1964, The regimes of eroding channels in cohesive materials: Soviet Hydrology, Selected Papers,No. 6, p. 551-579.

Krone, R.B., 1976, Engineering interest in the benthic boundary layer; in McCave, I.N., ed., The benthic boundary: London, Plenum Press, 323 p.

Leeder, M.R., and Bridges, P.H., 1975, Flow separation in bends: Nature, v. 253 (January), p. 338-339.

Leliavsky, S., 1966, An introduction to fluvial hydraulics: New York, Dover Publications, Inc., 257 p.

Leopold, L.B., and Emmett, W.W., 1976, Bed load measurements, East Fork River, Wyoming: Proceedings of the National Academy of Science USA, v. 73, p. 1000-1004.

Lewin, J., 1977, Channel pattern changes; in Gregory, K.J., ed., River channel changes: Chichester, John Wiley and Sons, Inc., p. 167-184.

_____, 1978, Meander development and floodplain sedimentation – a case study from Mid-Wales: Geological Journal, v. 13, p. 25-36.

Lo, K.Y., 1970, The operational strength of fissured clays: Geotechnique, v. 20, p. 57-74.

van Olphen, J., 1963, An introduction to clay colloid chemistry: London, Interscience Publishers (Wiley), 301 p.

Pickup, G., and Warner, R.F., 1976, Effects of hydrological regime on magnitude and frequency of dominant discharge: Journal of Hydrology, v. 29, p. 51-75.

Schumm, S.A., 1963, Sinuosity of alluvial rivers on the Great Plains: Geological Society of America Bulletin, v. 74, p. 1089-1100.

Sundborg, Å., 1956, The River Klarälven: a study in fluvial processes: Geografiska Annaler, v. 38, p. 125-316.

Task Committee on Sedimentation, 1966, Sediment transport mechanics; initiation of motion: American Society of Civil Engineers Proceedings, Journal of Hydraulics Division, v. 92, HY2, p. 291-314.

Terzaghi, K., and Peck, R.B., 1948, Soil mechanics and engineering practice: New York, John Wiley and Sons, Inc., 566 p.

Thorne, C.R., 1978, Processes of bank erosion in river channels [Ph.D. dissertation]: Norwich, University of East Anglia, 447 p.

Turnbull, W.J., Krinitzsky, E.L., and Weaver, F.J., 1966, Bank erosion in the soils of the lower Mississippi Valley: American Society of Civil Engineers Proceedings, Journal of Soil Mechanics and Foundations Division, v. 92, SM1, p. 121-137.

Twidale, C.R., 1964, Erosion of an alluvial bank at Birdwood, South Australia: Zeitschrift für Geomorphologie, NF8, p. 189-211.

White, C.M., 1940, Equilibrium of grains on the bed of a stream: Royal Society of London Proceedings, ser. A, v. 174, p. 322-338.

Wolman, M.G., 1954, A method of sampling coarse river-bed material: American Geophysical Union Transactions, v. 35, p. 951-956.

_____, 1959, Factors influencing erosion on a cohesive river bank: American Journal of Science, v. 257, p. 204-216.

Wolman, M.G., and Miller, J.P., 1960, Magnitude and frequency of forces in geomorphic processes: Journal of Geology, v. 68, p. 54-74.

Yalin, M.S., 1977, Mechanics of sediment transport, 2nd edition: London, Pergamon Press, 198 p.

EVENT FREQUENCY AND MORPHOLOGICAL ADJUSTMENT

OF FLUVIAL SYSTEMS IN UPLAND BRITAIN

A. M. HARVEY

Department of Geography
University of Liverpool
Liverpool, United Kingdom

D. H. HITCHCOCK

Geography Section
Central London Polytechnic
London, United Kingdom

D. J. HUGHES

Formerly Department of Geography
University of Liverpool
Liverpool, United Kingdom

ABSTRACT

Previous work on event frequency in fluvial systems has tended to focus on the frequency of the most effective event in terms of total sediment transport, or on the frequency of the event, often bankfull discharge, which expresses overall morphological adjustments. This paper attempts to identify process thresholds within fluvial systems and to consider the role of their frequencies in adjustments within dynamic equilibrium. Two main groups of event magnitude are identified: major events controlling overall morphology and moderate events causing adjustment within the overall forms produced by the major events. The discussion is based on field observations in three areas in N.W. England. The areas represent fluvial environments ranging from steep headwaters to braided and meandering river channels. Within these systems the major events occur from two to four times per year to once every two years. Moderate events occur from 14 to 30 times per year. The major events, although occurring with roughly the same frequency as is often quoted for bankfull discharge, do not equate with bankfull conditions. In this context bankfull discharge is not a very useful concept.

The period of field investigation coincided with the abnormally dry years of the early 1970's, during which time actual event frequencies were markedly less than expected. Although there were morphological adjustments to the changed relative frequencies of major and moderate events, the systems remained within dynamic equilibrium. There was no evidence of any approach to a major system threshold, nor evidence of any permanent change in the equilibrium.

A. M. Harvey, D. H. Hitchcock and D. J. Hughes

INTRODUCTION

Fluvial systems transfer water and sediment from source areas of a drainage basin through the channel network. System morphologies adjust to the prevailing rates of water and sediment transport. If these adjustments produce landforms whose morphology fluctuates around a mean form through a balance between erosion and deposition, then that form may be said to be in dynamic equilibrium. Dynamic equilibrium may exist between major system thresholds across which the relationships between rates of water and sediment transfer and the resulting morphology will be fundamentally different. Schumm (1977) has identified two types of geomorphic system threshold. One is extrinsic thresholds, which may be crossed as the result of external changes. The other is intrinsic thresholds, which may be crossed as a result of a dynamic equilibrium condition superimposed on a long term gradual change in the system structure (Schumm and Lichty, 1965).

Within fluvial systems, water and sediment supply are not continuous but take place as the result of discrete events. Any dynamic equilibrium form therefore appears to relate to events of a particular magnitude and frequency. Previous work on event frequency and the adjustment of fluvial systems has tended to focus on one of two aspects of these relationships: the frequency of the dominant events in the context of total sediment transport, or the frequency of the apparently dominant event in the context of landform adjustment. In regard to the former approach, Wolman and Miller (1960) demonstrated that many rivers transport at least 50 % of their total sediment loads during floods of a size which recur at least once every two to three years and 90 % during floods recurring at least once every five years. Tywoniuk and Cashman (1973) and Robinson (1972) suggested that for many Canadian rivers, up to 90 % of total suspended sediment may be carried by relatively frequent floods, occurring approximately twice per year.

The second approach deals with the frequency of those events which appear to control the channel morphology, including studies of bankfull frequency (e.g., Dury, 1961, 1973; Kilpatrick and Barnes, 1964; Woodyer, 1968). These studies suggested that in a variety of environments bankfull discharge has a recurrence interval, when calculated by the annual flood series (Langbein, 1949; Dunne and Leopold, 1978), of between 1.1 and 2 years. The indication is that alluvial channels tend to maintain a dynamic equilibrium form adjusted to approximately the annual event. There are some suggestions of departures from this condition and of variations in bankfull frequency (Harvey, 1969; Knowles, 1971; Williams, 1978).

The identification of one discharge (the bankfull discharge) as that to which channel systems adjust seems to be an oversimplification. A range of discharges to which different channel properties adjust has been identified (Pickup and Warner, 1976). Furthermore, meander geometry appears to adjust to discharges more frequent than bankfull (Carlston, 1965; Harvey, 1975). Recent studies have examined the relationships between events of different magnitudes and frequencies, particularly in relation to the recovery time following extreme events (Costa, 1974; Gupta and Fox, 1974; Wolman and Gerson, 1978; Anderson and Calver, 1977).

This paper examines the relationships between events of differing magni-

140

tudes and frequencies, not in the context of extreme events and recovery time but in the context of adjustments within dynamic equilibrium. Three areas in Northwest England have been studied (figs. 1 and 2). They represent a range of conditions found in the fluvial systems of upland Britain. Within each area process thresholds have been identified. These thresholds operate within dynamic equilibrium and control within-system morphological adjustment. They are not the geomorphic system thresholds that exist between dynamic-equilibrium states (Schumm 1973, 1977).

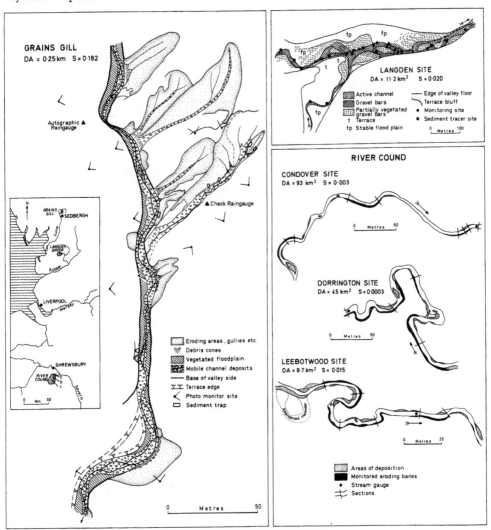

Figure 1. *The study sites – locations and general characteristics. Drainage area (DA) above study site (km^2). Approximate slope (S) of valley floor at study site.*

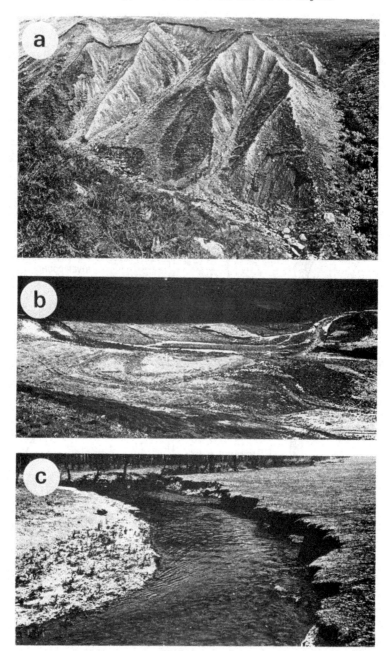

Figure 2.   The study sites; (a) Grains Gill gullies, (b) Langden Brook, (c) River Cound at Condover.

THE STUDY AREAS AND THEIR FLUVIAL SYSTEMS

Within each study area different aspects of the adjustment of fluvial systems have been monitored over various periods between 1969 and 1977. Grains Gill in Cumbria is a major sediment source area where active slope erosion supplies sediment to steep headwater channels. Langden Brook in Lancashire and the River Cound in Shropshire are main river channels. Langden Brook has a rapidly-changing braided channel and the Cound a more stable meandering channel. Each study catchment drains an upland area with steep slopes and maximum altitudes over 500 m and with sources of sediment from both bedrock and Pleistocene deposits. Vegetation is mostly open moorland but in the lower parts of the Cound basin is mixed farmland. Mean annual precipitation, which falls throughout the year, ranges between 1800-1900 mm over Langden, 1600-1700 mm over Grains Gill, and 700-1000 mm over the Cound basin. In winter, snowfalls occasionally but rarely persist on the ground for more than a few days. Except in the lower part of the Cound basin, soil moisture deficits are uncommon. This factor, together with the steep slopes and relatively impermeable substrates of the source areas, tends to result in rapid stream response to heavy precipitation. Rapid response, ample sediment sources, and steep headwater channels result in substantial sediment supply to the stream systems.

Grains Gill

The Grains Gill site (monitored by Harvey) is located midway along Grains Gill (fig. 1) and is representative of two aspects of upland fluvial systems: the supply of sediment from eroding slopes, and its subsequent movement through steep headwater channels. Grains Gill drains a steep catchment. The altitudinal range is 183-532 m, and the altitude of the study site is approximately 280 m. The site is developed on Silurian mudstones which in places are overlain by thick deposits of Pleistocene boulder clay and locally by coarse post-glacial gravels. The catchment lies wholly on unimproved land, predominantly grass moorland, but with heather-covered peat on the watershed.

Midway down Grains Gill the valley sides are deeply dissected by gullies cut into the boulder clay. These gullies are the major sediment sources for the stream. Upstream from the gullies there are minor sources of coarse sediment. However, the channel itself is either confined by solid rock or is narrow and stable and bordered by grass-covered alluvial gravels. Downstream from the gullies channel widths increase dramatically and show great local variability. The channel here takes on a braided character. In some places mobile gravels occupy the entire valley floor (Harvey, 1977). This abrupt change in channel morphology is clearly the result of massive contributions of sediment from the gullies.

Sediment production from the gullies takes place by a variety of processes (Harvey, 1974), major sediment production resulting primarily from heavy rainfall. Much of the sediment accumulates at the base of the gullied slopes in small fans and debris cones. During overland flow suspended sediment is carried directly into Grains Gill. Water with very high suspended sediment concentrations can be traced down Grains Gill into Carlingill and out into the River Lune. The debris cones at the base of the gullies consist of coarse fragments, together with fines when supplied by mudflows. The cones are a store in the sediment transfer system, awaiting removal by major floods. Such floods erode the cones and incorporate the debris into the bedload of Grains Gill. Reworking of the bed material during floods brings about changes in channel configuration. The major floods appear to control not only the channel morphology but also the rates of slope erosion and the morphology of the eroding slopes. If the debris cones are not removed by floods the slope processes continue to add material. As the base of the slope becomes buried, slope angles decrease, vegetation colonization begins, and the rate of sediment production then decreases. With the removal of the cones the gully channels are kept clear and the slopes are kept active, allowing maximum erosion rates to continue.

The dynamic equilibrium of the sediment production and transfer system depends not only on major runoff and flood events but also on the relation between sediment production and removal. Preliminary results of a study of this relationship have been published (Harvey, 1977). In the present study those data are extended to 1977 and include cover of the drought of 1975-1976 and the subsequent wet winter of 1976-1977.

Langden Brook

The Langden site (monitored by Hitchcock) is located midway along the main valley of the Langden catchment and is a major zone of sediment transport through an unstable braided channel (fig. 1). The catchment lies on Lower Carboniferous grits, with local patches of Pleistocene boulder clay in the main valley. The steep valley sides are mantled with coarse periglacial material, and peat covers the flatter upland surfaces. The vegetation is dominated by heather moorland. Altitude ranges from 164-520 m, with the study site at 225 m.

The primary sources of sediment to the stream system are (a) the gullies and scars cut into Pleistocene materials on the valley sides, (b) bedrock outcrops along some of the headwater streams, and (c) areas of peat erosion on the upland surfaces. During periods of high runoff much of the suspended sediment, which includes peat debris, moves rapidly through the system, but deposition of some of this material may take place on gravel bars and on the valley floor. The sediment load is dominated by a bedload of coarse sandstone cobbles which are stored within the channel system and which move in response to major floods, causing changes in channel morphology.

Channel changes may involve simply the movement of the mobile bed sediment, or additionally, the erosion of stabilized and vegetated parts of the floodplain. To a certain extent these two types of change result in differing mechanisms of channel pattern adjustment whereby the within-channel changes modify the form of lateral and braid bars and may lead to the formation of central shoals. The major changes may involve secondary anastomosis (Church, 1972) by the reoccupation of former channel lines within the flood plain and consequent scour of new channels (Hitchcock, 1977a).

Despite the rapid short-term changes and the unstable nature of the channel, maps dating back to 1848 and air photographs to 1947 suggest that the overall morphological characteristics have remained consistent (Hitchcock, 1977b) and that the short-term changes represent fluctuations about a dynamic equilibrium form. This equilibrium depends on the relationship between events producing within-channel changes and those producing major changes in channel configuration, in a manner not unlike that existing in Grains Gill.

River Cound

Three study sites on the Cound, at Condover (at 80 m), Dorrington (at 100 m) and Leebotwood (at 145 m) (fig. 1) (monitored by Hughes) like the Langden site are zones of sediment transfer through main river channels. However, the Cound sites are characterized by meandering rather than braided patterns, and the rates of sediment movement and channel change appear to be less than those at Langden.

The Counc catchment, altitudinal range 75-504 m, includes upland areas in the south developed on Pre-Cambrian slates and volcanic rocks and on Lower Paleozoic shales and sandstones. The northern, lower-lying part of the catchment is on Upper Carboniferous sandstones and shales, but these are extensively mantled by Pleistocene boulder clay and fluvio-glacial sands and gravels. The upland areas are grass moorland. The northern area is predominantly mixed farmland, the valley floors being pasture land.

The major sources of sediment to the Cound appear to be of two sorts: diffuse sources on agricultural land, supplying suspended sediment, and more localized sources adjacent to the river channel, supplying both coarse and fine material. The latter include areas where the channels are eroding bedrock, particularly in the steeper headwater reaches and areas where the streams have cut scars into Pleistocene deposits. Otherwise, much of the coarse sediment appears to be reworked floodplain material.

The main stream channels, away from the headwater reaches, are meandering except for occasional short divided reaches. Some divided reaches, such as one near Condover, are downstream of eroding scars. Although the channel is more stable than that of Langden Brook, there is evidence of channel change both by migration and by cutoff. Abandoned meanders can be seen at several locations on the floodplain. Maps dating from 1841 and air photographs from 1946 show clear evidence of meander migration. Migration takes place by erosion of stable, grassed floodplain sediments on the outer banks and by the formation of point bars, predominantly of gravel, on the inner banks of meander bends. Over the long term the channel shows aspects of dynamic equilibrium in that, despite the localized pattern changes evident from maps and air photographs, the overall channel geometry does not appear to have changed significantly. However, over the shorter period of field observation repeated cross section measurements (to be discussed later) suggest that there is a net loss of sediment from the study areas. It is possible that presently a dynamic equilibrium tendency is superimposed onto a slow progressive change in the system.

Adjustments take place within the system through two types of events: (a) lower-magnitude events which modify the configuration of the unconsolidated point bar or bed sediments and cause localized erosion of the banks, and (b) higher-magnitude events which cause widespread bank erosion of stabilized floodplain material and consequent changes in the plan form of the channel.

A. M. Harvey, D. H. Hitchcock and D. J. Hughes

MEASUREMENTS AND DATA COLLECTION

Grains Gill

Data on sediment production in Grains Gill have been derived from sediment traps. Two major traps were located at the bases of rill systems and three supplementary traps sited at the bases of open slopes (fig. 1). The data from the traps are augmented by observations of erosion pins and by repeated photography (Harvey, 1974). Sediment trap data relate to two periods, April 1971 - July 1972, and September 1975 - September 1977. Data on debris cone behavior and channel changes have been derived from repeated photography from fixed points during the period July 1969 - September 1977.

During the sediment-data periods autographic and check raingages were operated at the Grains Gill site. At other times data from the daily raingage at Sedbergh, 8 km to the south, have been used. Daily rains at Sedbergh correlate well with equivalent rains at Grains Gill. The data sets have a correlation coefficient of 0.90 and a regression equation so near linearity that the Sedbergh daily rain totals may reasonably be used as surrogates for the Grains Gill totals (Harvey, 1977). No data on streamflow at Grains Gill are available. However, Patrick and Gustard (1974) carried out a study of rainfall and runoff at Burnes Gill, 3 km to the northwest. They concluded that at Burnes Gill there is a linear relationship between storm rainfall totals and peak streamflow irrespective of season and antecedent soil moisture conditions.

Langden Brook

Sediment movement and channel change have been monitored at the Langden site over the period December 1972 - April 1975, through sediment tracer experiments at several locations and by repeated photography and visual monitoring of 20 sample positions (fig. 1).

Autographic streamflow data are available from a streamgage approximately 3 km downstream from the study site. Hitchcock (1977b) suggested that flood discharges at the study site are approximately 0.81 times those at the stream gage, and that those for that portion of the study site above the tributary junction are 0.50 times those at the gage. In this paper frequency is more important than at-site discharge; only the gaging station discharges from which the frequencies have been derived are quoted. Long term daily rainfall records are also available from the stream gage site. Event frequency has been estimated using both streamflow and rainfall records. Estimates of long term frequency on a basis comparable with the other study sites have been based ultimately on the long term rainfall data. Daily rainfall correlates with flood-peak discharge with a correlation coefficient of 0.71, from a sample of 65 daily rains in excess of 12 mm, irrespective of season.

River Cound

Channel changes at the Cound sites were monitored over the period December 1971 - October 1974, by measurements from pegs to the outer banks of 11 eroding arcs, to record rates of bank erosion, and by repeated cross section surveys at 18 sites. Some results from the Condover reach have been published (Hughes, 1977). This paper extends the coverage to the other two reaches and examines the frequency aspects of the system adjustment in relation to the other study areas.

Temporary stream gages were established at each study site. The gage at Condover was equipped with a flood peak stage recorder. No longer term stream gages exist on the Cound, and flow frequencies have had to be estimated by establishing a correlation between floodpeaks on the Cound and those on the neighboring Rea Brook. The Rea lies immediately to the west of the Cound draining a catchment of similar geology, relief and land use. The relationship, based on 24 floods, has a correlation coefficient of 0.80 and may be summarized by

$$Q_R = 2.75 \; Q_C^{1.00} \qquad\qquad (1)$$

where Q_R and Q_C = floodpeak discharge (m^3/sec), for the Rea, and Cound, respectively. There are good correlations between discharge at Condover and at the other sites (correlation coefficients 0.89 and 0.86, from samples of 29 and 19 events, respectively). However, because the primary concern of this study is with frequency, only the discharges at Condover are quoted on the assumption that individual floodpeaks have a similar frequency upstream and downstream.

Finally, in order to compare event frequency on the Cound with that in the other areas, estimates of long term frequency have been made. As at Langden, these estimates were obtained by using the flood frequency curve for the Rea and daily rainfall frequency from a nearby rainfall station. The only nearby rainfall station with an appropriate run of data is at Harnage Grange, 8 km southeast of Condover. Although there are only poor overall correlations between daily rains at Harnage Grange and floodpeaks on the Rea or the Cound (correlation coefficients 0.20 and 0.39), when the data are treated seasonally better correlations can be demonstrated. The correlation coefficients range, winter and summer, from 0.57 to 0.59 for the Rea and 0.77 to 0.76 for the Cound from samples of 27, 21, 13, and 7 events. All are significant at least at the 5 % level. Over the long term, as there is no marked seasonality of heavy rainfalls, estimates of event frequency derived from comparisons of the flood frequency curve with the longer period daily rainfall frequency curve appear reasonable.

THRESHOLD IDENTIFICATION

Grains Gill

In Grains Gill rainfall thresholds have been identified on the basis of gully erosion and downstream channel adjustment. (Thresholds here are defined as boundaries or limits between classes of events.) This paper refines the thresholds identified in earlier papers (Harvey, 1974, 1977) utilizing an extended run of data. Table 1 summarizes the event classification. The sediment-production classes of this table are based on the quantities of sediment collected in traps (Harvey, 1974) over the duration of 111 periods of about 10 days each. The channel-change classes of the table are based on repeated photography of fixed points (Harvey, 1977) and cover 99 periods of about 1 month each.

147

Table 1

Grains Gill Event Classification

Sediment Production Classes

(based on Harvey,1974, p. 52-53)

Class GS1   Major sediment yield, indicative of widespread surface erosion.

Class GS2   Transitional yield, indicative of very limited surface erosion, usually with heavy rill wash.

Class GS3   Minor yields, no surface erosion evident.

Channel Change Classes

(based on Harvey,1977, p. 308)

Class GC1   Major channel changes, substantial removal of material from all debris cones; major channel adjustment.

Class GC2   Important channel changes, involving at least some removal of material from debris cones and some modifications to channel patterns.

Class GC3   Minor channel changes; no substantial removal of material from debris cones, but minor changes to within-channel gravel bars.

Class GC4   No observable changes in channel form, but some movement of individual particles.

Class GC5   No gravel sediment movement observed.

The additional data have enabled several refinements to be made over the earlier work. In sediment-production class GS1, those events that resulted solely from heavy snowmelt without concurrent heavy rain (a total of six events during the three winters) have been omitted from the analysis. In regard to the channel changes, it has now become possible to treat class GC1 as distinct from class GC2. These classes were formerly treated together as class ABC (Harvey, 1977). The revised best-fit thresholds are given in table 2, which also summarizes the period maximum daily rainfalls for each class of event. In each case, the threshold daily rainfall has been defined as that which separates the event periods into classes containing the fewest exceptions to the thresholds (class limits). The definition has been further refined, except in the case of GC1 where too few data are available, by the use of the least squares discriminant method. With this method the threshold is defined by the value about which the sums of the squares of the exception values is least.

The additional data confirm the main sediment production threshold (GS1) and allow definition of a lower threshold, which appears to relate to the initia-

Table 2

Grains Gill Daily Rainfall Thresholds

Class	n	Daily rainfall (mm)					Best-fit threshold (class base)	Event Separations (a)		Exceptions (b)	
		Maximum	Period maximum					n(1)	n(2)	(1)	(2)
			UQ	Median	LQ	Minimum					
GS1 (c)	50	98	31.8	20.9	15.6	9.4	13	50	61	5	3
GS2	15	19.9	13.2	9.4	7.7	4.4	8.2	65	46	3	5
GS3	46	11.6	4.3	2.8	1.0	0					
GC1	4	Actual (d) 98, 57, 51, 31 (e),					51	4	95	1	1
GC2	6	values 60, 51, 44, 43, 42, 37					43	10	89	3	3
GC3	32	51 (f)	33	25	17	10.0	19	42	57	10	13
GC4	30	35	22	14.5	9.0	2.0	10	72	27	8	9
GC5	27	20	11.6	8.1	3.8	0.6					

(a) n(1) is the total number of event periods classified above the threshold for the class
n(2) is the total number of event periods classified below the threshold for the class
(b) (1) number of event periods not in this class which have values above the threshold for the class
(2) number of event periods in this class which have values below the threshold for the class
(c) excludes 6 snowmelt events, occurring without major rainfall
(d) actual rainfall values are given because of the small sample sizes for these classes
(e) 48 mm rainfall at Sedbergh over 2 days, maximum on one day (0900-0900 hrs) 31 mm
(f) no changes above Class GC3 detectable, but this event followed soon after a Class GC2 event. Next highest rainfall for a Class GC3 event was 47 mm

UQ and LQ are upper and lower quartiles respectively
n is number of event periods in each class

tion of sediment production processes. Apart from the six aberrant snowmelts mentioned above, these thresholds show no seasonal variation. The extra data have also allowed greater precision in defining the channel change thresholds, with some slight modifications. A differentiation between classes GC1 and GC2 at approximately 51 mm daily rainfall can now be suggested. Again, no seasonality is evident.

A. M. Harvey, D. H. Hitchcock and D. J. Hughes

Events of classes GC1 and GC2 are the major events controlling the morphological adjustment of the system. The relationship between these major events and events represented by the other classes (GS1, GS2, GC3, GC4) cause morphological variations within the dynamic equilibrium of the system.

Langden Brook

At Langden water-discharge and rainfall thresholds have been identified in relation to sediment transport and channel adjustment. The thresholds are based on tracer experiments and site observations at 20 locations over 18 periods during 1972-75. Table 3 gives the classification of events based on the most extreme change recorded. Hitchcock (1977a) summarized the period data. In this paper thresholds between classes are suggested, and are based on period floodpeak discharges and maximum daily rainfalls (table 4). In most cases the threshold was taken as the midpoint between the lowest discharge or rainfall in the upper class, and the highest from any lower class. Threshold LC3 was defined by the least squares discriminant method.

Table 3

Langden Brook Event Classification

(based on Hitchcock 1977a, 1976b)

Class LC1	Major channel changes, resulting from substantial movement of sediment of all sizes and formation of completely new gravel bars. Substantial erosion of floodplain material; completely new channel patterns may result from secondary anastomosis.
Class LC2	Considerable channel changes, resulting from transport of sediment of sizes over T N D 200 mm and formation of new gravel bars. Bank erosion widespread. Some spillage onto the lower parts of the floodplain, but not sufficient to cause fundamental shifts in channel position.
Class LC3	Cobbles larger than T N D 80 mm moved, some modification to the form of gravel bars. Some bank erosion occurs.
Class LC4	Gravels finer than T N D 80 mm moved; no bank erosion evident, minor collapse may be observed but this not due to hydraulic action.
Class LC5	Sand grades moved.
Class LC6	No sediment movement detected.

TND = True Nominal Diameter
    = cube root of product of axes of an ellipsoide

    = $\sqrt[3]{a \cdot b \cdot c}$ a = long axis, b = intermediate axis, c = short axis

Table 4

Langden Brook Thresholds for Discharge and Rainfall

Period Class	n	Floodpeak discharge (Q) (m³/sec)	Daily rainfall (P) (mm)	Approximate thresholds for class base (m³/sec)	(mm)
LC1	1	28.9	65.4	25	58
LC2	2	20.2, 18.6	46.5, 44.1	17.2	43
LC3	4	15.8, 12.1 12.1, 9,5	52.0, 41.1 36.8, 31.8	12.0 [a]	39 [b]
LC4	10	Max 13.7, UQ 11.5 Median 10.0 LQ 7.6, Min 3.4	Max 46.2, UQ 38.0 Median 33.8 LQ 27.0, Min 20.7	4.5	19
LC5	1	4.5	17.8	–	–
LC6	0				

(a) Determined by the least squares method, n(1)=4, n(2)=10, Exceptions(1)= 1, (2)=1.
(b) Determined by the least squares method, n(1)=4, n(2)=10, Exceptions(1)= 2, (2)=2.

UQ and LQ are upper and lower quartiles respectively,
n is number of periods

Events in class LC1 are the major ones to which the system adjusts. These events include overbank flow, secondary anastomosis, and the scouring of new channels. Events of class LC2 also involve flow through former channels on the floodplain but with less scour and therefore less radical alteration to channel configuration. Classes LC3 and LC4 primarily involve within-channel adjustment and are secondary in importance.

River Cound

On the Cound thresholds of water discharge were identified on the basis of bank erosion and channel adjustment. The thresholds are based on observations of bank erosion at 103 locations at Condover, observed on 21 occasions during 1972-74 (Hughes, 1977). These data were augmented by data from Dorrington (64 locations) and Leebotwood (40 locations), both observed on 11 occasions during 1973-74. The observation period data at each reach have been classified according to table 5. Appropriate thresholds, in terms of period peak discharges at Condover, are given in table 6. The classification of events and the definition of thresholds have been slightly modified from those based on Condover reach alone (Hughes, 1977), but there is no evidence to suggest dissimilar behavior between the three reaches. There are few

Table 5

River Cound Event Classification

(based on Hughes, 1977)

Class CC1	Major erosional event; more than 50 % peg sites in each reach affected by bank erosion; mean erosion at Condover at least 20 mm, at Dorrington and Leebotwood at least 10 mm. Major movement of bed sediment occurs.
Class CC2	Moderate erosional event; 15-49 % peg sites in each reach affected by bank erosion; mean erosion at Condover 1.5-19.9 mm, at Dorrington and Leebotwood 1.0-9.9 mm. Movement of bed sediment occurs.
Class CC3	Minimal erosional event; less than 15 % peg sites in each reach affected by erosion or mean erosion at Condover less than 1.5 mm, at Dorrington and Leebotwood less than 1.0 mm. Little movement of bed sediment occurs.

exceptions across the thresholds and only one of any importance. This is the event recorded in January, 1973, as a Class CC1 event at all three reaches, in spite of the fact that peak discharge at Condover was only 3.1 m^3/sec. This low flood occurred after a period of frost and snow when the stream banks were thoroughly wetted and disturbed by frost action, allowing a relatively small event to achieve considerable erosion. Similar behavior was described by Wolman (1959).

Events of class CC1 control the overall form of the channel although the threshold occurs at a discharge well below bankfull (see next section). The events of class CC2 cause minor adjustments within the channel.

EVENT FREQUENCY

The recurrence intervals (R.I.) of the thresholds identified in the previous section have been estimated from rainfall and streamflow data at nearby gaging stations. These thresholds were compared to the long term recurrence intervals of daily rains, derived by the partial duration series (Langbein, 1949; Dunne and Leopold, 1978) from data periods as near 40 years each as then-available data permitted (fig. 3). For rainfall thresholds this has been done directly at Grains Gill, using the Sedbergh records, and at Langden (table 7). At Langden and the Cound, long term frequencies for discharge thresholds have been estimated indirectly. Flood frequency curves (fig. 4) for the shorted periods of flood data were used to convert the shorter period rainfall frequencies (also shown on fig. 3) to the equivalent long-term frequencies. At Langden, where both methods have been used, the results are similar (table 7).

Table 6

River Cound Threshold Definitions

	Condover		Dorrington		Leebotwood	
	n	Q (m³/sec)	n	Q (m³/sec)	n	Q (m³/sec)
Class CC1	3	10.0, 8.9, 3.1	2	8.9, 3.1	1	3.1
Class CC2	11	Max 6.0, UQ 4.1 Median 3.3 LQ 2.3, Min 2.1	6	Max 6.0 Median 2.9 Min 1.49	4	8.9, 6.0 2.3, 2.1
Class CC3	3	1.49, 1.00, 0.87	2	2.3, 0.87	1	0.87

Approximate thresholds:

CC1 7 m³/sec at Condover,

    at Condover   n(1)=3, n(2)=14, Exceptions (1)1, (2)0.
    at Dorrington n(1)=2, n(2)=8, Exceptions (1)1, (2)0.
    at Leebotwood n(1)=1, n(2)=5, Exceptions (1)1, (2)1.

CC2 1.9 m³/sec at Condover,

    at Condover   n(1)=14, n(2)=3, Exceptions (1)0, (2)1.
    at Dorrington n(1)=8, n(2)=2, Exceptions (1)1, (2)1.
    at Leebotwood n(1)=5, n(2)=2, Exceptions (1)0, (2)0.

All discharge figures are at Condover.
UQ and LQ are upper and lower quartiles respectively,
n is the number of periods in each class.

Within each area the major thresholds (those defining classes GC1, GC2, LC1, LC2, CC1) control the overall adjustment of the system. The moderate thresholds (which define classes GS1, GC3, LC3, CC2, and possibly LC4) influence that adjustment. The minor thresholds (for classes GS2, GC4, LC4?) define the magnitudes of events below which little significant geomorphological activity takes place. The major thresholds are exceeded every 0.5 yr to 2 yr. These frequencies are generally similar to bankfull frequencies described in the literature, but the events are not necessarily comparable. Recent work by Williams (1978) questioned the concept of a common bankfull frequency. A study by Knowles (1971) on a typical upland British catchment demonstrated a more frequent bankfull discharge than is commonly cited in the literature. The return period of 0.24 yr (partial duration series) is reasonably consistent throughout the 19 sites investigated (drainage areas range from 0.5 km² to 98 km²).

Within our areas the major thresholds differ in their relationships to bankfull conditions. On Grains Gill the channel is steep, mobile, and ill defined, so

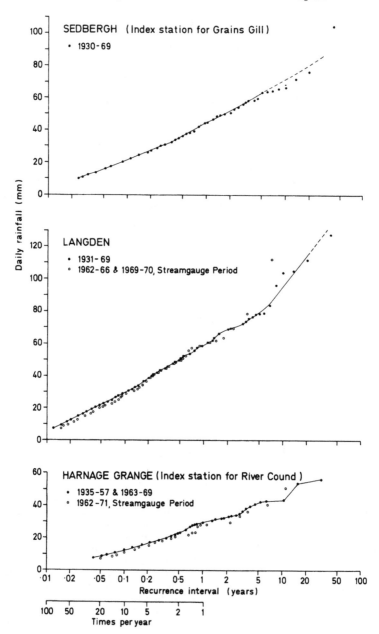

Figure 3. *Rainfall frequency curves for the index raingages, using the*
partial duration series for the long term daily rainfall data.

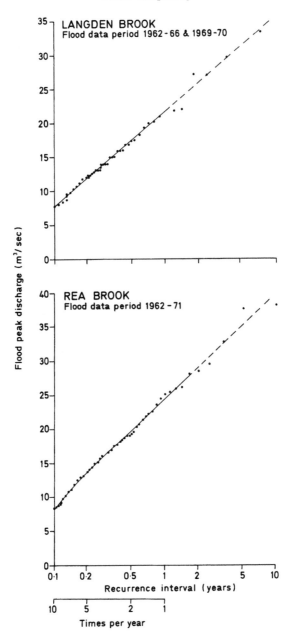

Figure 4. Flood frequency curves for Langden Brook and the Rea Brook, using the partial duration series for the short term periods of flood data availability.

155

Table 7

Threshold Frequency

	Threshold (base of class)		R.I. at gaging sta. [a]	Long term R.I. from index raingage	
	Daily rainfall (mm)	Discharge (m^3/sec)	(yr)	(yr)	(times/yr)
Grains Gill					
GS1	13	-	-	0.034	30 [b]
GS2	8.2	-	-	0.018	52 [b]
GC1	51	-	-	2.0	0.5
GC2	43	-	-	0.85	1.2
GC3	19	-	-	0.07	14
GC4	10	-	-	0.024	40
Langden Brook					
LC1	-	25	1.60	1.10	0.91
	58	-	-	0.90	1.11
LC2	-	17.2	0.50	0.25	4
	43	-	-	0.28	3.6
LC3	-	12.0	0.16	0.13	7.7
	39	-	-	0.20	5
LC4	-	4.5	0.06	0.048	20.8
	19	-	-	0.038	26.3
River Cound					
CC1	-	7 [c]	0.50 [d]	0.45	2.2
CC2	-	1.9 [c]	0.075 [d]	0.052	19

(a) Based on short term gaging-station records.
(b) Add to this figure approximately two events per year resulting from snowmelt alone.
(c) Discharge at Condover.
(d) R.I. based on Cound, Rea regression relationship (equation 1).

bankfull is not a very useful concept; its equivalent appears to lie between the thresholds for classes GC1 and GC2. At Langden, the threshold for class LC1 involves overbank flow, but the threshold for LC2 also involves some spillage across the floodplain into former channels. On the Cound the maximum flood during the field period (10 m^3/sec, long term R.I. estimated to be 1.2 yr) caused some spillage onto the lower parts of the floodplain. However, estimates of bankfull discharge at stable sections of the Cound, from stage-discharge relationships, are nearer 12 m^3/sec (long term recurrence interval estimated at about 2.8 yr). Quite clearly the threshold of class CC1 (discharge at Condover of 7 m^3/sec) occurs at stages well below bankfull on all three reaches of the River Cound. In each study area, major adjustment begins at stages below the bankfull equivalent.

Of the moderate thresholds that for class GS1 indicates substantial sediment supply, and those for classes GC3, LC3 and CC2 are of importance to within-channel adjustments. These events occur from 5 - 30 times a year. Minor thresholds which do not appear to have much morphological influence (classes GS2, GC4, LC4) occur from 25 - 50 times per year.

DISCUSSION - DYNAMIC EQUILIBRIUM

Within each of the areas dynamic-equilibrium is related not only to the major thresholds but also to the relationships between the major and moderate events. Field observations have permitted examination of two aspects of adjustments within dynamic equilibrium: (a) the morphological responses to particular sequences of major and moderate events, and (b) the sensitivity of the equilibrium in the context of the unusual climatic conditions of the early 1970's.

The seven-year period from September 1969 to August 1976, was the driest on record in many parts of Britain. In the study areas the period was the driest since at least 1930, and it culminated in two extremely dry summers (1975 and 1976). During only one of the seven years did the annual precipitation approach the long term mean annual figure at Sedbergh and Langden. None of these years produced rainfall near the average for Harnage Grange. The drought caused the marked discrepancies between the threshold frequencies predicted on the basis of the long term return periods and the actual occurrences during the field investigations (table 8). At Grains Gill and on the Cound these discrepancies are most marked in relation to the moderate and minor thresholds rather than the frequency of the major events. At Langden the discrepancy is greatest for the major events.

The sequence of events at Grains Gill is summarized by table 9. An illustration of the resulting morphological adjustments is given in figure 5. The data period (1969-77) can be subdivided on the basis of the occurrence of class GC1 events and the end of the 1975-76 drought. From 1969 to August 1971, although there were fewer events than predicted, there were at least the expected number of major events (classes GC1, GC2). Through this period the gullies maintained a dynamic-equilibrium form. Debris removal took place as a result of the major events, and the eroding slopes remained fully active. Slope activity is indicated by the vegetation cover, mostly grasses, on the gullies. From 1969 to 1971, there were large bare areas and

Table 8

Threshold Exceedence, Predicted and Actual

	Number of events (cumulated) exceeding each threshold											
	Grains Gill September 1969– September 1976						Langden [a] December 1972– April 1975				Cound January 1972– January 1974	
	GC1	GC2	GC3	GC4	GS1	GS2	LC1	LC2	LC3	LC4	CC1	CC2
Predicted [b]	4	8	98	294	203	392	2	10	19	50	5	48
Actual [c]	4	8	71	268	159	335	1	3	9	52	6	28

(a) On the basis of floodpeak discharges.
(b) Predicted on the basis of the long term recurrence intervals (table 10), for the duration of the field period in each study area.
(c) Actual occurrence, on the basis of rainfall or streamflow data for field period in each study area.

Table 9

Grains Gill Threshold Exceedence 1969-1977

	Number of rainfall events (cumulated) exceeding each class threshold					
	Classes					
Period to:	GC1	GC2	GC3	GC4	GS1	GS2
Nov. 1969 [1]	1 (0.2)	1 (0.4)	2 (5)	3 (14)	2 (10)	5 (19)
Aug. 1971	1 (0.9)	3 (2)	17 (24)	71 (74)	41 (51)	88 (98)
Sept.1975	1 (2)	3 (5)	45 (57)	163 (172)	98 (119)	208 (230)
Jan. 1976	1 (0.2)	1 (0.4)	3 (5)	12 (14)	10 (10)	18 (19)
Aug. 1976 [2]	0 (0.3)	0 (0.8)	4 (9)	19 (28)	11 (19)	19 (37)
Sept.1977 [3]	0 (0.6)	2 (1.3)	15 (15)	84 (46)	54 (32)	104 (62)

(1) Period begins September 1969.
(2) End of 1975-76 drought.
(3) End of observations.
Actual values for GC1 GC2 from field observations, the others estimated from rainfall data. Expected values, predicted from long term return periods, are given in parentheses.

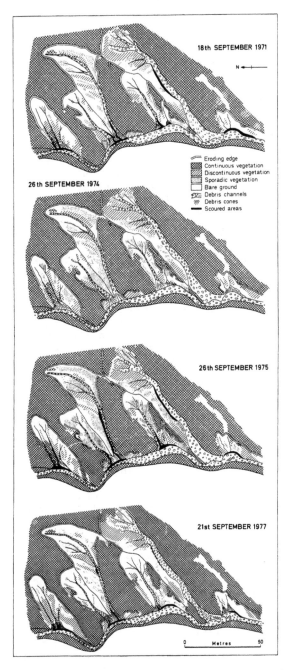

Figure 5. Grains Gill gullies, morphological sequence September, 1971 – September, 1977.

only small areas on the gullied slopes had even a sporadic plant cover (fig. 5).

No class GC1 events occurred for the next four years until September 1975, a longer period between GC1 events than appears to have occurred at any other time since 1930. There were also fewer GC2 events than expected. During this period the gullied slopes were partially stabilized by the growth of large debris cones and continued vegetation encroachment, especially during 1974 and 1975. However, the major flood of September 1975 removed much of the debris, trenched many of the gully channels, and re-activated the lower parts of the slopes (fig. 5).

During 1976 and 1977, despite the drought of 1976, the gullied slopes appear to have returned almost to their earlier form. Vegetation cover was reduced to a level similar to that of 1971. A GC1 event in January 1976, and two GC2 events during 1977 were sufficient in size to remove debris cones, thus maintaining active erosion on the slopes. The trend away from the earlier dynamic equilibrium towards stabilization that was apparent during 1974 and 1975 did not persist into 1976 and 1977.

At Grains Gill channel and gully morphology is controlled by major events of classes GC1 and GC2 and by their relationships to the moderate events, particularly to sediment production.

Adjustments at Langden can be illustrated by net changes in bed topography and by the response of the channel to the event sequence of the study period. Detailed surveys of the channel bed and gravel bar morphology were carried out in April 1973 and April 1975 (fig. 6). Although there was little change on some of the stable bar surfaces the channel shows an approximate balance between areas of erosion and deposition during the period. Despite the reduction in flood occurrence (table 8) the channel maintained its dynamic-equilibrium form. This is in contrast with Grains Gill, where for the same period (1973-75) a marked stabilization is apparent. However, Langden did experience several major flood events just before and during this period: class LC1 events just before the period began, in November 1972, then again in November 1973; and class LC2 events in April 1973 and January 1975. These events appeared to be sufficient to maintain dynamic equilibrium and prevent stabilization of the system.

The response of the channel to the sequence of events occurring during the field period is illustrated by the morphological change observed at a site approximately 1.5 km downstream from the main site (fig. 7). In December 1972, the main stream curved to the left towards the north side of the valley floor leaving an exposed gravel bar on the inside of the bend. Between December 1972 and April 1973, there were six LC4 and three LC3 events which caused only minor modifications to the form of gravel bars. However, in April 1973, there was a major flood of class LC2. At the upstream end of the reach the flood straightened the main channel and altered the form of the bars. Downstream the main channel continued to follow the inside of the left hand bend to impinge on the north side of the valley floor, as before. Until November 1973, there were only eight minor events (class LC4), but in November there was a major flood of class LC1. This flood caused major channel changes whereby the previous main channel on the inside of the bend was blocked by sediment and abandoned. Substantial flood-

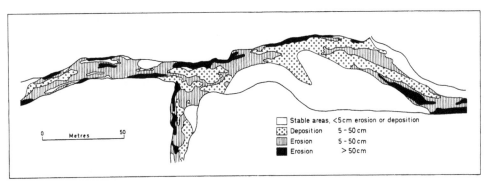

Figure 6. *Langden Brook net bed change at the study site, April, 1973 -*
April, 1975.

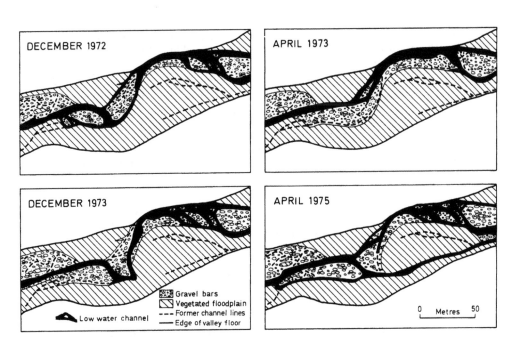

Figure 7. *Langden Brook, morphological sequence 1.5 km downstream of the*
main study site, December, 1972 - April, 1975. Flow is from
left to right.

plain erosion took place on the right bank sending the stream into a hairpin bend before entering the right hand bend on the north side of the valley floor (see map dated December 1973 in fig. 7). After this flood there were no major changes for 12 months despite the obvious instability of the hairpin bend. During this time 28 class LC4 and three LC3 events occurred. In January 1975, in response to a class LC2 flood, the hairpin was breached to the north by blockage of the channel and by spillage over the floodplain into a former channel that had last been occupied approximately 70 years ago. The head end of this new channel was scoured to depths of over 1 m. By April 1975, (fig. 7) after two more minor events, the reoccupied channel was taking most of the flow of the stream. Since then the new channel has widened, and the northern channel has been abandoned and has begun to be colonized by vegetation. This sequence of secondary anastomosis occurred in response to a class LC2 event, but only after major changes producing the hairpin bend resulting from the previous LC1 event.

Throughout the field period, despite the low occurrence of floods, there is no evidence of any progressive change. Dynamic equilibrium was maintained by the major floods that did occur and by minor changes in response to the other events.

On the Cound morphological adjustments can be identified at surveyed cross sections (table 10). At each site one section with stable banks, and several with eroding banks, have been monitored. The changes on the bed at the Condover stable section suggest that class CC1 events tend to be associated with scour and class CC2 events with aggradation. Net aggradation took place between May 1972 and May 1973, in response to a large number of CC2 events. Net scour occurred from May to October, 1973 and between October 1973 and February 1974. Both periods had few CC2 events relative to CC1 events. The shorter periods later in 1974 show little change, or minor aggradation, in response to CC2 events alone. The Dorrington and Leebotwood stable sections confirm this behavior. At these sections net scour occurred between May 1973 and February 1974, a period with five CC1 events, but there was net aggradation during 1974 in response to CC2 events alone.

This behavior is further confirmed by the bed changes at the eroding sites. At Condover during 1972-73 little net change occurred, but during 1973-74, with its higher proportion of CC1 events, there was net scour. Dorrington, and to a lesser extent Leebotwood, also show scour for 1973-74.

A net loss of bed sediment occurred during the field period but the field period shows a marked reduction in the occurrence of CC2 events (table 8). Over the longer term a dynamic equilibrium may be maintained in response to the relationship between CC1 and CC2 events.

A dynamic equilibrium in relation to bed sediment seems to be superimposed onto a slow progressive change in channel morphology. When bank sediment is taken into account (table 10) there is clearly a net loss of sediment, even when the erosion of low terrace material is discounted. At the eroding sections the channel appears to be increasing in width and the youngest parts of the floodplain are being built up to elevations lower than those being destroyed by erosion (fig. 8).

Table 10

River Cound, Sequence of Erosion and Deposition
at Surveyed Cross Sections 1972-1974

Stable bank sections

Dates		Event occurrence (totals in parentheses) CC1 CC2			Condover stable sect. Net bed change m^2	Dorrington stable sect. Net bed change m^2	Leebotwood stable sect. Net bed change m^2
8- 5-72	16- 5-73	1	10	(11)	+0.30		
16- 5-73	23-10-73	1	2	(3)	-0.64	+0.01	-0.07
23-10-73	21- 2-74	4	6	(10)	-0.99	-0.43	-0.09
21- 2-74	13- 3-74	0	1	(1)	-0.04		
13- 3-74	11- 7-74	0	2	(2)	+0.11	+0.15	+0.16
11- 7-74	31-10-74	0	1	(1)	+0.18	+0.12	-0.03

Eroding bank sections

Period and Reach	(n)	Net bed change mean m^2	Net bank change[a] mean m^2	Net bank change[b] mean m^2	Net section change[a] mean m^2	Net section change[b] mean m^2
1972-1973[c]						
Condover	7	-0.04	-0.11	-0.82	-0.15	-0.87
1973-1974[d]						
Condover	7	-0.53	-0.25	-1.13	-0.79	-1.55
Dorrington	3	-0.42	-0.14	-0.54	-0.55	-0.96
Leebotwood	5	-0.06	-0.05	-0.36	-0.11	-0.40

(n) Is the number of sections.
(a) Takes into account erosion and deposition below the surface of the
    modern floodplain.
(b) Takes into account all erosion and deposition, including erosion of
    material from low terraces.
(c) 7 September 1972 - 13 August 1973, includes 2 CC1 events and 9 CC2 events.
(d) 13 August 1973 - 20 July 1974, includes 4 CC1 events and 11 CC2 events.

163

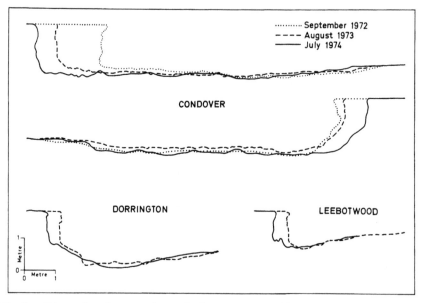

*Figure 8.   River Cound, examples of changes at eroding sections, September
1972, August 1973, and July 1974.*

Within each area there is evidence for dynamic-equilibrium, in the relation
of the relative frequencies of the major and moderate events. Changes in relative frequency could influence the stability of the system. A decrease in
stability could be brought about by lowering of an upper threshold due to the
cumulative effects of a number of moderate events. This has been observed
at Langden (Hitchcock, 1977a) where moderate events may influence the availability of sediment, or may cause changes in local flow directions triggering
localized instability and thus increasing the effectiveness of subsequent events.

An increase in stability may result from moderate events associated with
vegetation colonization and longer than usual period between major events.
This happened at Grains Gill during 1971-75 but the effects did not persist
into 1976-77, and dynamic equilibrium was restored. A similar tendency is
evident at Langden. Moderate and minor events deposit fine sand and peat
debris on gravel bars which may begin to stabilize by vegetation colonization:
however, the periods between major events are too short for this mechanism
to significantly influence the dynamic-equilibrium.

Finally it should be stressed that there is a certain random element within
the threshold frequencies identified. The random element is related not only
to a variable period between events but also to such aspects as the location
of the channel in relation to the debris cones at Grains Gill, or the location
of unstable flow conditions within the Langden and Cound channels.

CONCLUSIONS

The three study areas, representing different aspects of fluvial systems in upland Britain, all show characteristics of dynamic equilibrium. Morphology is adjusted to a range of event magnitudes, and particularly to the relationships between major and moderate events. The major, controlling events occur from two to four times per year to once every two years. The moderate events, influencing adjustments within the overall morphology created by the major events, occur about 14 to 30 times per year. The controlling events do not equate with bankfull conditions, and in this study the bankfull stage is not a very useful concept.

The stability of the equilibrium was viewed in the context of the dry spell of the early 1970's. At Langden, despite a very marked discrepancy between expected and actual event frequencies, there was no evidence of any departure from dynamic-equilibrium. At Grains Gill a temporary departure from equilibrium occurred in response to the reduced frequency of major events, but this departure was short-lived and the system appeared to return to equilibrium with the re-occurrence of major events. On the Cound fluctuations about equilibrium appear to result from variations in the relative frequencies of major and moderate events.

Over the longer term all three areas appear to operate within dynamic equilibrium. Even where slow progressive change may be taking place there is no evidence of any approach to a major system threshold. There is no evidence that any past events larger than those occurring during the study period have had any lasting influence on the contemporary morphology of the active landforms.

ACKNOWLEDGEMENTS

We are grateful to the Meteorological Office, Bracknell, and to the Northwest and Severn-Trent Water Authorities for permission to consult their records. Hitchcock and Hughes were supported by Natural Environment Research Council research studentships during their periods of field work. Field work at Grains Gill has been supported by University of Liverpool research grants. We are grateful to the photographic section of the Department of Geography, University of Liverpool for general photographic assistance and to the drawing office, particularly to Sandra, of the Department of Geography, University of Liverpool for producing the diagrams. We are also grateful to Professor K. J. Gregory of the Department of Geography, University of Southampton for his critical comments on the manuscript.

REFERENCES

Anderson, M.G., and Calver, A., 1977, On the persistence of landscape features formed by a large flood: Institute of British Geographers Transactions, v. 2, p. 243-254.

Carlston, C.W., 1965, The relation of free meander geometry to stream discharge and its geomorphic implications: American Journal of Science, v. 263, p. 864-885.

Church, M., 1972, Baffin Island sandurs: a study in Arctic fluvial processes: Geological Survey of Canada Bulletin 216, 208 p.

Costa, J.E., 1974, Response and recovery of a piedmont watershed from Tropical Storm Agnes, June, 1972: Water Resources Research, v. 10, p. 106-112.

Dunne, T., and Leopold, L.B., 1978, Water in environmental planning: San Francisco, W.H. Freeman and Co., 818 p.

Dury, G.H., 1961, Bankfull discharge: an example of its statistical relationships: Bulletin of the International Association of Scientific Hydrology, v. 5, p. 48-55.

_____, 1973, Magnitude frequency analysis and channel morphology; in Morisawa, M., ed., Fluvial geomorphology: Binghamton, [State University of New York], Publications in Geomorphology, p. 91-122.

Gupta, A., and Fox, H., 1974, Effects of high magnitude floods on channel form: a case study in Maryland piedmont: Water Resources Research, v. 10, p. 499-509.

Harvey, A.M., 1969, Channel capacity and the adjustment of streams to hydrologic regime: Journal of Hydrology, v. 8, p. 82-96.

_____, 1974, Gully erosion and sediment yield in the Howgill Fells, Westmorland; in Gregory, K.J., and Walling, D.E., eds., Fluvial processes in instrumented watersheds: Institute of British Geographers Special Publication 6, p. 45-58.

_____, 1975, Some aspects of the relations between channel characteristics and riffle spacing in meandering streams: American Journal of Science, v. 275, p. 470-478.

_____, 1977, Event frequency in sediment production and channel change; in Gregory, K.J., ed., River channel changes: New York, John Wiley and Sons, Inc., p. 301-315.

Hitchcock, D.H., 1977a, Channel pattern changes in divided reaches: an example in the coarse bed material of the forest of Bowland; in Gregory, K.J., ed., River channel changes: New York, John Wiley and Sons, Inc., p. 207-220.

_____, 1977b, Channel pattern changes in divided streams [Ph.D. dissertation] : Liverpool, University of Liverpool, 424 p.

Hughes, D.J., 1977, Rates of erosion on meander arcs; in Gregory, K.J., ed., River channel changes: New York, John Wiley and Sons, Inc., p. 193-205.

Kilpatrick, F.A., and Barnes, H.H., Jr., 1964, Channel geometry of piedmont streams as related to frequency of floods: United States Geological Survey Professional Paper 422E, 10 p.

Knowles, A.J., 1971, Channel morphology and flood hydrology of the River Greta, N. Riding of Yorkshire [M.Sc. thesis] : Liverpool, University of Liverpool, 68 p.

Langbein, W.B., 1949, Annual floods and the partial duration series: American Geophysical Union Transactions, v. 30, p. 879-881.

Patrick, C.K., and Gustard, A., 1974, Hydrology of upland catchments 1970-1974 (unpublished report): Transport and Road Research Laboratory, University of Lancaster, 23 p.

Pickup, G., and Warner, R.F., 1976, Effects of hydrologic regime on magnitude and frequency of dominant discharge: Journal of Hydrology, v. 29, p. 51-75.

Robinson, M., 1972, Sediment transport in Canadian streams: a study in measurement of erosion rates, magnitude and frequency of flow, sediment yields and some environmental factors [M.A. thesis]: Ottawa, Carleton University.

Schumm, S.A., 1973, Geomorphic thresholds and complex response of drainage systems, in Morisawa, M., ed., Fluvial geomorphology: Binghamton, N.Y., Publications in Geomorphology, p. 299-310.

_____, 1977, The fluvial system: New York, John Wiley and Sons, Inc., 338 p.

Schumm, S.A., and Lichty, R.W., 1965, Time, space, and causality in geomorphology: American Journal of Science, v. 263, p. 110-119.

Tywoniuk, N., and Cashman, M.A., 1973, Sediment distribution in river cross sections, in Fluvial processes and sedimentation, Proceedings of the Hydrology Symposium, Edmonton, University of Alberta, p. 73-95.

Williams, G.P., 1978, Bank-full discharge of rivers: Water Resources Research, v. 14, p. 1141-1154.

Wolman, M.G., 1959, Factors influencing the erosion of a cohesive river bank: American Journal of Science, v. 257, p. 204-216.

Wolman, M.G., and Gerson, R., 1978, Relative scales of time and effectiveness of climate in watershed geomorphology: Earth Surface Processes, v. 3, p. 189-208.

Wolman, M.G., and Miller, J.P., 1960, Magnitude and frequency of forces in geomorphic processes: Journal of Geology, v. 68, p. 54-74.

Woodyer, K.D., 1968, Bankfull frequency in rivers: Journal of Hydrology, v. 6, p. 114-142.

EFFECTS OF LARGE ORGANIC DEBRIS ON CHANNEL

FORM AND FLUVIAL PROCESSES IN THE

COASTAL REDWOOD ENVIRONMENT

EDWARD A. KELLER

Environmental Studies and Geological Sciences
University of California
Santa Barbara, California

TAZ TALLY

Geological Sciences
University of California
Santa Barbara, California

ABSTRACT

Large organic debris in streams flowing through old-growth redwood forests in California significantly influence channel form and fluvial processes in small to intermediate size streams. The role of large organic debris is especially important in controlling the development of the long profile and in producing a diversity of channel morphologies and sediment storage sites. The residence time for the debris in the channel may exceed 200 years.

The total debris loading along a particular channel reach represents a relation between rates of debris entering and leaving the reach. Loading is primarily a function of such interrelated variables as geology, valley-side slope, landslide activity, channel width, discharge, and upstream drainage area. Generally there is an inverse relationship between debris loading and stream size.

Large organic debris in steep mountain streams may produce a stepped-bed profile where a large portion of the stream's potential energy loss for a particular reach is expended over short falls or cascades produced by the debris. Approximately 60 % of the total drop in elevation over a several hundred meter second-order reach of Little Lost Man Creek is associated with large organic debris. The debris also provides numerous sites for sediment storage. Stored sediment covers up to about 40 % of the entire area of the active channel in the study sections. The sediment storage sites or compartments provide an important buffer system that regulates the bedload discharge.

The influence of large organic debris on channel form and process in low gradient stream reaches is less than in steeper channels. However, the debris still may affect development of pools and may help stabilize the channel banks. Root mats may armor banks and provide important fish habitats in the form of undercut banks. The stream channel of some low gradient reaches of Prairie Creek, California, may be quite stable. Lateral migration has only been one to two channel widths in the last several hundred years.

Management of streams in the coastal redwood environment so as to minimize adverse effects while maximizing anadromous fish habitat should consider the entire fluvial system. Managers should use natural stream processes to regulate channel conditions rather than strive for absolute control by artificial means.

INTRODUCTION

The virgin old-growth coastal redwood environment of Northern California is a unique forest system dominated by exceptionally large trees that grow to heights in excess of 100 m with diameters of several meters. They have an average life expectancy of about 1,000 yrs and may live more than 2,200 yrs (S.D. Veirs, 1978, personal communication). Even after death, large redwood trees resist decay and may remain on the forest floor or in stream channels for hundreds of years.

The range of ages of redwoods in old growth stands fits the concept of the "climax" forest. The density of trees in the studied area is approximately 50-125 trees per hectare, and the age distribution of the trees is uneven. Crown fires are rare because of the wet coastal climate, but surface fires have charred the bases of many trees. Mortality is dominantly a result of wind throw and landslides. The replacement rate for coastal redwoods is approximately 2-3 trees per 50 years per hectare (Veirs, 1978). This is a relatively low rate of reproduction compared to associated species in the forest such as Douglas fir, hemlock, spruce, and tan oak.

Large woody debris in the streams of the redwood forest can profoundly affect channel form and fluvial processes, particularly in small to intermediate sized streams. Old-growth forest along a small (first or second order) stream with abundant large organic debris is an exciting and aesthetically pleasing environment, providing a unique sense of permanency and perspective. In these small streams channel morphology is controlled by large (>10 cm in diameter) organic debris which is scattered where it fell in the channel. In larger (third and fourth order) streams the large organic debris tends to become concentrated in debris dams or jams that locally influence channel morphology and erosion-deposition patterns. However, the largest redwood debris (several meters in diameter) may remain where it fell for long periods of time. Large streams and rivers, such as Redwood Creek and the Klamath River, are capable of transporting even the largest redwood debris. Although it may be temporarily stored in mid- or side-channel bars, it is eventually transported to the sea. The influence of large organic debris on stream processes and form is thus most pronounced for small and intermediate sized streams. This generalization is consistent with studies in eastern hardwood or western Douglas fir forests (Keller and Swanson, in press).

The purpose of this paper is to discuss the role of large organic debris on channel form, fluvial processes, and development and maintenance of anadromous fish habitat for streams flowing through an old-growth redwood forest. Specifically we will consider: how the large organic debris affects channel width, depth, and slope; diversity of anadromous fish habitats such as pools, riffles and bars; areal sorting of bedload material; erosion-deposition patterns; and the instream residence time for large organic debris. We will also make preliminary recommendations for management of anadromous fish habitat in the coastal redwood environment.

STUDY AREAS

The study area is located in the Redwood Creek drainage basin, approximately 70 km north of Eureka, California (fig. 1). Two streams of contrasting environment -- Prairie Creek, a low gradient meandering stream, and Little Lost Man Creek, a steep mountain stream -- were studied in detail during summer low-flow conditions. Procedures included measurement and/or development of channel profiles and cross-sections, morphologic maps, sediment size distributions, debris loading, and minimum residence time of large organic debris in the channels. Channel characteristics and data on debris loading for the five study reaches on Prairie Creek and Little Lost Man Creek are summarized in table 1.

Prairie Creek, along most of the three reaches studied, is entrenched several meters into conglomerates and sandstones of the Plio-Pleistocene Gold Bluff Formation. Some of the entrenchment may be due to recent and ongoing tectonic activity.

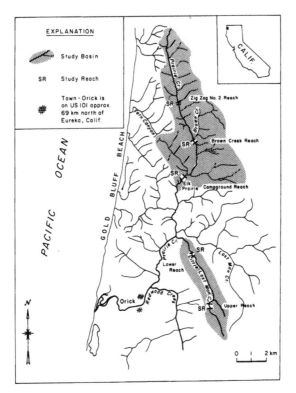

Figure 1. Index map of study areas.

Table 1

General Channel Characteristics for the Study Reaches of

Little Lost Man Creek and Prairie Creek

Reach	Thalweg Length (m)	Upstream area (km²)	Stream Order	Channel Width (m) (a)	Slope	Sinuosity	Debris Loading (kg/m²)	Pool to Pool Spacing (c)	% Channel area pool (e)	% Channel area riffle (e)	% Channel area stored sediment (e)	% Channel area under-cut bank (e)	Pool morph. influenced by debris (f) %	Debris controlled drop in elev. (g) %
Little Lost Man Cr.														
Upper	270	1.1	2	6.4	.033	1.1	141.6	1.9(d)	22	15	39	3	100	59
Lower	625	4.9	3	9.6	.048	1.1	49.0	1.8(d)	18	21	39	<1	90	30
Prairie Cr.														
Zig Zag No.2	220	5.7	2	6.7	.009	1.8	21.7	6.6	38	20	15	4	50	8
Brown Cr.	327	11.2	3	11.0	.010	1.2	84.8 (b)	6.0	26	18	29	<1	67	18
Campground	447	19.8	4	18.5	.005	1.3	19.6	4.0	25	25	13	1	50	<1

(a) Average width is the area of active channel divided by the reach length along the center line of the stream.

(b) Debris loading is calculated assuming density of wood averages 0.5 g/cm³. Debris loading is anomalously high for the Brown Creek reach. See text for explanation.

(c) Pool to pool spacing is measured in channel widths.

(d) Pool spacing in Little Lost Man Creek is due to spacing of large organic debris in the channel.

(e) Measured as percentage of total active channel area. Totals are less than 100 % because all stream environments are not included.

(f) Pools controlled or influenced by large organic debris as percentage of total number of pools in the study reach.

(g) Drop in elevation along the channel profile due to influence of large organic debris.

172

Little Lost Man Creek, a steep tributary of Prairie Creek, flows across steeply dipping sandstones, siltstones, shales and conglomerates of the Mesozoic Franciscan Formation. The local gradient of the stream is adjusted to the resistance of the various rock types.

LARGE ORGANIC DEBRIS: PROCESSES AND LOADING

The amount of large organic debris found in a channel reach is a function of the difference between the rate of debris entering the reach and that leaving. Large organic debris may enter a channel by landslides, blow-down of whole trees or portions of trees, bank erosion, and flotation of debris from upstream. Several of these processes may work in concert to deliver debris to stream channels. The dominant process depends upon local geologic conditions. For example, on steep sections of Little Lost Man Creek where the stream flows over resistant conglomerates and sandstone, the valley sides tend to be steep and landslides commonly deliver large organic debris to the channel. On the other hand, at locations where tributaries enter the stream, or along relatively low gradient sections where streamside trees are rooted in thick soils, undercutting of the stream banks may deliver most of the large organic debris to the channel.

Large organic debris, whether in the stream or on the banks, may be quite stable. For example, very large pieces of woody debris may be stabilized by having much of their mass resting out of the channel. Once in the channel, large organic debris may become stabilized by partial burial in sediment or by being wedged between other debris, boulders, or other obstructions. The debris may also be stabilized by the growth of nursed trees that send roots over the debris and into the soil, binding debris accumulations together.

Debris loading, for example the addition of one large root wad and trunk, may increase bank erosion which in turn is likely to undermine additional trees that subsequently will also fall into the channel. Furthermore, large debris in streams traps sediment and additional debris delivered from upstream by flotation (fig. 2). This process continues until the debris jam is removed by a combination of erosion, decay of supporting logs, and/or flotation during high flows.

Debris loading, measured in kilograms of vegetation per square meter of active channel (kg/m^2), was determined by measuring the length and diameter of all large organic debris with a diameter greater than 10 cm. Debris loading is calculated assuming the density of wood averages 0.5 gm/cm^3. This assumption facilitates comparison with other studies (e.g., Swanson and Lienkaemper, 1978; Keller and Swanson, in press) but may be questionable as the wet density of redwood may approach or even exceed 1.0 g/cm^3 (S.D. Veirs, 1978, personal communication). The value of 0.5 g/cm^3 is used cognizant of this limitation.

Because of the very large size of some redwood debris, a few large pieces may account for most of the debris loading in some reaches. For example, 60 % of the total loading along both the 400 m Upper Reach of Little Lost Man Creek and the 220 m Zig Zag No. 2 Reach of Prairie Creek consists of one redwood trunk (fig. 4 at DD6 and fig. 5 along sect. A-A').

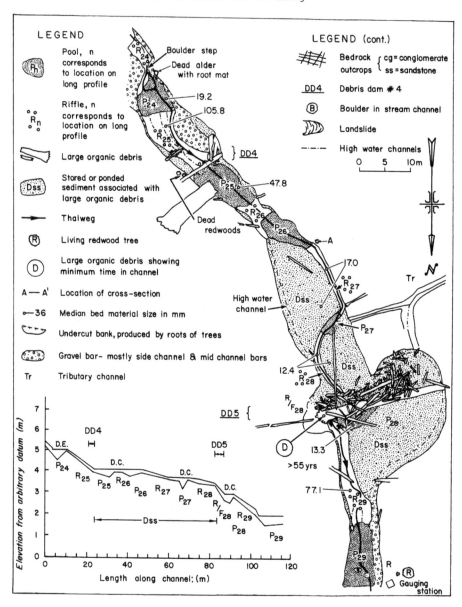

Figure 2. Morphologic map of a short section of the Lower Reach of Little Lost Man Creek. Notice the much wider channel at the site of debris dam (DD5), and the finer gravel which constitutes the debris-stored sediment. The location is downstream from the profile in figure 10. The symbol R/F indicates the location of a riffle-falls.

Redwood, Douglas fir, Sitka spruce, western hemlock, big-leaf maple and alder are the main contributors of large organic debris in the coastal redwood environment. However, because redwood debris tends to be very large and resistant to decay, it dominates the total loading.

Independent of other factors, in timber stands where all the trees are of about the same size, the debris loading should be greater in first and second order streams because the flow may not be able to float large debris. In larger streams, debris accumulates in distinct debris jams. With further increase in stream size, even the largest debris may be floated away. This is essentially the situation in Pacific Northwest streams that flow through Douglas fir forest (Swanson and Lienkaemper, 1978). In the coastal redwood environment, the situation is not entirely analogous because redwood debris occurs in a larger range of sizes. In spite of this, with the exception of the Brown Creek Reach of Prairie Creek, the general tendency of debris loading to decrease in the downstream direction as drainage area and channel width increase is apparent (table 1).

Debris loading in the study reaches of Prairie Creek varies from 19.6 to 84.8 kg/m². The highest measured loading occurs in the Brown Creek Reach and appears anomalously high when compared to the other two Prairie Creek reaches. Because most of the loading results from a few very large stems, perhaps the high loading represents a pocket of recurring blowdown. However, the origin of the high loading appears to be complex, involving several factors. Included is the long profile of Prairie Creek in the vicinity of the Brown Creek Reach. The profile is anomalous in that the study reach is located between two convex reaches along an otherwise uniform concave long profile (fig. 3). The origin of the convex reaches is not known but apparently is related to the geology and, hypothetically, to recent uplift. Admittedly dence for uplift is weak in that it is based primarily on anomalous topography including: local trellis drainage patterns, possible stream capture, and several meters of entrenchment of Prairie Creek in the vicinity of the convex profile.

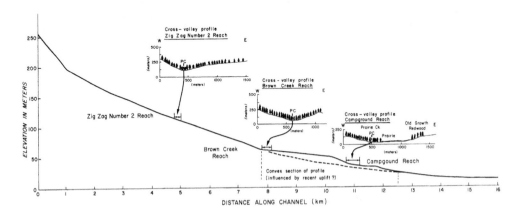

Figure 3. Long profile for Prairie Creek. Compiled from 1:24,000 scale topographic map.

175

Regardless of whether or not the entrenchment is due to recent tectonic activity, entrenchment influences debris loading by producing locally steep valley sides close to the channel. The cross-valley profiles of the three reaches (fig. 3) demonstrate that near Brown Creek the valley walls are steep and the valley width relatively narrow. Entrenchment is illustrated by the cross-channel profiles (fig. 8). The steep valley sides, more frequent entrance of tributary channels, and entrenchment increase the likelihood of large organic debris being delivered to the stream channel. Tributary junctions are especially significant because at these sites bank erosion acts along two fronts, thus increasing the prospects of a tree falling in the channel.

Measured debris loading in Little Lost Man Creek varies from 49 to 141.6 kg/m^2 (table 1). In the Lower Reach debris loading is lower. However, debris jams (or dams) are larger, more complex and spaced further apart than in the headwater areas (compare figs. 4 and 12 with figs. 13 and 14). A single large trunk generally supports the complex debris jams (see fig. 2, DD5 and fig. 4, DD3). Debris loading is locally higher along steep reaches where the frequency of landslides is relatively high (fig. 4).

Comparison of measured debris loading in Prairie Creek and Little Lost Man Creek with streams of similar size flowing through western Douglas fir forests shows interesting differences. For example, Mack Creek in the McKenzie River system, western Oregon, is a third-order stream with upstream drainage area of 6 km^2 and a channel width of 12 m; the debris loading is 28.5 kg/m^2 (Keller and Swanson, in press). By comparison the Lower Reach of Little Lost Man Creek, a third-order stream with upstream drainage area of 4.9 km^2 and a channel width of 9.6 m, has a debris loading of 49 kg/m^2. Furthermore, Devilsclub Creek and Watershed 2 Creek, also in the McKenzie River system, each with an upstream drainage area of less than 1 km^2, have debris loading of about 40 kg/m^2 (Keller and Swanson, in press). This is considerably less than the 141.6 kg/m^2 measured in the Upper Reach of Little Lost Man Creek, where the drainage area is 1.1 km^2. Debris loading in two other streams, Low Slope Schist Creek and Gans West Creek (both tributary to Redwood Creek) have also been measured. These small, steep streams, both with drainage area less than 1 km^2, flow through old-growth redwood forest and have debris loading of 51.9 kg/m^2 and 56.3 kg/m^2, respectively (F. Swanson, personal communication). These values are somewhat higher than the loading of small streams (for examples, Devilsclub and Watershed 2 Creeks) in the McKenzie River system flowing through western Douglas fir forest.

Additional data are needed before more meaningful comparisons of debris loading in streams flowing through different forest environments can be attempted. Nevertheless, the unusual size of the redwood debris demands that the general pattern of downstream decrease in debris loading be shifted downstream. Thus, in comparing two very similar second, third, or fourth order basins (one in redwood and the other in western Douglas fir forest), the debris loading at a particular location should be greater for the redwood basin.

CHANNEL FORM AND PROCESS

Large organic debris in forest streams may significantly influence channel form and fluvial process in small to intermediate size streams. This is especially true in the virgin coastal redwood environment where individual pieces

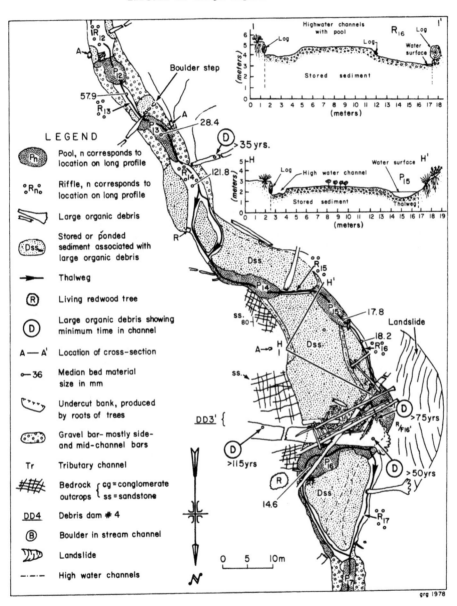

Figure 4. *Morphologic map of part of the Little Lost Man Creek, Lower Reach.*
The location is shown on figures 1 and 10. The landslide probably
started at least 75 yrs ago, which is the minimum residence time
for the large redwood trunk (D7). The pool, riffle and debris dam
numbers correspond to the long profile (fig. 10). The symbol R/F
indicates the location of a riffle-falls.

177

of organic debris can be extremely large. For example, a stream with an active channel width of about 5 m may have one entire bank, for a distance of several channel widths, completely formed and defended by a single downed redwood tree (see fig. 5, sect. A-A'). The role of large organic debris is especially significant in development of long profiles and of diverse channel morphologies and sediment storage sites.

Slope

Channel slope, one of the major dependent variables in the fluvial system, is a function of several independent variables including peak and mean annual water and sediment discharges, topography, size of bed material, bedrock type, and recent tectonic activity. All these factors, and others, interact to produce the long profiles of rivers. Furthermore, the long profile is adjusted so that there is a compromise between least work and equal work (Leopold and Langbein, 1962). Because channel slope tends to adjust over a period of years (Mackin, 1948), the long profile reflects recent development of the stream channel on a variety of scales. The patterns of erosion and deposition, as well as the effects of bedrock geology on profile development were examined.

The effect of varying rock type on channel slope was investigated along Little Lost Man Creek, where several sedimentary rock types occur. The long profile extending from the headwater reaches to a point several kilometers downstream through old-growth redwood forest, was surveyed with a hand level and tape during the summer of 1978. Figure 6 shows a 500 m segment of this survey. The slope is steepest where the stream flows over conglomerate and massive sandstone, intermediate over thin-bedded sandstones, and relatively gentle over shale. Thus, the profile clearly shows that slope is adjusted to geology. These data (fig. 6) are consistent with the conclusions of Hack (1957) on relations between bedrock resistance and channel slope.

Also shown on this profile are effects of debris dams. A good deal of the total drop in elevation occurs at sites of debris dams. Extensive stored sediment occurs just upstream from the larger accumulations of large organic debris. However, debris dams on some steeper sections associated with conglomerate are less discrete and less involved with profile control than are dams on the softer sandstone and shale sections. Bed material is much larger where the slope is steeper, and much of the large organic debris remains above the active stream channel, resting on large boulders. Being above the channel, the debris does not exert significant control on the channel-forming process.

Steep gradient sections of Little Lost Man Creek often contain waterfalls of various sizes interspersed with riffles. These sections are termed riffle/falls and are noted as R/F on our maps and profiles. The pools on these steeper reaches often form as plunge pools below the waterfalls.

Channel Depth

Variable channel depth (and width) is particularly important in producing diverse habitats for anadromous fish (primarily steelhead and salmon) as well as trout. A goal of this study was to evaluate fish habitat and how it may be managed and improved.

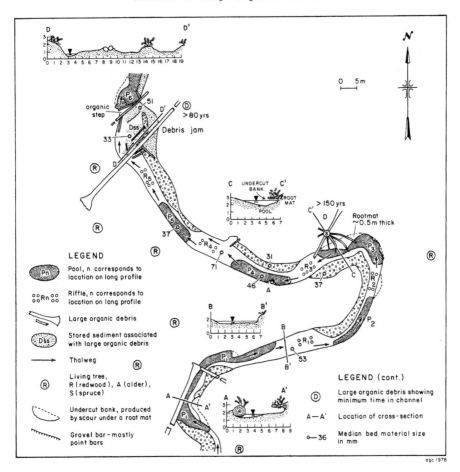

Figure 5. *Morphologic map for Zig Zag No. 2 Reach of Prairie Creek. Notice the large redwood trunk defending the bank at A-A', and the considerable area of undercut banks. The location of this reach is shown in figure 1. The pool and riffle numbers correspond to the long profile (fig. 7).*

Variation in channel depth is best analyzed on long profiles. Profiles of study reaches on Prairie Creek show well-developed pools and riffles (fig. 7). The upstream Zig Zag No. 2 reach has moderate debris loading (21.7 kg/m^2) and well-developed pools and riffles where large organic debris is locally very significant. At section A-A' in Zig Zag No. 2 Reach (fig. 5), the stream, for at least two channel widths, is defended by one large downed redwood tree. Undercut banks are common along this reach where the stream

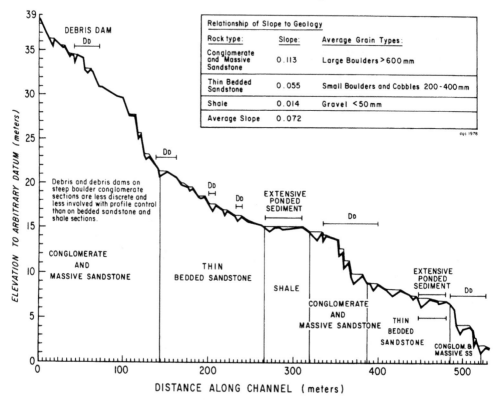

Figure 6. *Illustration of relationship between geology, channel slope, and bed material size for a 500 m reach of Little Lost Man Creek.*

scours beneath root mats. Undercut banks account for 4 % of the total channel area (table 1) and provide prime fish habitat. The Zig Zag No. 2 Reach has a relatively low gradient (0.009). Large organic debris account for only about 8 % of the drop in elevation along the 220 m reach. Nearly all of this is in one organic step at the downstream end of the reach. The phenomenon of organic stepping (Keller and Swanson, in press) is more important in relatively steep streams and will be discussed in the analysis of Little Lost Man Creek.

Figures 8 and 9 are morphologic maps along portions of the Brown Creek and Campground Reaches. These sites have well-developed pools and riffles. Particularly in the Brown Creek Reach, the organic debris loading is very high for Prairie Creek (84.8 kg/m^2; table 1) and the debris tends to form distinct debris jams. Most of these jams have been formed by one or two large trees that trap other debris derived in part by flotation from upstream. The debris dams are clearly associated with and facilitate the formation of major pools along the Brown Creek Reach and thus are important in providing valuable fish habitat. For the Brown Creek Reach, the debris loading is relatively high (84.8 kg/m^2), the major variations in depth of the channel are

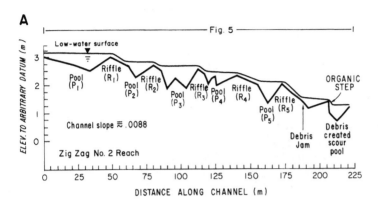

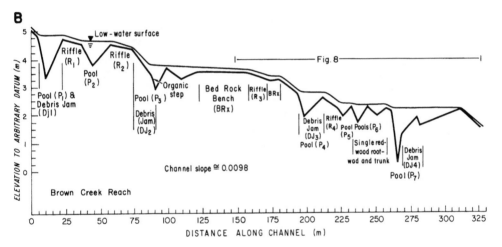

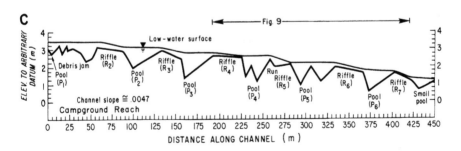

Figure 7.   Long profiles for the three study reaches of Prairie Creek; (a) Zig Zag No. 2 Reach, (b) Brown Creek Reach, (c) Campground Reach. Locations are shown on figure 1.

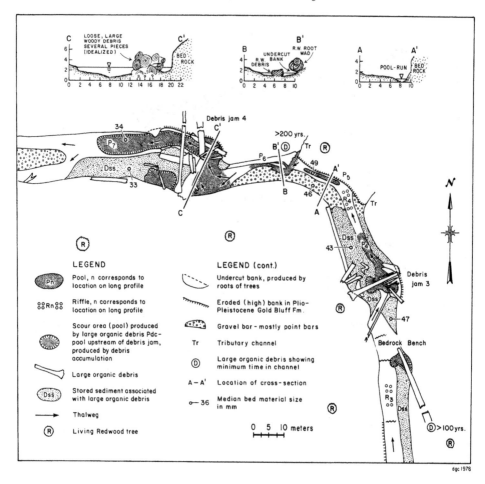

Figure 8. *Morphologic map of part of the Brown Creek Reach of Prairie Creek. Location is shown on figure 7. This reach shows distinct clumping of large organic debris and long residence time (>200 yrs) for one large redwood trunk and rootwad. The pool and riffle numbers correspond to the long profile (fig. 7).*

directly associated with (if not caused by) large organic debris, and 18 % of the total drop in elevation of the 327 m reach is at sites where large organic debris has accumulated.

Only one debris jam occurs in the 447 m Campground reach along Prairie Creek (fig. 9), but most of the major pools are associated with some large debris. A very large redwood log evidently causes differential scour. In fact, many pools are associated with such large debris. A definite cause-

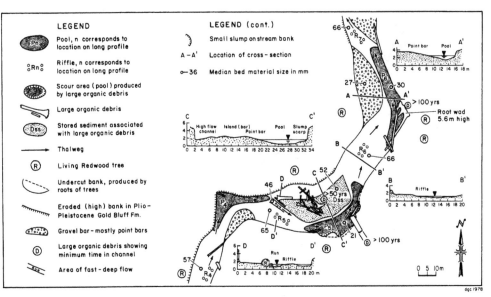

Figure 9. Morphologic map of part of the Campground Reach of Prairie Creek. Although debris loading is relatively low, large organic debris does significantly affect the development of channel forms such as pools. This site also has fairly good areal sorting of bedload material. Pool and riffle numbers correspond to the long profile (fig. 7).

effect relationship is difficult to demonstrate but the association is very strong. It is likely that large organic debris increases the variability of channel depth along the reach.

In Little Lost Man Creek the role of large organic debris in increasing the variability of channel depth is very pronounced. Figures 10 and 11 show profiles for the Upper and Lower Reaches and demonstrate some of the variability associated with large organic debris. In the Upper Reach the debris accumulations are discrete and more numerous. They provide many examples of organic stepping (Keller and Swanson, in press), which creates a variety of channel depths. Sediment is ponded above a debris dam and a pool is scoured immediately below the dam. The result of organic stepping is to produce a stream profile characterized by relatively long sections of stored sediment and low gradient alternating with short steep cascades or falls that terminate at a scour pool. The stepped profile is significant. Most of the downstream loss of potential energy takes place at the cascades or falls, thus reducing the energy available to erode the stream bed and banks. The pools provide temporary compartments for sediment storage. Furthermore, the steps produce a diversity of flow conditions and thereby create fish habitats (Swanson and Lienkaemper, 1978).

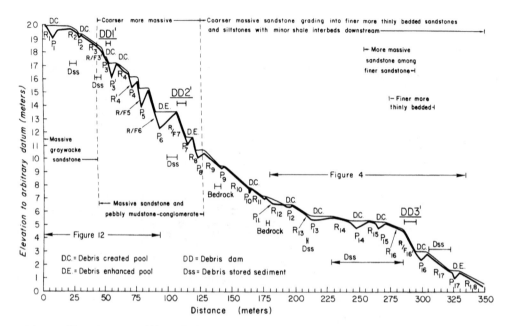

Figure 10. *Long profile of the upper 60 % of the Little Lost Man Creek, Lower Reach (location shown on fig. 1). The slope here is 0.056 compared to an average slope of .048 for the entire reach. The pool, riffle and debris dam numbers correspond to morphologic maps (fig. 4 and 12). The symbol R/F indicates the location of a riffle-falls.*

Channel Width

In some alluvial streams bankfull channel width is easy to define. However, in the coastal redwood environment trees and debris in the channel make it very difficult to define a characteristic channel width. In fact, the range of widths along a particular reach is often so great that one, or even several measurements, are nearly meaningless. Therefore, in this study the characteristic (average) width of active channel is defined as the area of active channel in a reach divided by channel length measured down the center line of the channel. The length of reach measured to calculate this width must be long enough to insure that several examples of various stream environments are included (pools, riffles, debris dams, etc.). Table 2 shows the variations of width for both Prairie Creek reaches and Little Lost Man reaches. With the exception of Brown Creek, the characteristic channel width is nearly the average of the channel width for riffles and pools (measured at their maximum width), exclusive of the debris dams. (Debris dams often have a channel width two or more times greater than the characteristic channel width.)

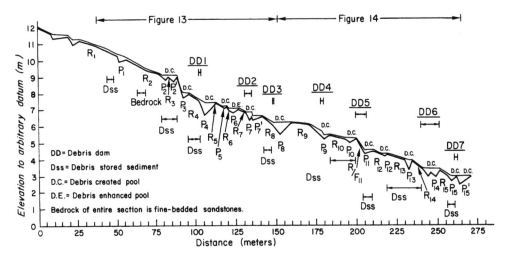

Figure 11.   *Long profile of the Little Lost Man Creek Upper Reach. The pool,
riffle and debris dam numbers correspond to figures 13 and 14.
Location of profile shown on figure 1. The symbol R/F indicates
the location of a riffle-falls.*

The configuration of a debris dam significantly affects both channel depth
and width. Some debris dams consist of a single log while others contain
several. Furthermore, some debris accumulations extend across the entire
channel, whereas others do not. Simple debris jams are common in the Upper
Reach of Little Lost Man Creek and many extend across the entire channel
(figs. 13 and 14). In the Lower Reach the debris accumulations are large
and only a few extend across the entire channel (figs. 2, 4, and 12).

Debris dams which extend across the entire channel in steep mountain
streams usually cause horizontal divergence of flow away from the center of
the channel. In turn this causes deposition of a mid-channel bar (stored
sediment) upstream from the dam. The bar diverts the flow toward the sides
of the channel, producing bank erosion and a locally wider channel. A
plunge pool may form at the base of the debris dam (see fig. 2, DD5). In
some instances debris dams that extend across the entire stream can also result in a more narrow channel width, by convergence of flow under the log
(see fig. 13, DD1). This type of dam often produces sediment storage upstream and a well-developed scour pool downstream.

Debris dams that extend across only part of the channel are also quite
common in Little Lost Man Creek. The dams divert the flow toward one bank
which may cause bank erosion and a wide channel. This may also cause local
scour (e.g., pool 5, fig. 13). In general these partial dams do not form the
pronounced organic steps associated with the large debris dams that block the
entire channel. However, they may influence the variability of channel width
and depth for tens of meters both upstream and downstream from the dam.

Table 2

Comparison of Some Channel Widths for Pools, Riffles and Debris Jams

Reach	Upstream drainage area (km^2)	Characteristic width (a) (m)	Mean pool width (b) (m)	Mean riffle width (m)	Mean width debris jam (m)	Range of width (m)
Little Lost Man Cr.						
Upper	1.1	6.4	7.1	6.6	8.1	2.4-15.6
Lower	4.9	9.6	10.8	11.4	17.2	2.0-24.0
Prairie Cr.						
Zig Zag No. 2	5.7	6.7	7.3	5.6	15.0	3.0-15.0
Brown Cr.	11.2	11.0	7.0	8.3	16.1	6.5-20.0
Campground	19.8	18.5	20.1	16.0	25.5	10.0-31.0

(a) The characteristic (average) width is the area of the active channel in the study reach, divided by the channel length.

(b) The mean widths of the pools, riffles and debris jams are the average of the widths measured at the location of maximum width for each pool, riffle or debris jam.

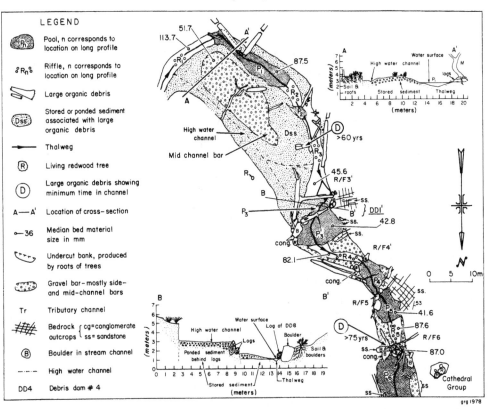

Figure 12. *Morphologic map of part of the Little Lost Man Creek Lower Reach. Location on long profile and corresponding pool, riffle and debris dam numbers are on figure 10. The symbol R/F indicates the location of a riffle-falls.*

Pools and Riffles

Observations of channel morphology were made at low water periods during summer. Little if any coarse bedload sediment is transported during these periods. Therefore observations were of channel features that developed in response to the higher flows during the winter months. Thus, the distribution of pools, riffles, and other channel features are, at times of low flow, relics of higher channel-forming stages.

Pools are topographic low areas of a channel produced by scour at higher flows. A variety of pools exists in the redwood coastal environment. Morphologic maps for Prairie Creek (figs. 5, 8, and 9) show locations of the major pools. In general, the spacing of pools varies from 4 to 6 channel widths, which is similar to the spacing of pools in other environments (Leopold and others, 1964; Keller, 1972). However, even in Prairie Creek there is consid-

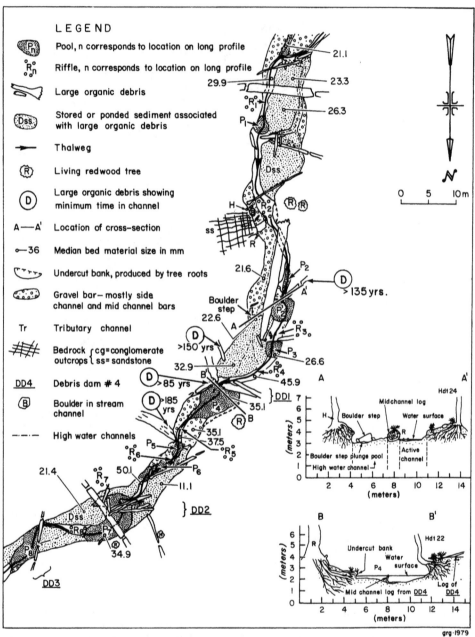

Figure 13. *Morphologic map of part of the Little Lost Man Creek Upper Reach. Location of long profile and corresponding pool, riffle and debris dams are shown on figure 11.*

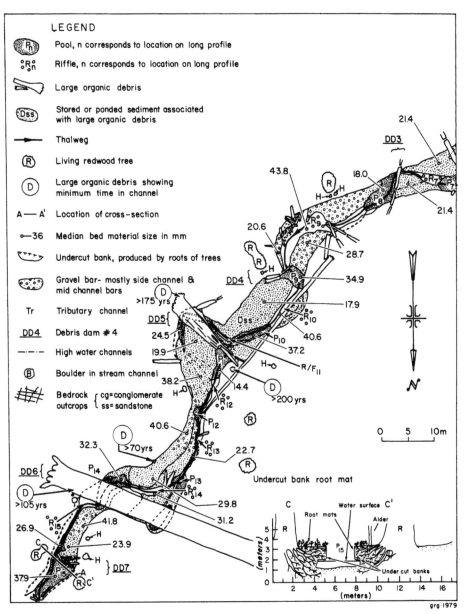

Figure 14.   *Morphologic map of part of the Little Lost Man Creek Upper Reach.
Location of long profile and corresponding pool, riffle and debris
dam are shown on figure 11. Notice that the large organic debris
is scattered along the channel rather than in distinct clumps or
groups. The symbol R/F indicates the location of a riffle-falls.*

erable variability in the development of the pools, and at least 50 % of them are associated with large organic debris if not directly formed by it. The effects of large debris are particularly pronounced in the Brown Creek Reach where all the major pools are directly associated with debris accumulations which store large volumes of sediment. Debris-stored sediment accounts for nearly 30 % of the total area of the stream in the Brown Creek Reach.

In the steeper Little Lost Man Creek, pool development is more directly related to the large organic debris. In the Upper Reach of Little Lost Man Creek (figs. 13 and 14) approximately 60 % of the total drop is associated with debris accumulations, whereas in the Lower Reach (figs. 4 and 12), only about 30 % is associated with the debris. The average spacing of pools in Little Lost Man Creek is about two channel widths, reflecting the spacing of large organic debris. With the exception of the Brown Creek Reach, approximately 50 % of the pools in Prairie Creek are similar to pools observed in meandering alluvial streams lacking significant large organic debris. Those in Little Lost Man Creek, on the other hand, are more characteristic of the mountain streams described by Heede (1972) and Swanson and Lienkaemper (1978). In these small mountain streams organic stepping is fully developed and most significant in controlling local erosion and depositional patterns.

The pool environment is particularly important to fisheries resources because the deep water provides cover and living space for young developing fish. Pools that have undercut banks provide additional habitat for fish. In Prairie Creek some of the pools have undercut banks that extend several meters beneath root mats. The area in undercut bank amounts to as much as 4 % of the total area of active channel (0.28 m^2 per meter of channel) in the Zig Zag No. 2 Reach of Prairie Creek compared to 3 % (0.22 m^2 per meter of channel) in the Upper Reach of Little Lost Man Creek (table 1).

Riffles are topographic high areas in the channel produced at relatively high flow by deposition. Riffles are particularly important in fisheries resources because spawning occurs in these locations. The head of a riffle is critical in fish production because the intergranular flow of water is greatest at this location, and it provides oxygen-rich water to developing fish eggs. In the Prairie Creek study reaches, 18-25 % of the total channel area is covered by riffles, with pools covering another 25-38 %. In Little Lost Man Creek the riffle area is approximately 21 % and 18-22 % for pools (table 1).

Sorting of Bed Material

Patterns of sorting were studied at approximately 100 locations by sampling the bed materials according to methods outlined by Wolman (1954). In Prairie Creek the sorting of bed material is best in the lower reaches where the hydrology of the stream is less affected by the organic debris. Here the largest bed material is characteristically found on riffles and finer material is found in pools (fig. 9). In the Brown Creek Reach, bed material is not as well sorted. Apparently this is due to the fact that large organic debris produces greater variability of the flow. In Little Lost Man Creek the sorting of bed material is also variable but finer gravels tend to be associated with sites of debris-stored sediment (fig. 2).

Lateral Migration

Although Prairie Creek has a relatively low gradient with well-developed pools and riffles and numerous meander bends, preliminary analysis suggests that in several locations little lateral migration has occurred during the last several hundred years. This was determined by mapping the distribution of large living redwood trees near the channel. Ages of these trees were estimated by first counting the rings of three large downed trees to determine the diameter-age relation in the environment and then by using this relationship to estimate the ages of living trees from their diameters. The diameters and ages of the 3 downed trees are: 1.9 m, 568 yrs; 2.2 m, 898 yrs; and 2.8 m, 1006 yrs respectively. Therefore living trees with diameters of 3-5 m are probably 800-1000 yrs old. The location of these trees suggests that along several reaches the lateral migration of the stream has been only one or two channel widths in the last several hundred years. However, this conclusion is based on limited available data and thus is tenuous. Furthermore, the apparent stability does not suggest that there is no change in position of the channel with time. Meander cutoffs have occurred at several locations along Prairie Creek, but these appear to be fairly rare events. Only two or three abandoned channels were observed along Prairie Creek in the areas of old-growth redwood stand.

Thus although large debris greatly affects sediment storage and sites of erosion within the channel, the stability provided by living rootmats on banks and large organic debris in the channel may restrict the lateral migration of the stream. However, all large organic debris does not increase channel stability. Debris that blocks or diverts the stream may cause bank and bed erosion, whereas trees that fall in the down valley direction may defend a considerable length of channel.

Sediment Routing and Storage

Large organic debris is important in the routing and storage of sediment. Studies by Megahan and Nowlin (1976) and Swanson and Lienkaemper (1978) have shown that for their study areas annual sediment yields in small forested drainage basins are generally less than 10 % of the sediment stored in the channels. In the Brown Creek Reach of Prairie Creek, the area of debris-stored sediment is approximately 30 % of the total channel area, whereas in Little Lost Man Creek the area of debris-stored sediment amounts to approximately 40 % of the total channel area (table 1).

A major effect of the large organic debris on sediment routing seems to be that of a buffer (Swanson and Lienkaemper, 1978). This is sufficient in the coastal redwood environment, particularly in the small streams where large organic debris is scattered along the length of the channel. Mass wasting processes produce most stream sediment (Janda, 1977). Were it not for the storage sites provided by the debris, the sediment would be transported quickly from the small tributaries to larger channels downstream. Thus the role of the debris in these small basins is to provide temporary storage sites and thereby lessen the impact of mass wasting.

E. A. Keller and T. Tally

RESIDENCE TIME OF LARGE ORGANIC DEBRIS

Large organic debris moves through the fluvial system primarily by flotation at high flows or perhaps, in very steep sections of the stream, by debris torrents (Swanson and Lienkaemper, 1978). Because stands of old-growth redwood include trees of varying sizes and ages, their debris is moved through the system during flows of varying magnitude. Very large debris in Prairie Creek and Little Lost Man Creek moves only very rarely. This was determined from the storage period of large debris in the channel. When a large redwood tree falls, nursed trees such as hemlock, spruce, and other redwood trees soon begin to grow on the downed tree. Coring of nursed trees provides a minimum estimate of the time that the debris has been in a particular location.

Some of the ages derived in this manner are shown on the morpohologic maps. Minimum residence time for 33 pieces of debris were determined. Only one is older than 200 yrs, but 17 are at least 100 yrs old (table 3).

Based on field observations of apparent residence times for fallen redwood trees, it appears that residence times of 500 yrs for large redwood debris in streams may be possible. Such potentially long residence times attest to the stability of the channel, which in turn is partly due to the large trees themselves. Of course, in the large streams, such as downstream reaches of Redwood Creek, there is sufficient water at high flow to float even the largest debris, so the residence time is shorter. Therefore, in these large streams, the role of large organic debris in influencing channel form and process is less than for smaller streams.

MANAGEMENT OF LARGE ORGANIC DEBRIS (PRELIMINARY STATEMENT)

Studies in Prairie Creek and Little Lost Man Creek are increasing our knowledge of the fluvial system in the coastal redwood environment. However, much remains to be done before comprehensive management criteria can be developed. Comprehensive studies by Janda (1977) in the Redwood Creek drainage established that the basin has been severely affected by clearcut logging. The logging is often associated with a greatly increased sediment load. Furthermore, loggers sometimes remove usable large redwood debris from stream channels. Therefore, stream basins that have been logged recently often have anomalously high sediment loads. Downstream channel aggradation is common and may result in bank erosion, producing a positive feedback that causes landslides. Thus still more material enters the stream channel. Sediment pollution and subsequent loss of habitat diversity in the channel have adversely affected anadromous fish populations.

Large organic debris in streams is sometimes assumed to be undesirable because it may block migration of fish and cause adverse channel erosion. In some cases this is true, but within limits large organic debris probably is necessary for a healthy stream environment (Sheridan, 1969; Meehan, 1974; Sedell and Triska, 1977; and Swanson and Lienkaemper, 1978). Debris helps produce diverse flow conditions that create a variety of stream habitats necessary for fish and other aquatic life.

How much large organic debris is necessary to produce the desired habitat? Stream cleanup operations may be "over zealous" in removing valuable

Table 3

Minimum Ages for Large Organic Debris in the

Study Reaches of Little Lost Man

Creek and Prairie Creek

Reach	Tree Type	Age (yrs)	Location/ Association	Environment
Little Lost Man Creek				
Upper	Hemlock	130	Not Shown	Partial D.D.[c]/B.D.Tr.[d]
	Hemlock	135	Fig. 13, P 2[a]	Partial D.D./B.D.Tr.
	Hemlock	150	Fig. 13, R 4[a]	B.D.Tr. on Debris Stored Sed.
	Hemlock	85	Fig. 13, P 4	D.D.[e]
	Hemlock	185	Fig. 13, P 5	Partial D.D./B.D.Tr.
	Hemlock	175	Fig. 14, P 11	D.D.
	Hemlock	200	Fig. 14, P 11	D.D.
	Hemlock	70	Fig. 14, P 13	M.C.B.[f] behind D.D.
	Hemlock	105	Fig. 14, P 14	D.D.
Lower	Alder	60	Fig. 12, R 3	M.C.B. behind D.D.
	Redwood	75	Fig. 12, R/F[b] 6	Log over stream
	Redwood	220	Not Shown	B.D.Tr. downed trunk
	Redwood	100	Not Shown	D.D.
	Redwood	80	Not Shown	Partial D.D.
	Alder	70	Not Shown	M.C.B.
	Sitka Spruce	22	Not Shown	Partial D.D./B.D.Tr.
	Sitka Spruce	35	Fig. 4, R 14	B.D.Tr.
	Sitka Spruce	115	Fig. 4, P 16	D.D.
	Redwood	75	Fig. 4, P 16	D.D.
	Redwood	50	Fig. 4, P 16	D.D.
	Sitka Spruce	65	Not Shown	D.D.
	Sitka Spruce	40	Not Shown	B.D.Tr. downed trunk
	Hemlock	20	Not Shown	B.D.Tr. downed trunk
	Redwood	55	Fig. 2, P 28	D.D.
Prairie Creek				
Zig Zag No. 2	Sitka Spruce	150	Fig. 5, P 3	B.D.Tr. with root mat
	Maple	80	Fig. 5, P 6	D.D.
Brown Creek	Redwood	160	Not Shown	D.D.
	Hemlock	100	Not Shown	D.D.
	Hemlock	100	Fig. 8, Side P	Partial D.D.
	Redwood	200	Fig. 8, P 6	B.D.Tr. downed trunk
Campground	Redwood	50	Fig. 9, P 5	M.C.B. after D.D.
	Redwood	100	Fig. 9, P 5	Partial D.D.
	Hemlock	100	Fig. 9, P 6	B.D.Tr. with root mat

Table 3 (continued)

(a) P = pool; R = riffle	(d) B.D.Tr. = bank defending tree
(b) R/F = riffle with falls	(e) D.D. = debris blocking entire channel
(c) Partial D.D. = debris dam blocking part of channel	(f) M.C.B. = midchannel bar

components of debris necessary for fish habitat (Swanson and Lienkaemper, 1978). The problem of too much large organic debris is likely to be encountered along streams that have been logged recently. Management of streams in areas previously logged may well involve removal of organic debris where it impedes or reduces the fish habitat. Or it may require placing a certain amount of large debris in other areas to help produce the desired habitat.

The best management procedures, from a philosophical standpoint, are those which utilize natural fluvial processes rather than traditional engineering approaches. The natural fluvial system must be worked with, rather than absolutely controlled.

A management program for small and intermediate size streams in the redwood coastal environment must necessarily involve a two-pronged attack. Consideration must be given to streams that have been logged and guidelines must be provided for future logging operations in virgin and second-growth timber. Management of streams affected by logging will be difficult. These streams, unless they have recovered morphologically, are seeking a new equilibrium in response to increased sediment and debris loads. The best approach might be to utilize sound conservation practices to reduce sediment production. The first concern should be the slopes adjacent to the streams. (This is essentially at the heart of the restoration work now being undertaken in the Redwood National Park.) After the sediment load has been reduced and the morphology of the streams is adjusted to the sediment load, stream habitats may be improved. The improvement can be achieved by (a) carefully evaluating the debris present and (b) adjusting debris loading to that expected under natural conditions.

Management of streams likely to be impacted by future logging must consider the possibility and likelihood of increased debris loading and sediment input as a result of the logging operations. This is just as important for streams flowing through advanced second-growth redwood stands as for old-growth, most of which is already logged. Greatest concern is with direct input of debris during logging, construction, and subsequent mass movement following logging. In most cases natural debris in the channels must be left in an undisturbed state, and steps should be taken to minimize the introduction of additional debris. It is assumed that logging regulations will continue to require a buffer strip along the stream. This should help minimize the impact of the logging on the stream. Lastly, conservation methods must be employed to minimize the occurrence of landslides and other mass wasting events. Each of these suggestions is aimed toward allowing the stream to exist in as natural an environment as possible. Even with this, logging in the vicinity of a stream channel is bound to disturb the fluvial system, but hopefully the effects can be minimized.

CONCLUSIONS

1. In the old-growth, redwood environment, the role of large organic debris is of primary importance in the development and maintenance of anadromous fish habitat in small and intermediate sized streams.

2. In Prairie Creek, at least 50 % of the pools in the low gradient study reaches are controlled or influenced by large organic debris. In the steeper reaches of Little Lost Man Creek more than 90 % of the pools are controlled by large organic debris.

3. Measured debris loading in kilograms per square meter of active channel varies from 19.6 - 84.8 kg/m^2 in Prairie Creek and 49 - 141.6 kg/m^2 in Little Lost Man Creek. These values are somewhat higher than those for streams of similar size flowing through western Douglas fir forest.

4. Generally there is an inverse relation between stream size and debris loading because small streams tend to have narrow valleys, steep valley slopes, relatively high frequency of landslides, narrow channels and small upstream drainage, all of which tend to increase the debris loading.

5. Large organic debris in the coastal redwood environment produces diverse flow conditions, directly causing or influencing the development of pools, riffles, and mid-and side-channel bars. The influence is greatest in small, steep streams, such as the Upper Reach of Little Lost Man Creek, where 60 % of the drop in elevation along the channel profile is associated with large organic debris. A stepped stream profile is produced that provides sediment storage sites covering approximately 40 % of the area of the active channel.

6. Debris is significant in the routing of sediment through the fluvial system by providing an upstream buffer during times of high sediment input from the basin.

7. Many residence times of large organic debris in stream channels of the coastal redwood environment are greater than 100 yrs. Maximum established residence time exceeds 220 yrs and may be much longer.

8. Locations of 800-1000 yr old redwood trees near the banks of the Prairie Creek study reaches suggest that channel configuration and position change little on a time scale of centuries.

9. Management of streams, for maximizing anadromous fish habitat must consider the entire fluvial system and utilize natural stream processes rather than strive for absolute control of the stream by artificial manipulation.

ACKNOWLEDGEMENTS

We gratefully acknowledge the financial support provided by the U.S. Forest Service, Redwood Laboratory at Arcata, California, and the Water Resources Center, University of California, Davis. The California Department of Parks and Recreation provided valuable logistical assistance at Prairie Creek State Park, as did the U.S. Geological Survey and the U.S. Park

E. A. Keller and T. Tally

Service at the Redwood National Park.

Critical review and suggestions for improvement of the manuscript by Thomas Lisle, Frederick Swanson and Steve Veirs are appreciated as are stimulating discussions of various aspects of the research with Richard Janda, Harvey Kelsey, Michael Nolan, and William Weaver.

Field assistance by Janet Levinson, Sharon Parkinson, Gary Lester, and Oscill Maloney is gratefully acknowledged. Without their able help little would have been accomplished.

REFERENCES

Hack, J.T., 1957, Studies of longitudinal stream profiles in Virginia and Maryland: U.S. Geological Survey Professional Paper 294B, p. 45-97.

Heede, B.H., 1972, Flow and channel characteristics of two high mountain streams: U.S. Department of Agriculture, Forest Service General Technical Report RM-96, 12 p.

Janda, R.J., 1977, Summary of watershed conditions in the vicinity of Redwood National Park, California: U.S. Geological Survey Open-File Report 78-25, 82 p.

Keller, E.A., 1972, Development of alluvial stream channels: a five-stage model: Geological Society of America Bulletin, v. 83, p. 1531-1536.

Keller, E.A., and Swanson, F.J., in press, Effects of large organic debris on channel form and fluvial processes: Earth Surface Processes.

Leopold, L.B., and Langbein, W.B., 1962, The concept of entropy in landscape evolution: U.S. Geological Survey Professional Paper 500A, 20 p.

Leopold, L.B., Wolman, M.G., and Miller, J.P., 1964, Fluvial processes in geomorphology: San Francisco, Freeman, 522 p.

Mackin, J.H., 1948, Concept of the graded river: Geological Society of America Bulletin, v. 59, p. 463-512.

Meehan, W.R., 1974, The forest ecosystem of southeast Alaska: fish habitats: U.S. Department of Agriculture, Forest Service General Technical Report PNW-15, 41 p.

Megahan, W.F. and Nowlin, R.A., 1976, Sediment storage in channels draining small forested watersheds in the mountains of central Idaho; in Proceedings of the Third Federal Interagency Sedimentation Conference, Denver, Colorado, p. 4-115 to 4-126.

Sedell, J.R. and Triska, F.J., 1977, Biological consequences of large organic debris in northwest streams; in Logging debris in streams, Workshop II, Corvallis, Oregon State University, 10 p.

Sheridan, W.L., 1969, Effects of log debris jams on salmon spawning riffles in Saginaw Creek: U.S. Department of Agriculture, Forest Service, Alaska Region, 12 p.

Swanson, F.J., and Lienkaemper, G.W., 1978, Physical consequences of large organic debris in Pacific Northwest streams: U.S. Department of Agriculture, Forest Service General Technical Report PNW-69, 12 p.

Veirs, S.D., Jr., 1978, Characteristics of old-growth redwood stands; in Proceedings of a workshop on: Techniques of rehabilitation and erosion control in recently roaded and logged watersheds, with emphasis to north coastal California: Resources Management Division, Redwood National Park, p. 48-49.

Wolman, M.G., 1954, A method of sampling course river-bed material: American Geophysical Union Transactions, v. 35, p. 951-956.

FOREST-FIRE DEVEGETATION AND DRAINAGE BASIN

ADJUSTMENTS IN MOUNTAINOUS TERRAIN

WILLIAM D. WHITE

STEVE G. WELLS

Department of Geology
University of New Mexico
Albuquerque, New Mexico

ABSTRACT

In 1977 the La Mesa fire burned forested portions of the Jemez Mountains and surrounding volcanic plateau in northcentral New Mexico. Fluvial systems in the devegetated watersheds are adjusting to different sediment and water discharge conditions. Instruments have been placed in six small watersheds to measure these adjustments in areas of different burn intensity (a qualitative estimate of devegetation). The amount of sediment delivered to the stream systems is influenced by: 1) the amount of hillslope devegetation; 2) seasonal variations in weathering and runoff; 3) protective post-fire forest litter; and 4) sediment production from burrowing animals. Sediment source areas in devegetated watersheds offer little resistance to erosional processes, and, as a result, threshold conditions are frequently attained. The primary source areas of individual watersheds are the basin headwaters and distal portions. Mid-basin regions of the watersheds studied underwent very little erosion during the period of measurement. Overland flow distances are greater in the mid-basin regions, requiring longer time periods to remove hillslope detritus. Extensive rilling in the basin headwaters and channel incision near basin mouths are post-fire fluvial adjustments. Cut and fill cycles in channel alluvium of the higher-order streams are related to seasonal variations in sediment supply. Accelerated mechanical weathering during freeze-thaw cycles and concomitant accelerated erosion increase the magnitude of sediment delivery to the streams in extensively devegetated areas. In areas of less devegetation, variables such as post-fire vegetal litter (needlecast) reduce rainsplash erosion and sheetwash erosion; however, accelerated sediment production by burrowing animals yields unexpectedly high erosion rates. Systematic adjustments between process and form in undisturbed watersheds of the western United States occur, in part, because geomorphic processes are adjusted to watershed vegetational characteristics. In devegetated watersheds sediment yield is influenced more by accelerated mechanical weathering or post-fire vegetal litter than by drainage basin size or slope.

A process-response model is developed to illustrate the complexities of fluvial adjustments in devegetated, mountainous terrain. This model differs significantly from the complex response model for alluvial valleys. In mountainous areas, tributary streams will adjust independently of trunk drainage lines where bedrock nickpoints separate the two. Thus, tributary streams can be semi-closed systems which supply sediment and water to trunk streams but do not adjust to any modifications made by the trunk streams. Revegetation and erosional stabilization of these semi-closed systems may occur before the stabilization of the trunk streams.

INTRODUCTION

Disturbances in the natural state of a drainage basin due to changes in vegetation commonly affect erosion and sedimentation of hillslopes and stream channels. The effects of vegetational change on hydrologic and geologic systems have been investigated for durations of geologic time (Schumm, 1968) and for shorter, historical periods (Dodge, 1902; Rich, 1911; Anderson and Trobitz, 1949; Storey et al., 1964; Dryness, 1965; Harris, 1977). The recent geomorphic studies deal primarily with sediment yield and runoff changes in response to land-use changes, such as logging, mining or devegetation (Harris, 1977; Brater, 1939). Catastrophic ecologic events, such as forest fires, can trigger complex geomorphic responses as portions of drainage basins react to changes in sediment source areas and to changes in sediment yields from these areas. This paper deals with forest-fire devegetation and geomorphic adjustments of small drainage basins in the Jemez Mountains, northcentral New Mexico (fig. 1). In June 1977, the La Mesa fire burned approximately 59.5 km^2 of forested land of which 41.5 km^2 are included in Bandelier National Monument. This forest fire presented an opportunity to evaluate quantitatively the types of fluvial adjustments made by small, headwater drainage basins to post-fire conditions.

The purposes of this paper are to (1) determine controls and rates of sediment supply to fluvial systems over short time periods, (2) delineate source areas for sediment contributed to fluvial systems, (3) determine the types of fluvial adjustments in response to accelerated erosion, and (4) compare established models of fluvial adjustments in undisturbed drainage basins to watersheds which differ in the amount and extent of fire devegetation, hereafter called burn intensity.

STUDY AREAS

To accomplish these objectives, measuring instruments were placed in six drainage basins of the Frijoles Canyon watershed (fig. 1), three months after the fire. The basins are located in two areas which differ in vegetation and climatic settings. The Burnt Mesa study area is at an elevation of approximately 2195 m and is characterized by a mosaic of Ponderosa pine (Pinus ponderosa) in dense young stands, open park like stands of mature trees, and open meadows of grasses. The Apache Springs area, at an elevation of about 2530 m, is a mixed-conifer forest comprised of Ponderosa pine (Pinus ponderosa), White fir (Abies concolor), Douglas-fir (Pseudotsuga menzieisii), and aspen (Populus tremuloides). Precipitation in the Burnt Mesa area for October 1977-November 1978 was about 825 mm. Most of the precipitation fell during the winter and spring. No accurate precipitation records exist for the Apache Springs area.

In each study area three low-order drainage basins were selected on the basis of the degree of burn intensity:

intense = complete destruction of all ground and tree foliage, leaving only charred trunks and branches (fig. 2A);

Figure 1. Study areas within Bandelier National Monument, New Mexico.

moderate = destruction of the ground foliage, leaving only the crowns of
tall trees (fig. 2B);

light = destruction of the ground foliage and small shrubs; larger shrubs
and trees not destroyed.

These watersheds carry water only during precipitation events or snowmelt.
Streams draining these watersheds are tributary to the major perennial, base-
level stream, Rito de los Frijoles. Rito de los Frijoles is incised into the
Bandelier Tuff, a welded ash-flow deposit consisting of rhyolite ash and
pumice. These volcanic rocks are derived from the Pleistocene Valles Caldera.
All the study watersheds are developed on the Bandelier Tuff. Apache
Springs basins are on the caldera rim, and Burnt Mesa watersheds are on the
eastward-sloping volcanic plateau that flanks the caldera.

The fluvial systems consist of long, parallel trunk streams which discharge into the Rio Grande Valley. Three types of watersheds contribute

A. Headwater area of the intense-burn watershed, Burnt Mesa. Note the lack of tree crowns and forest litter covering the colluvium.

B. Headwater area of the moderate-burn watershed, Burnt Mesa. Note the amount of surviving tree crowns and presence of forest litter.

Figure 2. Examples of burn intensity in Jemez Mountains.

runoff and sediment to the stream system: (1) steep, headwater basins developed on the caldera rim (Apache Spring area); (2) steep, side-valley watersheds developed on the Frijoles Canyon walls (not studied); and (3) moderate relief watersheds developed on the mesa tops above Frijoles Canyon (Burnt Mesa area). All these watershed types were devegetated by forest-fires. Mesa-top watersheds have larger drainage areas than inner canyon watersheds, and more sediment and runoff are provided from these areas to Rito de los Frijoles.

Low-order drainage basins of Burnt Mesa and Apache Springs differ in their geomorphic character (table 1). Drainage basins of the Burnt Mesa area are typically elongate due to their development on gently sloping mesa tops. The Apache Springs basins are characteristically circular due to their development on valley side-slopes of incised trunk-drainage lines (table 1). The largest drainage basin investigated is an intense-burn third-order basin of Burnt Mesa which covers 4116.5 m^2 (fig. 3). The smallest basin investigated is the light-burn, first-order basin of the Apache Springs area which covers 280 m^2. Drainage basins of the Apache Springs area are steeper and shorter than those on Burnt Mesa. Relief in the drainage basins is similar in both areas (table 1).

GEOMORPHIC MAP OF INTENSE-BURN WATERSHED, BURNT MESA

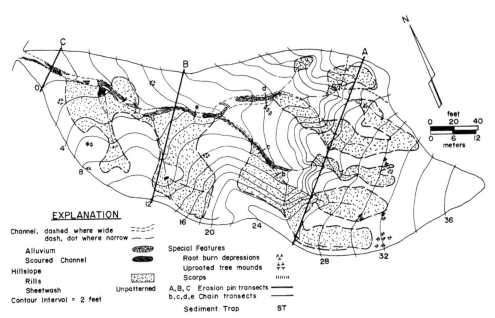

Figure 3. Geomorphic map of the intense-burn watershed, Burnt Mesa. The map shows the extent of rilling versus sheetwash as of July, 1978. Also shown are the erosion-pin transect and sediment-trap locations.

Table 1

Geomorphic Characteristics of Instrumented Watersheds in
Bandelier National Monument, New Mexico

Basin	Area (m²)	Aspect	Order	Relief (m)	Length (m)	Perimeter (m)	Basin Circularity	Basin Slope
Burnt Mesa								
Intense	4116.5	WNW	3rd	11.86	129.97	298.09	.582	.091
Sediment Trap	92.3	WSW	-	2.72	12.98	41.15	.685	.210
Moderate	813.8	S	1st	8.65	67.06	146.30	.478	.129
Light	1426.0	S	1st	5.57	70.84	177.39	.570	.079
Apache Springs								
Intense	1431.5	S	2nd	11.16	51.51	143.26	.876	.217
Moderate	677.7	S	1st	8.12	49.07	124.05	.553	.166
Light	380.4	ESE	1st	5.41	25.97	67.36	.776	.208

Bedrock outcrops are not extensive in the drainage basins; rather, varying thicknesses of colluvium, water-worked colluvium, and alluvium cover the basins. Grain-size characteristics for the streams and hillslopes of the watersheds have been analyzed by sieve analysis. Grain-size distributions for the intense-burn watershed are summarized in figure 4. Commonly, grain-size distributions near the hillslope divides are unimodal, whereas a bimodal distribution is typical for downslope sections (fig. 4). Changes in the distributions suggest reworking and sorting of the hillslope sediment with a depletion in the 2 to 3 ϕ size (sand). Grain-size distributions of the channel alluvium show increases in the 1 to 3 ϕ size downchannel. This suggests transport of sand grains from the hillslopes to the channels.

MEASUREMENTS

Hillslope sections and channel reaches are instrumented in each basin to monitor the amount, distribution, and seasonal variations in sediment contribution to the stream channel. Modifications of channel morphology are measured with erosion chains (table 2), and the amount of runoff was estimated by slope-area techniques. To measure the hillslope sediment contribution, two types of instruments are employed: (1) erosion pins, and (2) a movable-contour-plotting frame which permits 25 elevation measurements over a 1 m^2 area (Campbell, 1974) (table 2). Validation of erosion measurements determined from the erosion pins and movable-contour-plotting frame is accomplished

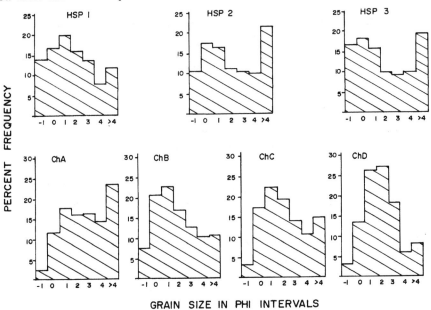

GRAIN SIZE IN PHI INTERVALS

Figure 4.   Grain-size distributions of hillslope (HSP) and channel (Ch) samples for the intense-burn watershed, Burnt Mesa. HSP 1 is at hillcrest divide and HSP 3 is at base of slope; CHA is at headwater and CHD is near basin mouth.

Table 2

Summary of the Instrumentation of each Study Watershed,

Bandelier National Monument, New Mexico

Study Area	Watershed	Channel Scour & Fill Chain Transects	Hillside Erosion Pin Transects	Hillside Erosion Contour Plots
BURNT MESA	INTENSE	4 transects	3 transects/100 m length 145 pins/0.6-0.9 m spacing	2-1 m^2 plots
	MODERATE	none	3 transects/50 m length 41 pins/0.9 m spacing	1-1 m^2 plot
	LIGHT	none	3 transects/65 m length 55 pins/1.2 m spacing	1-1 m^2 plot
APACHE SPRINGS	INTENSE	2 transects	3 transects/65 m length 86 pins/0.9 m spacing	none
	MODERATE	none	3 transects/33 m length 46 pins/0.9 m spacing	none
	LIGHT	none	2 transects/17 m length 37 pins/0.6 m spacing	none

by comparing sediment collected in traps. The sediment trap consists of a series of three drums (0.5 m^3 total capacity) which collect runoff and sediment from a 92 m^2 headwater portion of the intense-burn watershed, Burnt Mesa (fig. 3).

Another technique for monitoring geomorphic processes and landscape changes in the devegetated basins is terrestrial photography (Malde, 1973). In the present study, the camera has been used in two ways: (1) as a portable and visual note-taking method, and (2) for repetitive photography at 5 permanent camera stations to monitor changes over extended time periods.

SEDIMENT SOURCES

The lengths and types of surface runoff influence the location of sediment supply in the devegetated watershed. Extensive rill systems develop on the steep slopes near the watershed headwaters (fig. 5). Erosion is greater in regions undergoing rill erosion than in areas eroded by sheetwash. In the distal portion of the drainage basins, drainage lines incised. Channel incision caused accelerated hillslope erosion near the basin mouths.

The length of overland flow (maximum transport distance on the hillslopes) is greater in the mid-basin regions than in the headwaters of watersheds for the intense- and moderate-burn watersheds but is greatest in the headwaters in light-burn watersheds (table 3). Sediment is removed from hillslopes in the upper and lower basin areas faster than from the mid-basin because of shorter transport distances. Overland flow distances in the headwaters continue to decrease as rill development progresses. Photography was used to document the decrease in area between rills with time. Some headwater overland-flow distances have been reduced by one half their original distance. Increased channelways, in proportion to interfluve area, results in increased efficiency of sediment removal from the watersheds.

During the forest fire, impervious hydrophobic layers developed on the colluvial and alluvial cover of the deforested areas. Hydrophobic layers are chemically and physically altered zones (Krammes and DeBano, 1965; DeBano et al., 1970; Savage et al., 1972). The infiltration potential of soil and colluvial cover was diminished because grain sizes were reduced by the fire (Tarrant, 1956). Fines moved downward through the profile and filled voids.

Figure 5. Rill development on the ashy, hydrophobic layer in the mid-basin area of the intense-burn watershed, Burnt Mesa.

Table 3

Average Length of Overland Flow on Interfluves
within the Burnt Mesa Watersheds

Watershed	Location	Mean Overland Flow Distance (m)
Burnt Mesa, Intense	Headwaters	5
	Mid-basin	16
	Basin mouth	10
Moderate	Headwaters	10
	Mid-basin	12
	Basin mouth	8
Light	Headwaters	26
	Mid-basin	14
	Basin mouth	6

Overland flow is more effective on the hydrophobic layers. After the fire, hydrophobic layers were eroded and the underlying unconsolidated material was available for transportation. Hydrophobic layers on steep slopes were eroded by rilling as flow was concentrated into minute channels (fig. 5). Sheetwash erosion and rilling destroyed the impervious layers on gentle and steep slopes, respectively. Fire-induced grain size reduction most likely contributed to the supply of easily-erodible material.

EROSION IN DEVEGETATED WATERSHEDS

In undisturbed watersheds, the sediment supply to the fluvial systems is sporadic through time and is commonly dependent upon geomorphic events which attain threshold conditions (Schumm, 1973; Scott and Williams, 1974; Schick, 1974). Sediment sources, such as hillslopes and valley margins, are stable over longer time periods because threshold conditions are infrequently attained. Stress applied to undisturbed watersheds must be of higher magnitude than that applied to devegetated watersheds if a given amount of erosion is to occur. Drainage basin devegetation results in an increase of sediment delivery because lower threshold conditions are required to produce and to entrain sediment. In the forest-fire devegetated watersheds of the Jemez Mountains, movement of sediment supplying fluvial systems has increased in frequency and has become more dependent upon seasonal variations in weathering and runoff.

After one year of field instrumentation, generalized relationships can be drawn between the amount and location of erosion within a single watershed. The amount of erosion is greater in the headwaters and near-mouth portions of the watershed than in the mid-basin regions. Comparison of November 1978, measurements for each transect (headwater, mid-basin, and basin mouth) exemplifies this relationship (fig. 6). This relationship held true throughout the period of measurement except for the light-burn basin, which was influenced by animal activity.

Seasonal Variations in Erosion

In October more erosion occurred in the Apache Springs watersheds than in the Burnt Mesa drainage basins (table 4). Average erosion along each transect was less than 1 mm in Burnt Mesa but averaged 3 mm in Apache Springs. These differences in erosion are controlled, in part, by the rate of destruction of the hydrophobic layer. Apache Springs watersheds have steeper slopes (table 1), and therefore overland flow can attain greater velocities. Consequently, most of the hydrophobic layers were removed from the Apache Springs area. Longer time periods are needed to erode the

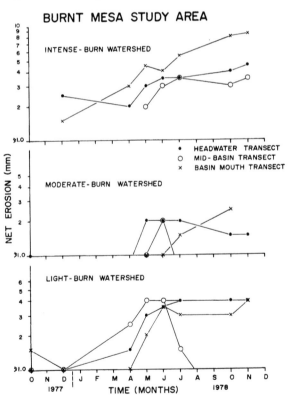

Figure 6. *Semi-logarithmic plots of seasonal variation in erosion for the Burnt Mesa watersheds.*

Table 4

Average Erosion Values for each Transect in the Burnt Mesa and Apache Springs Study Areas. Net Loss of each Transect is Averaged over one year.

Net surface elevation change averaged for each transect (mm)

Transect #		1977 Oct.	1977 Dec.	Apr.	May	June	July 1978	Oct.	Nov.	NET LOSS/TRANSECT (mm/year)
Burnt Mesa										
Intense	A	-0.5	-2.5	-2.0	-3.0	-3.5	-3.5	-4.0	-4.5	-4.0
	B	0.0	+1.0	-0.5	-2.0	-3.0	-3.5	-3.0	-3.5	-3.0
	C	-0.5	-1.5	-3.0	-4.5	-4.0	-5.5	-8.0	-8.5	-7.3
Moderate	A	-1.0	0.0	-0.5	-2.0	-2.0	-2.0	-1.5	-1.5	-1.3
	B	-0.5	+1.0	-0.5	-1.0	-2.0	0.0	-0.5	-1.0	-0.8
	C	-0.5	0.0	+2.5	-1.0	-1.0	-1.5	-2.5	-1.0	-0.8
Light	A	-1.0	-1.0	-1.5	-3.0	-3.5	-4.0	-4.0	-4.0	-3.4
	B	-1.0	-1.0	-2.5	-4.0	-4.0	-1.5			--
	C	-1.5	0.0	-1.0	-2.0	-3.5	-3.0	-3.0	-4.0	-3.4
Apache Springs										
Intense	A	-3.0	--	-2.0	--	-2.5	-2.5	-3.0	--	-2.6
	B	-1.0	--	-2.0	--	-2.0	-1.0	-1.0	--	-0.8
	C	-2.0	--	-2.0	--	-3.0	-3.0	+4.0	--	-3.4
Moderate	A	-4.0	--	-5.5	--	-5.0	-4.0	-6.0	--	-5.1
	B	-2.0	--	0.0	--	-2.0	-2.0	-2.0	--	-1.7
Light	A	x	x	+2.5		-1.5	-2.0	-2.0	-2.5	-1.7
	B	x	x	-3.0		-3.5	-4.0	-3.0	-2.5	-2.6

hydrophobic layer on the gentler slopes of Burnt Mesa study area, and more extensive remnants of the layer are still preserved on the hillslopes there (fig. 5).

In the late fall of 1977, another change in ground conditions and weathering processes resulted in an increase in sediment availability and erosion. Freeze-thaw cycles began in November 1977 as available soil moisture was subjected to diurnal fluctuations in surface temperature. Freezing and thawing disrupted the previous ground-surface texture (figs. 7A and 7B). Frost heaving resulted in the upward movement of fine particles relative to coarser ones. Loose, fine-grained sediment was concentrated at the ground surface where minor runoff events could easily remove it. A three- to eight-fold increase in erosion occurred between October and December, 1977, in response to this surface alteration (fig. 8).

Differential rates and amounts of erosion over the watershed are most likely related to differences in soil moisture and consequently, frost-heaving intensity. Variations in erosion on the hillslopes are common along erosion-pin transects; however, all transects showed a dramatic increase in erosion during the freeze-thaw cycles (fig. 8). The intense-burn watershed of Burnt Mesa experienced the greatest amount of erosional response to frost heaving (fig. 8). This is attributed to (1) hillslope aspect (orientation) and moisture regimes, (2) relative proportion of fine sediment in the colluvial cover, and (3) lack of extensive vegetational cover.

Seasonal variations in the erosion of the Burnt Mesa watersheds are given in table 4 and figure 6. Erosion rates increased significantly during late fall and spring. These seasons coincide with periods of accelerated mechanical weathering (frost heaving). Erosional stability and periods of low sediment production occur when the watersheds are blanketed with snow (fig. 6).

Other Factors Controlling Watershed Erosion

Erosion measurements given in table 4 suggest that factors other than mechanical weathering influence sediment supply. For example, erosion rates are greater and more variable in the Burnt Mesa area than in the Apache Springs area (table 4). Surface denudation in the Apache Springs area is distributed more uniformly through time than in Burnt Mesa (fig. 6). Rates of revegetation of the watersheds are partially responsible for these variations in erosion. Grass quickly re-established cover over a greater part of the surface in the Apache Springs area than in the Burnt Mesa area. Revegetation reduced the amount and variation of surface erosion over time because mechanical weathering was less effective in the vegetated zones.

Another vegetational factor influencing the variability of erosion through time is the post-fire needlecast. A needlecast is a thin blanket of pine needles covering the ground surface which forms when partially-burned trees drop dead needles to the ground (fig. 2B). Needlecasts are common in moderate and light burns. Their occurrence depends on the density of standing conifers, and of dead, unburned needles. This type of forest litter protects the recently devegetated colluvial cover and reduces erosion (Connaughton, 1935; Megahan and Molitor, 1975). Needlecasts break the impact of raindrops and distribute rainfall over the surface, allowing slow in-

A. Compaction and cracking of the ground surface in the headwaters of the intense-burn watershed, Burnt Mesa, September, 1977.

B. Textural change due to frost heaving, November, 1977.

Figure 7. Seasonal variations in ground surface texture.

INTENSE BURN,   BURNT MESA

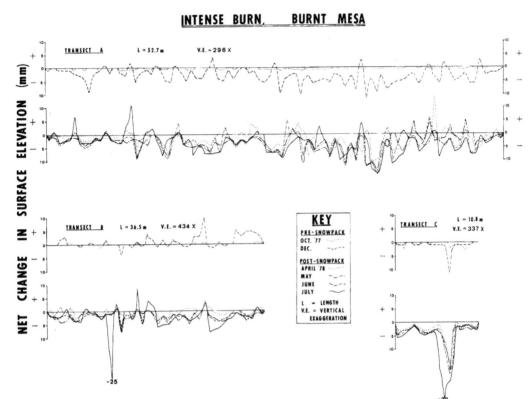

Figure 8. *Erosion-pin data for transects in the intense-burn watershed, Burnt Mesa. The plot represents changes in ground surface elevation before and after the 1977-78 snowpack. The horizontal line represents datum; deviation above the line depicts gain, deviation below the line depicts loss.*

filtration. In addition, needlecasts trap small particles which are transported by overland flow and exert some influence on the microclimate of the soil by decreasing insolation (Megahan and Molitor, 1975). Decreased insolation influences ground temperature and frost heaving of the soil. The influence of needlecast on erosion is demonstrated, in part, by comparing erosion measurements in the intense- and moderate-burn watersheds of Burnt Mesa (fig. 6; table 4). The moderate-burn watershed has an extensive cover of pine needles, while the intense-burn essentially has no needlecast. Erosion rates in the intense-burn watershed are almost double those in the moderate-burn basin. Similar relationships occur in the Apache Springs area.

Burrowing animals also influence the amount of erosion over time. The light-burn watershed of Burnt Mesa has higher erosion rates than the moderate-burn watershed (fig. 6) because of a large and wide-spread gopher population in the light-burn watershed compared to the other watersheds. Gopher rehabitation and subsequent burrowing in this area after the fire produced unexpectedly high volumes of easily-transported sediment. Thus erosion may not be directly proportional to the extent of devegetation because other activities, such as animal burrowing, may produce significant amounts of sediment.

Sediment Delivery Rates

The rate of movement of individual grains on hillslopes can be studied by tracer (colored) sand. This technique was used to gather data on the rate of sediment transport across the interfluves. Tracer sand (0 to 1Φ grain size) was placed along transect A on a hillcrest in the intense-burn watershed of Burnt Mesa on May 11, 1978. One month later (June 14, 1978) distribution of the tracer sand showed the effects of raindrop impact. Some particles were transported at least 0.11 m downslope from the drainage divide. Two months later (July 14, 1978) the sand was transported by sheetwash during a summer thunderstorm. Some particles were moved 0.77 m downslope from the hillslope divide. Whether this transportation occurred during a single runoff event or several is not known. By August 1, 1978, the tracer sand was difficult to locate because of dispersion on the hillslope; however, grains were found at 0.82 m downslope of the hillslope divide. One grain was found only 0.08 m from the initial starting point on the divide, indicating selective or random transport of similar sized grains. An approximation of the rate of sediment transport by overland flow can be derived from this test. The mean rate of transport is approximately 3.0 m/yr. Obviously, sediment movement on hillslopes is not constant. For example, faster movement should occur on steep reaches and slower movement on gentle slopes. However, 3.0 m/yr may be representative of the mean transport rate for the intense-burn watershed.

FLUVIAL ADJUSTMENTS TO FOREST-FIRE DEVEGETATION

Types of Fluvial Adjustments

Adjustments of fluvial systems to changes in geomorphic variables have been described in terms of complex responses (Schumm, 1973, 1977; Womack and Schumm, 1977). In the complex-response model, variations in sediment supply from tributary streams to trunk streams cause modifications of the larger fluvial systems which are, in turn, transmitted back through the tributaries. Adjustment of Rito de los Frijoles to increased sediment load has resulted in frequent overbank events as well as channel modifications. These adjustments are not transmitted back to the tributary watersheds on the mesas because 20 m-high bedrock nickpoints separate the two systems. Thus, the mesa top watersheds can be considered semi-closed systems which progressively adjust to a single change in a drainage basin variable (devegetation).

Post-fire adjustments of fluvial systems on the mesa tops involve alterations from stable channel forms (shallow, parabolic cross-section) to unstable, degrading or aggrading channels. In general, low-order channels of Burnt Mesa are scouring as the high-order channels are cutting and filling (fig. 9); however, net filling is typical in the higher-order streams (fig. 9). Most

214

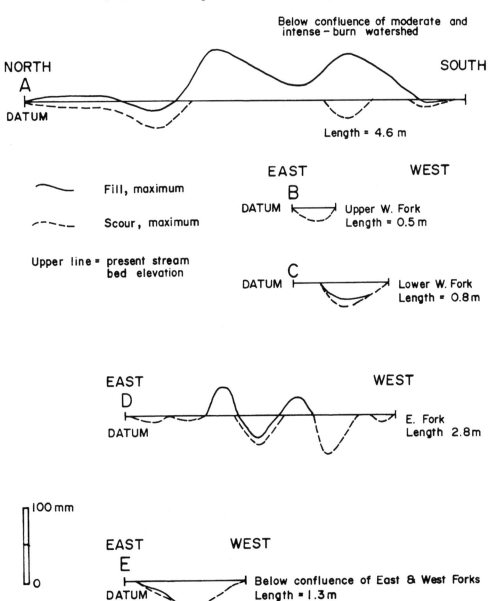

Figure 9. Cross-sectional profiles of channels illustrating amount of scour
and fill (see figure 3 for locations). Cross-section A is the
fifth order channel below the confluence of the intense- and moderate-burn watersheds on Burnt Mesa. Locations B (headwaters)
through E (near mouth) are within the intense-burn watershed.

channels aggraded immediately after the fire in response to summer precipitation events. Channel widening occurred as point bars and channel bars formed. Some channels doubled in width during this initial response. Small alluvial fans formed at the junctions of low-order drainage lines.

Field observations indicated that sediment transport was dominantly bedload. Between September and November 1977, high-order channels carried less bedload because coarse sediment supplies from the hillslopes declined. In response to decreasing sediment supply and lower magnitude runoff events, the channels began scouring.

Rejuvenation in the sediment supplies from the hillslopes occurred during freeze-thaw cycles in fall, 1977. The stream channels were able to transport this second sediment pulse without significant change in channel width; however, channel scour did continue (fig. 10). Fine-grained sediment produced by the second pulse was transported as suspended load. Little channel widening is necessary for transporting high silt-clay loads (Schumm, 1960). Incision of most streams during this period is related to frost heaving in the channel bottoms. Moisture in the alluvium repeatedly froze and uplifted fine material which could then be easily transported. This process accounts for increased channel incision during freeze-thaw cycles (fig. 10).

Following these two sediment pulses, the winter snowpack prevented extensive erosion (table 4) and stabilized channels. During the spring 1978 melt and runoff, channel incision recommenced and continued until mid-summer,

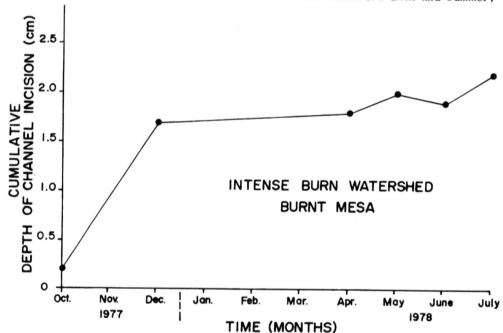

Figure 10. *Rate of channel incision at the mouth of the intense-burn watershed, Burnt Mesa.*

216

1978. During the summer, vertical accretion on stable channel bars caused net deposition in the higher-order channels (fig. 9). Filling of these channels occurred in response to revegetation along the margins of these streams (fig. 11). Tall grasses which grew on stable bars trapped fine sediments during summer runoff events. After summer channel filling, channel scouring dominated through the fall, 1978. Dormancy of channel margin vegetation during the fall resulted in increased erodibility of the stream banks. Furthermore, because stream transport power is typically lower during fall and spring runoff events, commonly only suspended sediments are transported.

Process-Response Model in Devegetated Watersheds

The ephemeral fluvial systems in mountainous terrain show complex responses to forest-fire devegetation. These complexities are related to variations in amounts and source areas of sediments. Sediment delivered from mesa-top fluvial systems to the trunk drainage, Rito de los Frijoles, caused channel geometry modifications. Alterations in the Rito de los Frijoles geomorphology cannot be transmitted back to the mesa-top tributaries because of the bedrock nickpoints at the canyon walls. Only tributary streams topographically lower than the bedrock nickpoints (within the canyon walls) can react to changes induced by Rito de los Frijoles.

A process-response model for fluvial systems in bedrock terrain and in forest-fire devegetated areas is given in figure 12. The model shows the importance of seasonal variations in sediment delivery from the mesa-top tributaries. In figure 12 geomorphic responses and sediment transport are illustrated by the arrows, and the volume of sediment storage on hillslopes, in channels, and in valleys is indicated by the relative size of the boxes. The model (1) illustrates the continuous decrease in sediment supply from the mesa-top systems to the Frijoles canyon, (2) illustrates the continued response of fluvial systems in the canyon to devegetation after geomorphic stability is attained on the mesa tops, and (3) emphasizes that tributaries in a fluvial system may adjust dependently or independently of the master drainage line. Complex responses (Schumm, 1977) are typical in streams with alluvial valleys, whereas in mountainous terrain the bedrock nickpoints act as barriers to interactions between components of the fluvial systems and reduce adjustments between tributaries and trunk streams.

DRAINAGE BASIN ADJUSTMENTS

Altered runoff and sediment yield illustrate drainage basin adjustments to varying degrees of devegetation. These two geomorphic parameters have been measured in numerous southwestern drainage basins and have been compared to the drainage basin geomorphology (Hadley and Schumm, 1961; Strand, 1975; Schumm, 1977). Such comparisons show high correlations between these drainage basin processes and morphologic elements. However, these drainage basins represent undisturbed conditions, and in devegetated watersheds factors other than basin morphology influence runoff and sediment yield.

Runoff

Water discharge in the instrumented watersheds of Burnt Mesa was measured for single, peak runoff events. Runoff during the summer of 1978

A. Post-fire devegetation of channel margin in early fall, 1977.

B. Revegetation along channel margins in summer, 1978.

Figure 11. Revegetation along channel margins in the Burnt Mesa study area.
Note that these are repeat photographs of exactly same area.

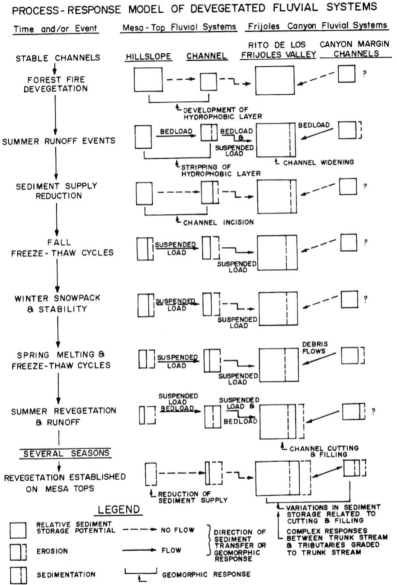

Figure 12. *Process-response model for fluvial systems in Jemez Mountains. Model illustrates seasonal variations in fluvial processes due to devegetation. Boxes represent sediment-storage capabilities of the fluvial system, and arrows represent types and directions of sediment transfer in the fluvial system. Note that Mesa-top streams are independent of subsequent adjustments made by Frijoles Canyon streams.*

was measured in each type of burn-intensity area. The amount of runoff per unit area increased as the devegetation increased (fig. 13). Runoff in the intense-burn basin may be 60 times greater than in the light-burn area. Additionally, figure 13 and table 5 indicate that runoff is not directly proportional to drainage basin size, because the moderate-burn watershed is smaller than the light-burn watershed. In undisturbed watersheds, runoff as a dependent variable bears a direct power relation to drainage basin size. Relief and drainage density are the geomorphic parameters which influence the runoff in devegetated drainage basins. Table 5 compares selected basin parameters to runoff and sediment yield and shows that runoff increases with increasing relief and drainage density. Drainage density is given in qualitative terms because the density changes with time due to headward growth of drainage lines.

Sediment Yield

Sediment yield from the instrumented watersheds is determined from erosion-pin data, erosion-contour plot data, and sediment-trap data. Calculated sediment yields from the pin- and contour-plot data were compared to the volume of sediment trapped in buried drums. Sediment collected in the trap installed in the intense-burn basin of Burnt Mesa gave a sediment yield of 0.7 m^3/yr over a 92 m^2 area. Using the erosion rate of 0.5 cm/yr, determined from erosion-pin data, the computed sediment yield is 0.5 m^3/yr over a 92 m^2 area. The two types of measurements are similar, so the use of erosion-pin data may provide minimum estimates of sediment yields.

In the Burnt Mesa watersheds the intense-burn basin has the highest sediment yield, but the light-burn basin has a higher sediment yield than the moderate-burn watershed due to animal activity (table 5). Hadley and Schumm (1961) demonstrated that sediment yield increases with increasing relief-ratios in small drainage basins in the western United States. In the devegetated watersheds of the Jemez Mountains, sediment yield is not as sensitive to basin slope relationships as to types of mechanical weathering. In this area other drainage-basin variables affect

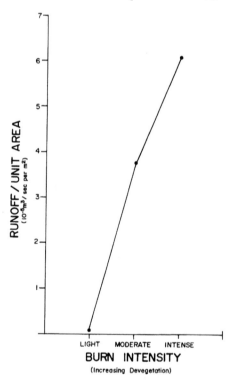

Figure 13. Comparison of runoff per unit area and burn intensity, Burnt Mesa study area.

Table 5

Comparison of Sediment Yield and Runoff from Instrumented

Watersheds and Selected Geomorphic Parameters of

the Drainage Basins in Burnt Mesa Area

Drainage Basin	Area (m^2)	Relief (m)	Relief-Length Ratio	Drainage Density	Runoff ($10^{-5} m^3/$ sec/m^2)	Sediment Yield (m^3/yr)
Intense-burn	4116.5	11.86	0.09	high	6.2	19.4
Moderate-burn	813.8	8.65	0.13	medium	3.8	0.8
Light-burn	1427.0	5.57	0.08	low	0.1	4.5

erosional processes and the supply of sediment to fluvial systems. For example, lower erosion rates of the moderate-burn basins correlate with the extensive needlecast and pine canopy which reduce raindrop impact and hillslope runoff. Also, higher-than-expected erosion rates occur in the light-burn watershed in response to sediment production by burrowing animals.

Another difference between fluvial adjustments in undisturbed drainage basins and devegetated watersheds is illustrated by the relationship between sediment yield and drainage basin size. Sediment yield per unit area decreases with increasing drainage basin size for undisturbed basins in the southwestern United States (Strand, 1975). This relationship is quantitatively expressed as $Q_s = 2.4 \, A^{-0.229}$ where Q_s is sediment load in AF/mi^2/yr, and A is drainage area in mi^2. Larger drainage basins are capable of storing more sediment than smaller drainage basins (Schumm, 1977). In larger drainage basins, sediment is stored on hillslopes and in valleys for longer time periods. For example, given a single event, a higher magnitude of event would be necessary to carry the material out of a large basin, relative to a small basin. However, for the devegetated watersheds of Burnt Mesa sediment yield increases with increasing drainage basin size (table 5). In these disturbed watersheds, vegetation reduction decreases the sediment-storage potential of larger drainage basins and increases the sediment removal efficiency. Net channel incision, which is common to the streams of the intense-burn watersheds (fig. 9) is another indication of decreased sediment-storage potential. Sediment storage provided by hillslope and channel vegetation is important in these small, mountainous watersheds, as demonstrated by the reappearance of summer vegetation in 1978 and the subsequent aggradation of higher-order channels.

CONCLUSIONS

Devegetated mountainous watersheds have relationships between drainage processes and form which differ from those for undisturbed drainage basins in northern New Mexico. Systematic relationships between runoff and sediment yield in undisturbed watersheds occur, in part, because the processes are adjusted to watershed vegetational characteristics. Changes in the vegetation of watersheds cause other geomorphic parameters to become influential in controlling sediment and water discharge. In devegetated watersheds, runoff depends on the amount of devegetation, basin relief, and drainage density; sediment yield is dependent upon complex factors which control sediment production and availability. Parameters such as slope and basin size are not as important process controls as needlecast cover or burrowing-animal activity. Quantitative expressions of drainage basin adjustments in undisturbed watersheds often do not apply to forest-fire devegetated watersheds.

ACKNOWLEDGEMENTS

The authors express their thanks to Burchard Heede, Wayne Lambert, and Thomas Gardner for reviewing the manuscript. Special thanks are given to Raymond Ingersoll, Milford Fletcher, and Ro Wauer for their comments and criticism on the paper and to Sandra Anderson for her field assistance. Photographs in the paper were provided by Rick Dingus (Fine Arts Dept., UNM); thanks are given to Rick's artistic and scientific contribution. This paper and research is supported by Grant No. PX7029-7-0809 of the National Park Service (Southwest Regional Office), U. S. Department of Interior.

REFERENCES

Anderson, H.W., and Trobitz, H.K., 1949, Influence of some watershed variables on a major flood: Journal of Forestry, v. 47, p. 347-356.

Brater, E.F., 1939, The unit hydrograph principle applied to small watersheds: American Society of Civil Engineers Proceedings, v. 65, p. 1191-1215.

Campbell, I.A., 1974, Measurements of erosion on badlands surfaces: Zeitschrift für Geomorphologie, suppl. bd. 21, p. 122-137.

Connaughton, C.A., 1935, Forest fire and accelerated erosion: Journal of Forestry, v. 33, p. 751-752.

DeBano, L.F., Mann, L.D., and Hamilton, D.A., 1970, Translocation of hydrophobic substances into soil by burning organic litter: Soil Science Society of America Proceedings, v. 34, p. 130-133.

Dodge, R.E., 1902, Arroyo formation [abs.]: American Geologist, v.29, p. 322.

Dryness, C.T., 1965, Erosion potential of forest watersheds; in Sopper, W.E., and Lull, H.W., eds., Forest hydrology: New York, Pergamon Press, p. 599-611.

Hadley, R.F., and Schumm, S.A., 1961, Sediment sources and drainage basin characteristics in upper Cheyenne River basin: U.S. Geological Survey Water-Supply Paper 1531B, p. 137-196.

Harris, D.D., 1977, Hydrologic changes after logging in two small Oregon coastal watersheds: U.S. Geological Survey Water-Supply Paper 2037, 31 p.

Krammes, J.S., and DeBano, L.F., 1965, Soil wettability: a neglected factor in watershed management: Water Resources Research, v. 1, p. 283-286.

Malde, H.E., 1973, Geologic bench marks by terrestrial photography: U.S. Geological Survey Journal of Research, v. 1, p. 193-206.

Megahan, W.F., and Molitor, D.C., 1975, Erosional effects of wildfire and logging in Idaho: American Society of Civil Engineers, Watershed Management Symposium, Logan, Utah, Aug. 11-13, 1975, p. 423-444.

Rich, J.L., 1911, Recent stream trenching in the semi-arid portion of southwestern New Mexico, a result of removal of vegetation cover: American Journal of Science, v. 32, p. 237-245.

Savage, S.M., Osborn, J., Letey, J., and Henton, C., 1972, Substances contributing to fire-induced water repellency in soils: Soil Science Society of America Proceedings, v. 36, p. 674-678.

Schick, A.P., 1974, Formation and obliteration of desert stream terraces-- a conceptual analysis: Zeitschrift für Geomorphologie, suppl. bd. 21, p. 88-105.

Schumm, S.A., 1960, The effect of sediment type on the shape and stratification of some modern fluvial deposits: American Journal of Science, v. 258, p. 177-189.

_____, 1968, Speculations concerning paleohydrologic controls of terrestrial sedimentation: Geological Society of America Bulletin, v. 79, p. 1573-1588.

_____, 1973, Geomorphic thresholds and complex responses of drainage systems, in Morisawa, M., ed., Fluvial geomorphology: Binghamton, State University of New York, Publications in Geomorphology, p. 299-311.

_____, 1977, The fluvial system: New York, John Wiley and Sons, 338 p.

Scott, K.M., and Williams, R.P., 1974, Erosion and sediment yields in mountain watersheds of the Transverse Ranges, Ventura and Los Angeles Counties, California -- Analysis of rates and processes: U.S. Geological Survey Water Resources Investigations 47-73, 66 p.

Storey, H.C., Hobba, R.L., and Rosa, J.M., 1964, Hydrology of forest lands and range lands in Chow, V.T., ed., Handbook of applied hydrology: New York, McGraw Hill, p. 22-1 to 22-52

Strand, R.I., 1975, Bureau of reclamation procedures for predicting sediment yield: Agricultural Research Service, ARS-S-40, p. 10-15.

Tarrant, R.F., 1956, Effect of slash burning on some physical soil properties: Forest Science, v. 2, p. 18-22.

Womack, W.R., and Schumm, S.A., 1977, Terraces of Douglas Creek, northwestern Colorado: an example of episodic erosion: Geology, v. 5, p. 72-76.

SLACK-WATER DEPOSITS: A GEOMORPHIC TECHNIQUE FOR THE INTERPRETATION OF FLUVIAL PALEOHYDROLOGY

PETER C. PATTON

Department of Earth and Environmental Sciences
Wesleyan University
Middletown, Connecticut

VICTOR R. BAKER

R. CRAIG KOCHEL

Department of Geological Sciences
University of Texas at Austin
Austin, Texas

ABSTRACT

During large floods overbank sedimentation is greatest in slack-water areas associated with channel expansions and the mouths of tributaries which are either back-flooded or hydraulically dammed. Slack-water deposits in eastern Washington and central and west Texas provide information on the magnitude, frequency and areal extent of past floods.

The distribution of the slack-water facies of the Pleistocene Lake Missoula Flood deposits throughout the Channeled Scabland of eastern Washington defines the area covered by those floods. The sediments accumulated in basins created by downstream flow constrictions and in back-flooded valleys adjacent to the major channelways. The rhytmically layered slack-water sediments, in many ways analogous to turbidity current deposits, probably resulted from discharge surges during a few great floods.

The lack of soil-profile development on central Texas slack-water deposits is evidence that the accumulation of alluvium at these sites is related to the present hydrologic regimes of the streams. The elevations of the deposits were used to estimate the maximum stage of flooding. The discharge values calculated from the stage estimates approximately agree with the maximum flood of record on these streams.

Tributary canyons of the Pecos and Devils rivers in west Texas are filled with slack-water deposits. The Pecos River deposits are composed of sediment deposited during flooding back up the tributaries from the main stream. In contrast, slack-water deposits in the adjacent Devils River basin are silts deposited during flood surges up the tributaries, interbedded with gravel deposited during floods down the tributary canyons. The more complex hydrologic record in the Devils River slack-water deposits is probably the result of the greater drainage efficiency of its tributary canyons.

The paleohydrologic record derived from studies of slack-water alluvial stratigraphy is useful in assessing the rates of geomorphic change in river channels that result from infrequent processes that cannot be measured directly.

INTRODUCTION

Paleohydrologic investigations of ancient and modern river systems provide insight to the long-term hydrologic conditions that controlled the development and evolutionary adjustments of stream channel systems. Basically, two approaches are used in fluvial paleohydrology. The first approach concerns the mean hydrologic conditions that existed during the formation of identifiable paleochannels. Such studies have concentrated on alluvial stream channel systems and have based their hydrologic analysis on the morphology and sedimentology of the preserved channels (Dury, 1965; Schumm, 1968; Baker and Penteado-Orellana, 1978). Estimates of stream discharge are obtained by comparing the geomorphic characteristics of the paleochannels to analogous modern rivers. This approach yields general estimates for relatively high frequency runoff events such as bankfull discharge. This information can then be used to infer the mean characteristics of the prevailing climate at the time of channel formation and thus to estimate the magnitude of paleoclimatic fluctuations (Schumm, 1965).

The other paleohydrologic approach attempts to determine the characteristics of discrete hydrologic events, usually high-magnitude floods. The method is based on identifying indirect evidence of the stage and frequency of flooding. Because river stage estimates can be converted to discharge, the method can, in theory, be used to construct unusually long flood-frequency records. In contrast to the paleochannel-morphology method, the paleoflood analysis provides insight to the frequency of extreme events within a climate or range of climates, as opposed to the mean state.

Estimates of the frequency of occurrence and the discharge magnitudes of ancient floods are important in geomorphic studies of channel evolution. Several studies have shown that in certain physiographic and climatic regions the results of a single large flood can dominate the fluvial landscape (Hack and Goodlett, 1960; Baker, 1973, 1977; Williams and Guy, 1973; Patton and Baker, 1977). Morphological elements of these stream systems may be formed and controlled by rare floods. For example, on small central Texas streams floodplains are formed from bars deposited during the major floods (Baker, 1977; Patton and Baker, 1977). Frequently occurring flows in these streams have little or no impact on these deposits. Conversely, other aspects of central Texas stream systems may slowly recover to their preflood morphology. For example, a regular pool and riffle pattern that was obliterated or enlarged during a flood flow may be reestablished. In this example, the higher-frequency flows are the most important part of the discharge spectrum, and they control the channel morphology. In both instances, data on the number and magnitude of infrequent floods would be useful in assessing the relative importance of rare floods in controlling the development and adjustment of fluvial landforms, and in learning more about the recovery time of different stream systems.

From an engineering and land-use perspective, more precise evidence of large floods that occurred prior to historical records and more accurate flood-frequency determinations of large historical floods are necessary to estimate accurately the flood potential of a basin. This is increasingly important because economic losses from flooding are continuing to rise, and it has been predicted that the greatest amount of future flood damage will result from extreme catastrophic floods (U.S. Water Resources Council, 1968).

Indirect evidence of the frequency of individual prehistoric flood events can be obtained by the following methods: (a) botanical studies of floodplain vegetation (Sigafoos, 1964; Helley and LaMarche, 1973); (b) analysis of the degree of soil development of floodplains (Cain and Beatty, 1968), flood bars (Bretz, Smith, and Neff, 1956; Baker, 1973; Patton and Baker, 1978), and on geomorphic surfaces truncated by floods (Costa, 1978); (c) radiocarbon dating of buried organic debris within flood deposits (Fryxell, 1962; Helley and La-Marche, 1973) and within organic rich floodplain soils (Patton and Baker, 1977); and (e) paleostratigraphy of flood deposits (Patton and Dibble, 1978).

Although there are numerous methods for determining flood frequency, there are only a few methods for estimating the stage and probable discharge of ancient flood flows. When Baker (1973) analyzed the paleohydrology of the Lake Missoula flooding in eastern Washington, he used the elevation of drainage divide crossings, the highest ice-rafted erratics, and the highest eroded scarps to interpret the ancient water-surface profile. While useful in investigating the extreme conditions related to scabland flooding, these criteria are generally not applicable to the analysis of most river flooding. However, he also noted, as did Bretz (1929), that fine-grained sediment had back-flooded into tributaries and formed deposits whose highest elevation approximated the elevation of the flood waters in the main channels. These deposits have been termed "slack-water sediments" because they accumulate in overbank areas where current velocity is reduced. They may be found adjacent to a main stream and also along back-flooded tributaries.

The purpose of this paper is to describe the results of recent and ongoing research on the utility of slack-water deposits for paleohydrologic analysis. This paper is a progress report. Investigations of slack-water stratigraphy are continuing in eastern Washington, south-central Utah, central and west Texas, and southern New England.

PREVIOUS INVESTIGATIONS

In addition to Bretz's (1929) observations of Missoula Flood slack-water deposition, several researchers have noted that such deposits are formed during floods on major river systems. The tendency for sediment accumulation to be greatest in tributary streams has been documented during Skagit River floods in Washington (Stewart and Bodhaine, 1961), Ohio River basin floods of 1937 (Mansfield, 1938), Connecticut River floods of 1927, 1936, and 1938 (Jahns, 1947), and the 1973 Agnes Hurricane floods in Pennsylvania (Moss and Kochel, 1978) and Maryland (Costa, 1974). McKee (1938) also described the occurrence of small terraces at the mouths of tributaries of the Colorado River in the Grand Canyon of Arizona. These terraces, which are found above the floodplain or "beach" of the river, are formed by reverse surges of sediment into the tributary from the main stem. McKee's (1938) interest in these terraces was strictly related to their internal sedimentary structures. He made no attempts to interpret their stratigraphy or to estimate the probable flood stages based on the elevation of the terraces above the modern Colorado River.

Stewart's work (Stewart and Bodhaine, 1961) was one of the first attempts to use slack-water deposits to reconstruct flood magnitudes. He determined the stages of floods that had occurred about 1815 and 1856 on the Skagit River from water marks on trees and canyon walls and from the height of sediment lodged in the bark of trees and deposited in crevices in the canyon sides. He observed that the sediment which backfilled tributary gulches was at an elevation consistent with his other high-water-mark data, and he incorporated this information into his stage estimates.

Jahns (1947) was able to identify stratigraphic layers in slack-water silts along the Connecticut River. He related them to the 1927, 1936, and 1938 floods, and used the elevation of the deposits and their stratigraphic relationships to infer the relative frequency of the floods. Sediment from the 1938 flood occurred at the highest elevation and covered the low terrace (Terrace III) of the Connecticut River. Because the flood sediment buried a fine-grained sedimentary sequence typical of an open flood-plain environment, he reasoned that the 1938 flood must have been the largest flood to occur since Terrace III had formed. He estimated that Terrace III was the active river floodplain level approximately 2500 to 6000 years B.P.

SLACK-WATER DEPOSITS RELATED TO LAKE MISSOULA FLOODING

Bretz (1929) described fine-grained sediments infilling 41 valleys on the eastern margin of the Cheney-Palouse tract of the Channeled Scabland in eastern Washington and western Idaho. The sediments were deposited by back-flooding of the tributaries during the catastrophic Pleistocene floods caused by the rapid drainage of glacial Lake Missoula. These floods were also responsible for the erosional scabland topography of the main channelways (Bretz, 1929). Bretz supported his hypothesis for the flood origin of the tributary deposits with the following lines of evidence. He observed that the slack-water deposits begin at the mouths of the tributaries where the scabland topography ends. The terraced accumulations are thickest near the mouth of the tributary and become thinner upstream. The elevations of the terraces coincide with water-surface elevations in the main channels, which he had determined from other field evidence. The grain size of the slack-water sediments decreases upstream, the internal sedimentary structures indicate upstream flow, and the composition of the sediment requires a source from outside of the drainage basin. Bretz (1929) reasoned that the slack-water sands and silts could not have been deposited in marginal lakes because the elevations of the terraces coincide with the water-surface gradient of flooding down the Cheney-Palouse channelway, Bretz (1929) also described upstream dipping slack-water deposits on the Snake River above Lewiston-Clarkson, the massive deposits in the Tucannon Valley opposite the Palouse-Snake River confluence, and the thick sequences deposited in the Walla Walla basin and in the Yakima River basin.

Bretz (1929) was impressed with the range of grain sizes and the diversity and complexity of the primary sedimentary structures within various slack-water deposit exposures. He later wrote that the mechanics of slack-water deposition were poorly understood but that the deposits themselves offered potentially the best opportunity for precisely determining the chronology of scabland floods (Bretz, 1969, p. 151).

Baker (1973) investigated and described in detail the slack-water deposits in the Tucannon Valley. The Tucannon slack-water deposits are interesting because the mouth of the valley is almost directly opposite the confluence of the Snake and Palouse rivers (fig. 1). The immense discharge of the Cheney-Palouse scabland system, which overtopped the pre-flood Palouse River divide and carved Palouse Canyon, surged directly up the Snake River and into the Tucannon Valley. Baker (1973) measured sections of the slack-water deposits from the proximal facies of coarse sand and gravel at the bar that blocked the mouth of the valley (fig. 2A) to the distal (upvalley) facies of numerous rhythmically bedded sand and silt units (fig. 2B). Each rhythmite ideally consists of the following vertical sequence:

(1) A basal layer of structureless upward fining coarse sand and gravel, only present in the more proximal rhythmites.

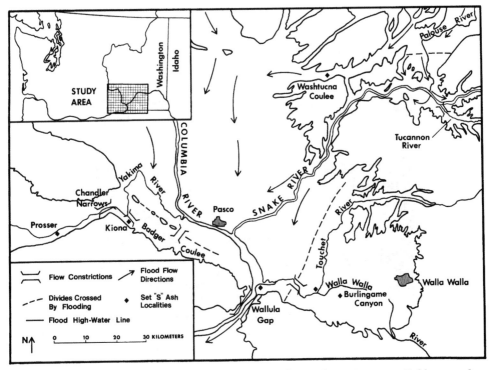

Figure 1.   *Location map showing the Pasco Basin region, Tucannon Valley, and locations of late Pleistocene flood slack-water deposits containing the Mt. St. Helens set "S" tephra.*

A. *Proximal facies deposit in the eddy bar at the junction of the Tucannon and Snake rivers. The upvalley flood surges moved from left to right. Note the large "rip-up" silt boulders, the largest of which is about 1.5 m wide.*

B. *Distal facies rhythmites exposed 10 km upvalley from site "A" (Baker, 1973, section 7, p. 43). Note the prominent clastic dikes.*

Figure 2.   Typical Missoula Flood slack-water deposits exposed in the Tucannon River valley.

(2) Horizontally stratified medium sand.

(3) Ripple-drift-laminated fine sand.

(4) Parallel lamination in very fine sand and silt.

The basal layer of the rhythmite sequence is commonly graded and is characterized by load casts on the interface with the underlying laminated silts. In more extreme cases of soft-sediment deformation, convolute laminations and slumps are present on this interface. As Baker (1973) noted, these deformation structures were created as a result of increased pore pressure during loading of the overlying basal graded unit on the water-saturated silts. The ripple-drift lamination of zone (3) in an ideal rhythmite frequently displays an upward progression from "type A" to "type B" climbing ripples (Jopling and Walker, 1968), overlain by "type S" (sinusoidal) ripples. Jopling and Walker (1968) concluded that sinusoidal ripples result if the rate of suspended-load deposition on the ripple exceeds the rate of traction transport. Experiments by J.C. Boothroyd and J.B. Southard (personal communication, 1978) show that A-B-S ripple drift structure develops very rapidly (1-2 hours) from flows containing extremely high concentrations of suspended sand and silt.

Baker (1973) stated that the sequences of slack-water sedimentary structures and textures are somewhat analogous to those of turbidity currents (Bouma, 1962). Thus, zones (2), (3), and (4) are similar to Bouma zones b, c, and d. According to this hypothesis, the Missoula Flood was phenomenally charged with suspended silt, sand, and gravel because of its extremely high stream power (Baker, 1973) and the extensive erosion of the sedimentary mantle that blanketed the Columbia Plateau prior to flooding. The slack-water deposits were created by upvalley surging and subsequent settling of this sediment-charged water in the tributary mouths. The tributary mouths acted as "stilling basins" adjacent to the main scabland flood channelways.

The turbidity-current analogy should be applied with caution to these fluvial deposits. The precise process of turbidity-current initiation and flow is still subject to considerable debate (Middleton, 1970), and both the scabland sediments and the flow geometries differ markedly from marine turbidite environments. The overall proximal to distal variation is, however, analogous. The deposits at the mouth of the Tucannon Valley are coarse-grained, crudely-graded, and irregularly bedded with large-scale forset cross-stratification. Large rip-up clasts are common (fig. 2A). Gravel beds are commonly amalgamated, indicating non-deposition or erosion of associated finer-grained components (Walker, 1967). These characteristics are typical of proximal marine turbidites. A short distance upvalley the rhythmites are relatively thin bedded, with cross-bedded sands and well-graded basal units (fig. 2B). The rhythites lack scour features, have a lower sand/mud ratio, and beds of sand are rarely amalgamated. Again, from Walker's criteria, these deposits would be analogous to marine turbidites occurring on the distal portions of submarine fans. The progression for the Tucannon rhythmites seems to be (a) deposition of a few rhythmites, predominantly from traction transported sediment, at the mouth of the valley and (b) upvalley increasing sedimentation from suspension as the velocity of the density current rapidly waned, creating multiple rhythmites.

P. C. Patton, V. R. Baker and R. C. Kochel

The slack-water facies of the Missoula flood deposits are the current subject of an intense study of the number and timing of late glacial floods in the Columbia River system. The upper portions of some slack-water sequences contain tephra layers that correlate to the "set S" pumice deposits of Mount St. Helens (Moody, 1978; Mullineaux et al., 1978). This correlation indicates a limiting upper age of about 13,000 yrs B.P. for the last major episode of catastrophic flooding in the Columbia Basin of eastern Washington (Mullineaux et al., 1978). The regional distribution of this tephra (fig. 1) makes it an important time-stratigraphic marker in the slack-water sequence. Its discovery has refocused attention on the origin of the flood slack-water sediments.

Richard B. Waitt, Jr., (1978, written communication) studied the remarkably thick late-glacial rhythmite sequences of the Walla Walla basin and the lower Yakima Valley (fig. 1). He found evidence for subaerial exposure following floodwater emplacement of some of the rhythmites. At the Burlingame Canyon section (fig. 1), Waitt counted at least 38 well-formed distal facies rhythmites. Each rhythmite may possibly represent a separate catastrophic flood event (Richard B. Waitt, 1978, written communication). However, the proximal facies sediments for each of these inferred floods have not yet been discovered. Thus, the possibility remains that single great floods can produce multiple distal rhythmites, and the Burlingame section may represent a somewhat smaller number of late Wisconsin flood events than given by the rhythmite count.

The geometries and locations of the Walla Walla and Yakima valleys (fig. 1) would have made these sites ideal "stilling basins", adjacent to the main flood paths through the Pasco Basin. The water surface of any one flood was probably unsteady, or fluctuating in elevation, because of the extremely complex flow geometry of the Channeled Scabland. Thus, in this conception, a single catastrophic flood would not result in a simple rise and subsequent fall of water in the Pasco Basin. Some volumes of water might arrive sooner by flowing along more direct channelways, while other volumes might be delayed by their longer flow routes. The development of divide crossings, the recession of cataracts, and perhaps even the jamming of berg ice would also contribute to fluctuating water levels in the Pasco Basin. Because this water would contain extremely high concentrations of suspended sand and silt eroded from the Palouse loess and Ringold Formation on the Columbia Plateau, any water surface fluctuation in the Pasco Basin could generate an upvalley surge of sediment-charged water into the Pasco and Yakima valleys. Sediment would gradually settle from suspension in these stilling basins until the cycle repeated with a new surge from a subsequent main-channel water-surface fluctuation.

Neither the one-rhythmite-for-each-flood hypothesis or the several-rhythmites-for-each-flood hypothesis has been proven. Definitive breaks, such as buried soils, or materials for radiocarbon dating have not yet been uncovered at the tops of each rhythmite. Obviously, further work on the Columbia Basin slack-water deposits is necessary to precisely interpret the exact sequence of late Pleistocene floods.

SLACK-WATER DEPOSITS RELATED TO HOLOCENE FLOODS

Central Texas

The studies of the Lake Missoula Flood slack-water deposits suggest that similar deposits on modern rivers might be used to estimate ancient flood stages. The stage data might then be converted to discharge estimates to extend the Holocene hydrologic record in a given basin.

Central Texas (fig. 3) was chosen to test this idea because it is a region characterized by intense rainfalls and large floods (Baker, 1977; Patton and Baker, 1977). Many of the high-intensity rainfalls, for durations up to about 24 hrs, and the resulting peak flows are records or near records for the United States (Baker, 1977). The region has also been identified as one of high flash flood potential (Beard, 1975). Both the geology and physiography of the region contribute to the magnitude of the floods resulting from these rainfalls. The Edwards Plateau of central Texas is underlain by Cretaceous limestone. Because of the thin, stony soils and steep slopes developed on this limestone plateau, overland flow is enhanced. The high densities of rill networks on the hillslopes further illustrate the pronounced surface runoff (Baker et al., 1975; Patton and Baker, 1976).

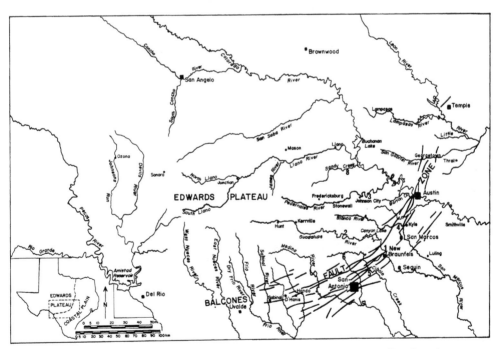

Figure 3. Location map for central Texas streams. Numbers refer to stream sites noted in table 1.

Depositional Setting

During large floods in the confined bedrock valleys of the Edwards Plateau, flood waters completely cover the floodplain of the valley and impinge on the bedrock valley walls. Figure 4 is a cross-section of the Pedernales River in Pedernales State Park. The low flow channel of the Pedernales is lined by bald cypress trees. The intermediate water line is based on a trash line left by a flood in 1973. The highest water line is the estimated stage of the September 1952 flood, the greatest recorded flood on the Pedernales River. During floods of this magnitude the tributaries are dammed by the high stages in the main channel, and reverse surges are common. In addition the flood-producing storms are often quite localized and, as the flood wave moves downstream through portions of the basin that are not in flood, back-flooding of the tributaries is common. Most of the slack-water deposits in central Texas are located at tributary junctions, but several have been found in marked channel expansions and where zones of flow stagnation might be expected during high stages. A striking example of such deposition was observed in August, 1978, during the flooding that resulted from tropical storm Amelia (fig. 5).

Slack-water deposits were found on nearly all of the major tributaries of the Pedernales River in Pedernales Falls State Park. A prominent deposit occurs at the mouth of Mescal Creek (fig. 6). The deposit is a small flat terrace on both sides of the creek approximately 18 m above the bed of the river. Unlike the scabland deposits, the alluvium is unstratified medium-to-fine sand. At a lower elevation on the floodplain an alternating sequence of sand and mud is exposed. Each couplet probably represents individual floods of lesser magnitudes than the floods that produced the slack-water deposit. This stratification is not present in the highest alluvium. The Mescal Creek slack-water deposit contains abundant feldspar, biotite, and quartz grains, although the drainage basin is entirely underlain by carbonate rocks. The

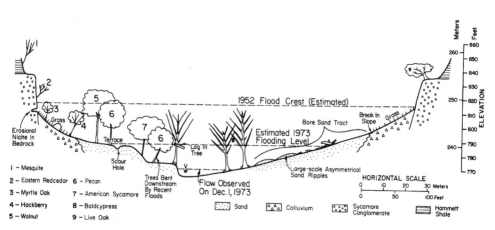

Figure 4. *Cross-section of the Pedernales River valley at Trammel Crossing, Pedernales Falls State Park, Texas. Flood crests, vegetation associations, and geomorphic features are shown schematically.*

A. Aerial photograph taken at maximum flood stage, 3:00 P.M.,
August 3, 1978. Note the prominent slack-water zone and eddy
at "S", where the flow expanded below the falls. Water stage
was approximately 10 m.

B. Aerial photograph taken on August 17, 1978, showing fresh accumulation of slack-water sand at position of flood eddy.

Figure 5. Formation of a slack-water deposit at Pedernales Falls during tropical storm Amelia flooding in central Texas.

nearest potential source for sediment of this composition is North Grape Creek, an upstream tributary of the Pedernales River which is incised into granite. This mineralogical evidence indicates that the sediment in the Mescal Creek flood terrace was derived from back-flooding of the Pedernales River, i.e. from flow moving up Mescal Creek during the flood.

No soil has developed on the upper 50 cm of the Mescal Creek slack-water alluvium. The alluvium caps an organic-rich buried A horizon 10-15 cm thick, which was the surface of the deposit prior to the last major flood (fig. 7A). Given the fresh appearance and the lack of any soil development on the upper 50 cm of the alluvium, it is likely that this sediment was deposited during the 1952 flood.

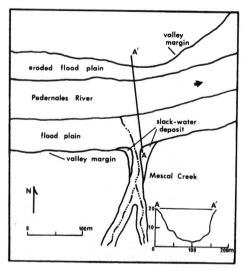

Figure 6. Plan view of the confluence of Mescal Creek with the Pedernales River in Pedernales Falls State Park. Surges up Mescal Creek from the Pedernales River have created a slack-water deposit at this stream junction.

On Barton Creek near Bee Caves, Texas, a slack-water deposit has formed as a result of a channel expansion. The channel of Barton Creek abruptly widens downstream from the State Highway bridge and then narrows about 300 m further downstream (fig. 8). The expansion is created by an abandoned meander scarp cut into the limestone valley walls. This expansion is filled with fine-grained sediment and forms a 6-meter-high terrace parallel to the stream. The alluvium is predominantly carbonate silt and fine sand. There are no internal sedimentary structures and no evidence of soil development in the profile (fig. 7B). The appearance of adventious roots on several small trees growing on the terrace is further proof of recent sedimentation (Sigafoos, 1964).

A third style of slack-water deposition is illustrated in a reach of Cibolo Creek near Boerne, Texas. Approximately 1.5 km downstream from Curry Creek Road (fig. 9) elongate scour holes and trash piled up against trees on the southern floodplain suggest that during major floods large volumes of water flow across the inside of an abrupt meander bend. The bedrock-incised meander probably causes stagnation of the flow and reduced velocity along the left bank of the channel. A large slack-water deposit has formed along this bank. The sediment consists mainly of carbonate sand and silt. A stratigraphic break occurs at about 50 to 65 cm below the surface. Charcoal fragments were found at this contact. The uncorrected radiocarbon date for the charcoal is 610 ± 90 years B.P. (Univ. of Texas at Austin Radiocarbon Laboratory Sample TX-2348). This date sets a maximum age for the subse-

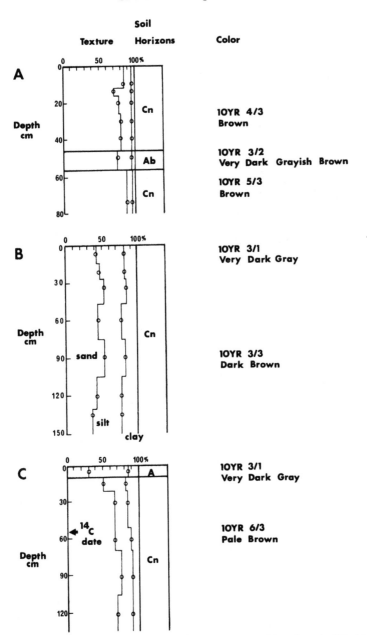

Figure 7. *Soil profiles developed on slack-water sediments at (A) the con-*
fluence of Mescal Creek and the Pedernales River, (B) Barton
Creek near Bee Caves, and (C) Cibolo Creek near Boerne. Note
the position of the radiocarbon dated horizon in the Cibolo Creek
profile.

237

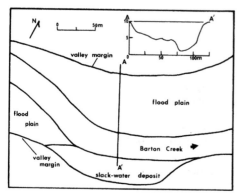

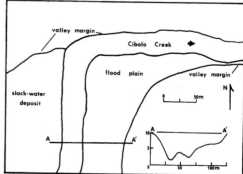

Figure 8. Plan view of Barton Creek near Bee Caves, Texas. A slack-water deposit has formed in a channel expansion which is probably an abandoned meander scarp.

Figure 9. Plan view of Cibolo Creek near Boerne, Texas, illustrating the dispositional setting of the slack-water deposit formed along the left bank of the stream.

quent deposition and pedogenesis of the overlying deposit. The alluvium has a distinct A horizon but neither an argillic or cambic B horizon (fig. 7C). The degree of soil development and the radiocarbon age can be used as a reference point for relative dating based on soil development of other flood-plain alluvium of similar parent material in central Texas.

Discharge Calculations from Slack-Water Elevations

One potential value of slack-water deposits is that they can be used as approximate water stage indicators for prehistoric floods or for poorly documented historical floods. This technique is especially appropriate in Texas, where gaging records are inadequate for prediction of the potential for high-magnitude floods. The conversion of indirect stage data to actual runoff values may be fairly accurate in these narrow bedrock valleys, because a small increase in discharge will result in a relatively large increase in stage.

Before using slack-water deposits as high-water markers to estimate peak flood discharge several qualifying assumptions must be considered. First, if significant aggradation or degradation of the stream channels has occurred since the flooding that produced the deposit, the accuracy of using the deposits and the present stream bed to estimate the cross-sectional flow area of the flood would be questionable. There is little detailed information on current aggradation or degradation of central Texas streams. Shepherd (1975) documented the aggradation of Sandy Creek in the Llano region of Texas but could not assign a precise age to the period of aggradation. All streams in central Texas, when considered over geologic time periods, are downcutting. But evidence of erosion on streams flowing across limestone bedrock, such as Cibolo Creek and Blieders Creek, indicates that downcutting occurs as individual scour holes (Baker, 1977; Patton and Baker, 1977). This process is related to large infrequent floods (Patton and Baker, 1976). There-fore the mean rate of downcutting, although difficult to determine, must be relatively slow. Other evidence also supports this hypothesis. There is an exposure of gravel, 1.2 m thick, in the vertical bank of Cibolo Creek near Bulverde, Texas. The gravels are well-cemented with calcium carbonate and are capped by sand and silt on which a well-developed soil has formed. Be-cause the gravel layer is only 0.3 m above the present stream bed, only minor downcutting could have occurred during the relatively long time interval repre-sented by the soil formation. Similar evidence on other reaches indicates a greater amount of downcutting. On one reach 2 to 3 m of erosion may have taken place. Slack-water deposits were not sampled on river reaches which appeared to be abnormally scoured or where evidence indicates recent down-cutting.

A second assumption is that no significant scour and fill occurred during the flood. If, during the high-water time of the flood, the bed had scoured and subsequently refilled, the actual cross-sectional flow area was greater than the post-flood survey would indicate. Consequently, the peak discharge actually would have been higher than estimated. Scour and fill is probably minimal along bedrock-floored streams in central Texas.

A third key assumption, at least if discharge is to be estimated, is that the top of the slack-water deposit corresponds to the top surface of the water during the flood. Most likely the water surface was higher than the top of the deposit. In some cases (e.g., for wide, shallow flows) the diffe-rence probably would be insignificant, but in other cases the difference could be considerable. The estimated cross-sectional flow area of the flood therefore should tend to be smaller than the actual flow area, to an unknown extent, when the top of the deposits is assumed to be the high-water level.

A fourth assumption in using slack-water deposits as water-stage indi-cators, is that the sediment resulted from floods which occurred during the present hydrologic regime of the stream. Evidence must be accumulated to prove that deposition is still occurring on these sites or has occurred within the recent past.

Each slack-water site investigated was searched for datable archeological artifacts, wood, or charcoal. Datable material was found in the slack-water deposit at Cibolo Creek near Boerne, Texas. The radiocarbon age of 610 years B.P. indicates that the deposition of the sediment has taken place in the present climate and stream regime at this location.

The degree of soil development was measured on slack-water deposits to estimate the relative age of the flood sediment. Surface and buried A horizons (recognized by low soil color values and chromas) are weakly devel-oped or non-existent on these soils (fig. 7). The soils have neither argillic or cambic B horizons and are classified as fluvents (Soil Survey Staff, 1960).

The soils have weak profile development because they are cumulative soils, formed when the influx of parent material is more rapid than the rate of soil formation. Similar examples of cumulative soils are those forming on active alluvial fans (Birkeland, 1974, p. 148) and in regions of active loess deposition (Rieger and Juve, 1961). In these soils, textural variations between each increment of fluvial parent material may be great enough to mask any incipient B horizons that may be present. The alternating sequence of fine and coarse-grained floodplain sediment on the Pedernales River is an example of textural variability. Although the degree of visible stratification seen in the Pedernales example is rare in central Texas slack-water deposits, textural analyses often revealed subtle variations in grain size throughout the profile (fig. 7B).

The number of buried soils and their degree of development in cumulative profiles provide information on the frequency of the process responsible for the sediment deposition. In areas of active loess deposition, for example, soil development is enhanced by distance from the source area and decreasing rates of deposition (Rieger and Juve, 1961). In the farthest downwind areas soils may be buried by periodic influxes of loess. In a similar manner, soil development on slack-water alluvium is related to flood frequency. If flooding is frequent, only minimal soil formation will occur. If flooding is infrequent, long periods of non-deposition will be represented by buried soil horizons. Thus a qualitative estimate of flood frequency can be made based on the absence or presence of buried soils and on the stage of development of the buried soil horizons.

The soils formed on slack-water deposits contrast sharply with soils formed on older terrace deposits. Although the terraces are frequently inundated, suspended sediment is transported across them because they are flat and hydraulically smooth. (This observation is consistent with other measurements of sediment accumulation made after large floods (Moss and Kochel, 1978). As a result, the terrace soils have well developed A horizons and argillic B horizons. This contrast in soil development allows discrimination of depositional floodplain sites from those sites which are stable and not aggrading.

To test the hypothesis that the elevations of slack-water deposits are useful estimators of water stage during high-magnitude floods, discharge estimates were made at eight different slack-water sites and compared with known historical record flows. The sites were either near long-term gaging stations or near the locations of previous indirect measurements of historical maximum floods.

Cross-sections were measured with a level and stadia rod. Discharges were calculated using the slope-area method (Dalrymple and Benson, 1967). The variables for the calculations are listed in table 1. The estimates of roughness values are one of the greatest sources of error in these calculations. Manning n values used are in the proper range for bedrock lined channels (Chow, 1959) and conform to values used by the U.S. Geological Survey for these streams. The reported magnitude of the greatest historical flood at the site nearest to each slack-water deposit location is listed in table 2. Two historical values are listed for the Guadalupe River near Kendalia and for the Blanco River upstream from Kyle because there is a significant difference between values recorded at upstream and downstream sites. The slack-water discharge estimates are approximately equal to the historical maximum flood on these streams. Certainly, within the error range of the slope-area method for discharge estimation the results are good.

Table 1

Measured, Estimated, and Calculated Variables for Slope-Area
Discharge Calculations, Central Texas Slack-Water Sites

Location	Sec-tion	Sub-sec-tion	n	A (m^2)	R (m)	K (m^3)	S	Q m^3/sec
Guadalupe	1	1	.100	705.2	4.36	18,481.4	.0022	3,000
River (at		2	.040	365.6	10.31	41,463.7		
Elm Creek)		3	.060	88.6	4.94	4,177.9		
Guadalupe	1	1	.100	141.8	4.76	3,932.6	.0011	3,280
River (near		2	.040	359.8	13.85	50,555.1		
Kendalia)		3	.080	172.6	9.12	9,197.4		
		4	.060	443.8	8.78	30,742.3		
		5	.100	207.1	3.90	5,047.1		
Pedernales	1	1	.060	120.8	3.72	4,742.2	.0021	12,170
River (at		2	.075	549.1	12.47	38,447.0		
Pedernales Falls		3	.035	567.6	17.35	123,327.1		
State Park)		4	.050	285.0	16.87	36,514.0		
		5	.060	339.9	8.48	60,975.9		
Blanco River	1	1	.070	100.1	3.05	2,979.6	.0038	2,770
(at Hallifax		2	.035	361.1	7.04	37,217.5		
Ranch)		3	.070	168.3	2.74	4,634.7		
Cibolo Creek	1	1	.050	20.7	1.65	571.7	.0027	1,270
(near Bracken)		2	.030	281.3	3.99	23,221.0		
		3	.100	48.0	1.40	598.2		
Cibolo Creek	1	1	.100	73.9	2.65	1,343.6	.0036	1,840
(near Boerne)		2	.050	347.6	6.83	24,490.0		
		3	.070	171.6	3.23	5,268.9		
	2	1	.100	123.5	2.41	2,204.6		
		2	.045	246.8	7.50	20,557.8		
		3	.060	245.4	2.74	7,905.0		
Barton Creek	1	1	.050	7.9	.55	104.1	.0028	2,090
(near Bee		2	.040	246.1	6.10	20,132.6		
Caves)		3	.045	300.2	3.48	15,043.7		
	2	1	.045	300.2	4.24	18,511.0		
		2	.040	261.0	7.78	25,031.3		
		3	.050	46.0	1.95	1,414.7		
Sandy Creek	1	1	.100	111.8	1.83	1,645.9	.0022	4,415
(10 km upstream		2	.035	772.0	8.33	88,489.5		
from Lake LBJ)		3	.050	53.2	3.51	2,410.2		
	2	1	.050	75.8	1.13	1,621.9		
		2	.035	890.3	6.98	90,947.7		
		3	.100	118.7	3.48	2,675.8		

Symbols: A, cross-section area; n, Manning roughness coefficient; R, hydraulic
radius; K, channel conveyance; S, channel slope; and Q, peak discharge.

Table 2

Discharge Estimates Calculated for Slack-Water Sites Compared
to Historical Maximum Discharges

Location of slack-water deposit	Slack-water discharge (m³/sec)	Nearest gaging station	Historical maximum discharge (m³/sec)	Exceedance probability	Standard deviation of the logarithms of the annual peaks
Guadalupe River at Elm Creek	3,000	New Braunfels	2,860	.020	.53
Guadalupe River near Kendalia	3,280	Comfort Spring Branch	5,150 3,420	.020 .014	.57 .48
Pedernales River at Pedernales Falls State Park	12,170	Johnson City	12,480	.005	.48
Blanco River at Hallifax Ranch	2,770	Wimberley	3,200	.030	.60
Cibolo Creek near Bracken	1,270	indirect measurement at Bracken	1,550	—	—
Cibolo Creek near Boerne	1,840	indirect measurement at Curry Creek Rd.	1,670	—	—
Barton Creek near Bee Caves	2,090	indirect measurement at Bee Caves	1,115	—	—
Sandy Creek 10 km upstream from Lake LBJ	4,415	indirect measurement near Lake LBJ	4,610	—	—

242

Frequency of Estimated Floods

The frequency of the floods represented by the slack-water deposits is not known. Table 2 lists the approximate exceedance probabilities for these flood magnitudes, based on a log Pearson III analysis. Unfortunately, because of the short historical record, the annual series frequency distributions have high standard deviations. These values are not therefore good statistical estimates of exceedance probability (Beard, 1975). The A horizon on the Cibolo Creek deposit is at least 600 yrs old. The soil profile on the same sediment is by far the best developed of all of the sites sampled. This would suggest that the floods which are presently topping the slack-water sites have frequencies significantly less than 600 yrs, perhaps on the order of a hundred years.

West Texas

Continuing research on slack-water deposits is concentrating on the applicability of this approach to problems of Holocene paleohydrology in the southwestern United States. West Texas was selected for study because a well stratified slack-water deposit near the mouth of the Pecos River had been investigated previously (Dibble, 1967; Patton, 1977). This deposit, a thick accumulation of flood sediment incorporating archeological habitation layers in the Arenosa Shelter (fig. 10, site A) offers a starting point for the paleohydrologic reconstruction of this region (Patton and Dibble, 1978). In addition, the narrow canyons of west Texas provide excellent settings for the accumulation of sediment delivered by back-flooding.

Canyons cut by tributaries of three major rivers in southwest Texas contain well-preserved slack-water deposits near their junctions with the main stream. The slack-water accumulations are best developed along the lower 60-70 km of the Pecos River and Devils River, along the Rio Grande from Big Bend eastward to Amistad Reservoir, and at the confluence of the Rio Grande and Pecos (fig. 10). Tributaries of the lower Pecos River and Devils River are deeply incised into the Cretaceous limestone units of the western Edwards Plateau. Although each tributary canyon contains slack-water sediment accumulations, not all are well enough preserved or exposed for detailed study. Tributary canyon characteristics which appear to permit excellent preservation and access to the stratigraphy of the deposits include: (1) an intermediate-size drainage area and relatively low gradient, to prevent frequent flushing of canyon floor sediments during tributary flooding (large tributaries may have shallow gradients but may collect too much runoff to prevent erosion of back-flooding deposits, while small tributaries invariably have steep gradients and fewer protected sites for back flood deposition); (2) protected accumulation sites such as the lee of bedrock protrusions, rock shelters (shallow caves) in bedrock walls, or on the inside of meander bends; (3) junction angles with the main stream sufficient to permit substantial back-flooding from the main stream; (4) minimal vegetative cover established on the deposits, to reduce bioturbation and to allow access for trenching of the site.

Lower Pecos River tributary canyons have extensive slack-water sediment accumulations preserved as terraces along the margins of the canyons. The terrace sediments are composed predominantly of quartz sand. This sand could be introduced only by back-flooding from the Pecos River. The well-stratified sequence of deposits consists of layers of structureless and horizontally laminated silt, sand, and organic debris (fig. 11). Horizontal lami-

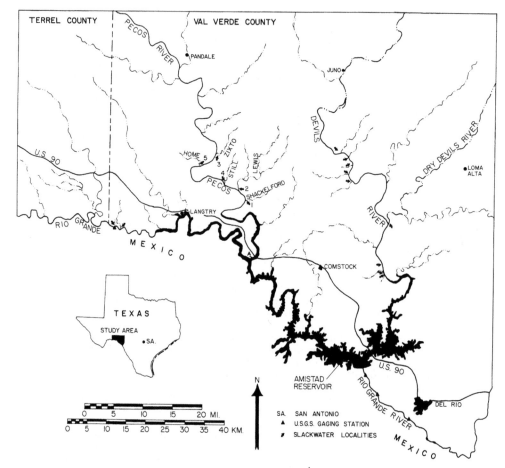

Figure 10. Index map of the west Texas study area showing the location of major rivers and the slack-water sites referred to in the text. The letter "A" refers to Arenosa Shelter, now inundated by Amistad Reservoir.

nation is most common in the sandy layers. Logs and a variety of fine-grained organic materials, including leaves and twigs, are commonly interbedded with the sediments. Radiocarbon ages for these materials are currently being determined (fig. 12). All of the deposits rest on limestone bedrock and are floored by as much as several meters of cobbles, boulders, gravel, and sand. In addition to the fine quartz sand and silt derived from the Pecos River, occasional thin layers of dark brown, loamy, snail-rich colluvium derived from adjacent limestone slopes are present in some of the sections (fig. 12). Average grain size of slack-water deposits at all sections decreases with increasing distance from the Pecos River (table 3). In addition, sorting

Figure 11. *Pecos River slack-water terrace exposed in Zixto Canyon. Light-colored sediments are silt and sand. Darker zones have large amounts of incorporated organic debris. Downstream is to the right. Note the log jutting out of the deposit to the right of the cleared profile. Rule is approximately 1.8 m long.*

appears to improve slightly with increasing distance from the Pecos River. Mean grain size (Mz of Folk, 1974) is variable within a single flood unit. Most commonly, there are no significant changes in grain size. However, in some deposits upward-fining and upward-coarsening sequences are present (fig. 12). Apart from occasional horizontal laminations, sedimentary structures are noticeably absent. In a few layers, small-scale cross-bedding displaying both up-canyon and down-canyon transport directions have been observed. The absence of primary sedimentary structures may indicate very rapid fallout from suspension during back-flooding surges.

Evidence of minor soil development is present in several of the flood deposits and in the colluvial sediments buried by subsequent flood deposits. Buried soil horizons indicate that relatively great periods of time may have elapsed between the deposition of the sediment on which the soil is developed and the next flood of comparable or greater magnitude. Evidence of soil formation in these deposits includes root mottling, oxidized horizons, formation of sand-sized nodules of silt and clay particles cemented by calcium carbonate, vertical color variations, and evidence of incipient textural B horizons. Parametric plots of various grain size generally discriminate between soils developed on colluvium versus the soils developed on alluvial flood sediments.

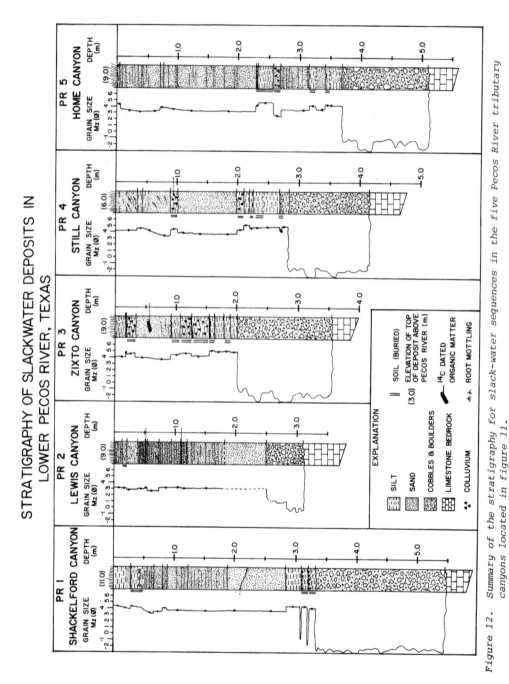

Figure 12. Summary of the stratigraphy for slack-water sequences in the five Pecos River tributary canyons located in figure 11.

Table 3

Trends of Average Grain Size and Sorting in Relation to
Distance up a Tributary Canyon from the Pecos River

	Canyon [a]				
	Lewis (2)	Home (5)	Shackelford (1)	Zixto (3)	Still (4)
Distance from Pecos River (m)	50	250	290	320	380
Average sediment size (M_z [b] in $\emptyset$ units)	3.01	3.80	3.94	4.26	4.44
Inclusive graphic standard deviation (σ_I [b] in $\emptyset$ units)	0.97	0.89	0.84	0.82	0.76

(a) Number refers to location on figure 10
(b) M_z, σ_I from Folk, 1974

Preliminary analysis of stratigraphic sections in five Pecos tributary canyons (fig. 10) indicates that there are 10 to 12 distinct layers (sedimentation units) on each terrace which may correlate with separate floods (fig. 12). No erosional unconformities have been found in the sequences. Initial correlations can be made readily between the 5 slack-water profiles (fig. 12) on the basis of buried soils, colluvial layers, grain size, and thickness of sedimentation units. The best correlations occur between canyons PR1, PR3, PR4, and PR5. Lewis Canyon (PR2) was sampled very close to the Pecos River and the sequence may have been significantly affected by smaller-magnitude floods. Radiocarbon dates will be used to refine these correlations.

The record of flood sedimentation preserved in tributaries of the adjacent Devils River basin is completely different. In the Devils River basin, coarse bouldery deposits emanating from the mouths of tributaries are interbedded with fine-grained carbonate silts of probable back-flood origin. In many cases, the carbonate silts are massive, and no stratigraphic breaks have been detected in these layers. Therefore the precise number of floods recorded in each deposit is not known. Nevertheless, the first reconnaissance of these tributary valleys left the impression that the flooding history of each tributary was drastically different. Each deposit appears to have a distinct sequence of interbedded slack-water silts and flood bars (fig. 13).

The difference in the flood stratigraphy can perhaps be explained in terms of the relative efficiency of the tributary networks in the two basins. The lower Pecos River basin is characterized in part by poorly developed tributary

247

Figure 13. *Grey carbonate silt slack-water deposits interbedded with gravel at the mouth of an unnamed tributary of the lower Devils River. Apparently periods of tributary flooding and gravel deposition are interspersed with periods of low tributary runoff and Devils River back-flooding. Entire exposure is about 7 m thick.*

stream networks and broad flat undissected interfluves. As a result, flood runoff is dominated by the main stem, and the influence of the downstream tributaries is probably minor. The slack-water deposits all contain approximately the same number of flood layers, indicating that the flooding of the lower Pecos River is probably uniform over this reach. In the Devils River drainage the tributary basins have much greater drainage densities, and the tributaries apparently respond much more rapidly to floods. Tributary contribution to flood runoff must be an important feature of the flooding regime of this river because large volumes of water and sediment flow out of the tributaries into the main stem. In addition, small thunderstorms can probably produce substantial runoff in some tributaries while the others are unaffected. As a result of such localized flooding the tributaries develop out of phase with one another. Therefore, the flood record in each tributary will produce a unique history of tributary flooding and perhaps indicate different frequencies of main-stem floods, depending on the spatial distribution of rainfall.

SUMMARY

Overbank sedimentation in eddies and backwaters during large floods creates deposits which are useful in the evaluation of the present and past hydrologic regimes of streams. These slack-water deposits are most common at the mouths of tributary valleys where hydraulic damming and reverse surges deposit thick accumulations of fine-grained alluvium. The deposits can be used to estimate the magnitude, frequency, and areal extent of discrete flood events.

Extensive slack-water sediments were deposited during the Pleistocene Missoula Floods in the Channeled Scabland of eastern Washington. Bretz (1929) used the areal distribution of the slack-water facies as evidence for the geographic extent of the flooding. Slack-water sediments are well exposed in the Tucannon River valley, which flood waters entered as they surged up the Snake River. The deposits grade from a coarse-grained proximal facies at the mouth of the valley to fine-grained rhythmically bedded sand and silt units upvalley. The sequence of sedimentary structures present and the repetitious beds are analogous to those reported for turbidites. The Tucannon valley slack-water deposits appear to be the result of traction deposition near the mouth of the valley, grading upvalley into sequences deposited primarily from suspension. Whether the rhythmites are the result of numerous floods (Richard B. Waitt, Jr., 1978, written communication) or discharge surges during a single flood has not been clearly determined. Mount St. Helens "set S" ash has been found in the slack-water facies at several localities (Moody, 1978; Mullineaux et al., 1978). The ash places an upper limit of 13,000 years B.P. on the occurrence of the last major scabland flood (Mullineaux et al., 1978).

Slack-water deposits have been identified along the margins of the major stream valleys in central Texas. These deposits lack well-developed soils and rarely contain buried soils. Soil evidence and one radiocarbon date from charcoal found in a slack-water deposit indicate that sedimentation is continuing at these sites and that these deposits are related to the modern stream regime. The elevations of the deposits were used as measures of water stage. Slope-area discharge estimates were made based on these approximate stages. The calculated discharges agree closely with, or exceed, the recorded maximum historical flood discharges on these stream reaches. Therefore, these slack-water deposits are useful in estimating the flood potential in these basins.

Tributary canyons of the Pecos River and Devils River all contain well-developed terraces composed of back-flooded sediment. Slack-water deposits in Pecos River tributaries are well stratified. They contain layers of structureless and horizontally laminated silt, sand, and organic debris. Ten to 12 distinct layers are present in each slack-water terrace. Radiocarbon dating of the organic debris within these deposits will be used to correlate the sedimentary sequence between tributary canyons, and with the sequence known from Arenosa Shelter at the mouth of the Pecos River (Patton, 1977). The adjacent Devils River basin has a more complicated flood history. The slack-water silts from the main stream are interstratified with gravel bars deposited by the tributary streams. Each Devils River tributary appears to have a different flood history. The difference in the flood history of the lower reaches of these two basins may be explained by the greater drainage efficiency of the Devils River tributaries.

Paleohydrologic studies are important in determining the magnitude and rate of stream channel adjustments through time. In analyzing most alluvial stream channels, correlations between stream morphology and hydrologic variables with high frequencies of occurrence explain most of the channel adjustments. However, in those fluvial systems where the morphology of the stream channels is related to infrequent large events, it is necessary to have an understanding of the magnitude and frequency of floods over time spans which greatly exceed the historical record. The hydrologic information derived from slack-water deposits can be used to extend the flood record over geologic time spans. This paleohydrologic record can in turn be used to evaluate the geomorphic changes in river channels which result from processes that occur too infrequently to be measured directly.

ACKNOWLEDGEMENTS

This research was supported in part by the Division of Earth Sciences, National Science Foundation Grant EAR 77-23025, The Geology Foundation of the University of Texas at Austin, and The Geological Society of America through a Penrose Research Grant no. 1947-75. S.D. Hulke, E. Patton, and R.G. Shepherd assisted with the field work in central Texas. We thank R.G. Shepherd for his review of the manuscript.

REFERENCES

Baker, V.R., 1973, Paleohydrology and sedimentology of Lake Missoula flooding in eastern Washington: Geological Society of America Special Paper 144, 79 p.

_____, 1977, Stream channel response to floods with examples from central Texas: Geological Society of America Bulletin, v. 88, p. 1057-1071.

Baker, V.R., Holz, R.K., Hulke, S.D., Patton, P.C., and Penteado, M.M., 1975, Stream network analysis and geomorphic flood-plain mapping from orbital and suborbital remote sensing imagery; application to flood hazard studies in central Texas (final report): National Aeronautics and Space Administration Report EREP no. 064B, Contract no. 9-13312, 187 p.

Baker, V.R. and Penteado-Orellana, M.M., Adjustment to Quaternary climatic change by the Colorado River in central Texas: Journal of Geology, v. 85, p. 395-422.

Beard, L.R., 1975, Generalized evaluation of flash-flood potential: University of Texas Center for Research in Water Resources Technical Report CRWR-124, 27 p.

Birkeland, P.W., 1974, Pedology, weathering, and geomorphological research: New York, Oxford University Press, 285 p.

Bouma, A.H., 1962, Sedimentology of some flysch deposits: a graphic approach to facies interpretation: Amsterdam, Elsevier, 168 p.

Bretz, J.H., 1929, Valley deposits immediately east of the Channeled Scabland of Washington [Parts 1 and 2]: Journal of Geology, v. 37, p. 393-427, p. 505-541.

_____, 1969, The Lake Missoula floods and the Channeled Scabland: Journal of Geology, v. 77, p. 505-543.

Bretz, J.H., Smith, H.T.U., and Neff, G.E., 1956, Channel Scabland of Washington: new data and interpretations: Geological Society of America Bulletin, v. 67, p. 957-1049.

Cain, J.M., and Beatty, M.T., 1968, The use of soil maps in the delineation of flood plains: Water Resources Research, v. 4, p. 173-182.

Chow, V.T., 1959, Open-channel hydraulics: New York, McGraw-Hill, 680 p.

Costa, J.E., 1974, Response and recovery of a piedmont watershed from tropical storm Agnes, June, 1972: Water Resources Research, v. 10, p. 106-112.

_____, 1978, Holocene stratigraphy in flood-frequency analysis: Geological Society of America Abstracts with Programs, v. 10, p. 383.

Dalrymple, T., and Benson, M.A., 1967, Measurement of peak discharge by the slope-area method: U.S. Geological Survey Techniques of Water-Resources Investigations, Book 3, Chapter A2, 12 p.

Dibble, D.S., 1967, Excavations at Arenosa shelter, 1965-1966 (preliminary report): National Park Service, Texas Archeological Salvage Project, 85 p.

Dury, G.H., 1965, Theoretical implications of underfit streams: U.S. Geological Survey Professional Paper 452C, 43 p.

Folk, R.L., 1974, Petrology of sedimentary rocks: Austin, Texas, Hemphill Publishing Co., 182 p.

Fryxell, R., 1962, A radiocarbon limiting date for scabland flooding: Northwest Science, v. 36, p. 113-119.

Hack, J.T., and Goodlett, J.C., 1960, Geomorphology and forest ecology of a mountain region in the central Appalachians: U.S. Geological Survey Professional Paper 347, 66 p.

Helley, E.J., and LaMarche, V.C., 1973, Historic flood information for northern California streams from geological and botanical evidence: U.S. Geological Survey Professional Paper 485E, 16 p.

Jahns, R.H., 1947, Geologic features of the Connecticut valley, Massachusetts, as related to recent floods: U.S. Geological Survey Water Supply Paper 996, 158 p.

Jopling, A.V., and Walker, R.G., 1968, Morphology and origin of ripple-drift cross-lamination, with examples from the Pleistocene of Massachusetts: Journal of Sedimentary Petrology, v. 38, p. 971-984.

Mansfield, G.R., 1938, Flood deposits of the Ohio River, January-February, 1937: in Grover, N.C. and others, U.S. Geological Survey Water Supply Paper 838, p. 693-736.

McKee, E.D., 1938, Original structures in Colorado River flood deposits of Grand Canyon: Journal of Sedimentary Petrology, v. 8, p. 77-83.

Middleton, G.V., 1970, Experimental studies related to problems of flysch sedimentation: Geological Association of Canada Special Paper 7, p. 253-272.

Moody, U.L., 1978, Microstratigraphy, paleoecology, and tephrochronology of the Lind Coulee Site, central Washington [Ph.D. thesis]: Pullman, Washington State University, 273 p.

Moss, J.H., and Kochel, R.C., 1978, Unexpected geomorphic effects of the Hurricane Agnes storm and flood, Conestoga drainage basin, southeastern Pennsylvania: Journal of Geology, v. 86, p. 1-11.

Mullineaux, D.R., Wilcox, R.E., Ebaugh, W.F., Fryxell, R., and Rubin, M., 1978, Age of the last major scabland flood of the Columbia Plateau in eastern Washington: Quaternary Research, v. 10, p. 171-180.

Patton, P.C., 1977, Geomorphic criteria for estimating the magnitude and frequency of flooding in central Texas [Ph.D. thesis]: Austin, University of Texas, 222 p.

Patton, P.C. and Baker, V.R., 1976, Morphometry and floods in small drainage basins subject to diverse hydrogeomorphic controls: Water Resources Research, v. 12, p. 941-952.

_____, 1977, Geomorphic response of central Texas stream channels to catastrophic rainfall and runoff; in Doehring, D.O., ed., Geomorphology in ared regions, Binghamton, State University of New York, Publications in Geomorphology, p. 189-217.

_____, 1978, New evidence for Pre-Wisconsin flooding in the channel scabland of eastern Washington: Geology, v. 6, p. 567-571.

Patton, P.C., and Dibble, D.S., 1978, Archeological and geomorphic evidence for the paleohydrologic record of the Pecos River in west Texas: Geological Society of America Abstracts with Programs, v. 10, p. 469.

Rieger, S., and Juve, R.L., 1961, Soil development in recent loess in the Matanuska Valley, Alaska: Soil Science Society of America Proceedings, v. 4, p. 285-291.

Schumm, S.A., 1965, Quaternary paleohydrology: in Wright, H.E., and Frey, D.G., eds., The Quaternary of the United States, Princeton, N.J., Princeton University Press, p. 783-794.

_____, 1968, River adjustment to altered hydrologic regimen, Murrumbidgee River and paleochannels, Australia: U.S. Geological Survey Professional Paper 598, 65 p.

Shepherd, R.G., 1975, Geomorphic operation, evolution, and equilibria, Sandy Creek watershed, Llano region central Texas [Ph.D. thesis]: Austin, University of Texas, 209 p.

Sigafoos, R.S., 1964, Botanical evidence of floods and floodplain deposition: U.S. Geological Survey Professional Paper 485A, 35 p.

Soil Survey Staff, 1960, Soil classification, a comprehensive system, 7th approximation: Washington, D.C., U.S. Department of Agriculture, Soil Conservation Service, 265 p.

Stewart, J.E., and Bodhaine, G.L., 1961, Floods in the Skagit River basin, Washington: U.S. Geological Survey Water Supply Paper 1527, 66 p.

United States Water Resources Council, 1968, The nation's water resources: Washington, D.C., U.S. Government Printing Office, 32 p.

Walker, R.G., 1967, Turbidite sedimentary structures and their relationship to proximal and distal depositional environments: Journal of Sedimentary Petrology, v. 37, p. 25-43.

Williams, G.P., and Guy, H.P., 1973, Erosional and depositional aspects of Hurricane Camille in Virginia, 1969: U.S. Geological Survey Professional Paper 804, 80 p.

RIVER CHANNEL AND SEDIMENT RESPONSES TO BEDROCK

LITHOLOGY AND STREAM CAPTURE, SANDY CREEK

DRAINAGE, CENTRAL TEXAS

RUSSELL G. SHEPHERD

Willard Owens Associates, Inc.
7391 West 38th Avenue
Wheat Ridge, Colorado 80033

ABSTRACT

Sandy Creek watershed is approximately 70 km west of Austin, Texas. Sixteen percent of the drainage area (now 1,025 km²) of Sandy Creek was added suddenly by the stream capture of an adjacent drainage. The main channel of Sandy Creek as well as that of the capturing tributary (Crabapple Creek) adjusted dramatically as channel and sediment responses were relayed through the system.

Today the channel of Sandy Creek fails to conform to the usual hydraulic geometry relations. In a downstream direction channel-bed width and channel-fill depth decrease while flood depth increases anomalously, and gradient, grain size, and sorting remain nearly constant.

Sandy Creek fails to conform to the usual relationships of hydraulic geometry because it has aggraded from upstream. The aggradation is due to successive responses to 1) stream capture of limestone drainage, which abruptly increased sediment-deficient discharges of water, producing 2) degradation in Sandy and Crabapple Creeks, producing 3) rejuvenation of their tributary networks, producing 4) aggradation due to increased sediment loads from the rejuvenated, dominantly granitic drainages.

Investigation of the controls of pre-capture bedrock channel and sediment characteristics revealed that the type and abundance of watershed bedrock exert the most important influence on the adjustments. A spectrum of bedrock channel types was documented. Channels dominated by limestone or schist bedrock characteristically have the thinnest, worst-sorted, and coarsest channel-bed sediments, the lowest width-depth ratios, and pool-riffle thalweg patterns. Channels dominated by granite characteristically have the best sorted and thickest sediments, the highest width-depth ratios, and braided thalweg patterns at low flow. Channels dominated by gneiss or sandstone are transitional.

This spectrum is exemplified by six major channels tributary to Sandy Creek. They exhibit a trend, from east to west, of suspended-load to bedload bedrock channels. Commensurate with the trend is a decrease in granite and gneiss abundance, described quantitatively by their percentages of watershed area. The spectrum of channel types is analogous to that for alluvial channels.

INTRODUCTION

The importance of stream capture in geomorphology has long been recognized (Gilbert, 1877) and continues to receive attention (Woodruff, 1977). While mapping the bedrock geology of the Crabapple Creek quadrangle in the Llano Region of Central Texas (fig. 1), Barnes (1952) noted that Crabapple Creek, a tributary of Sandy Creek and the Colorado River, had captured the headwaters of Willos Creek (fig. 2), a tributary of the Pedernales River. In addition to observing a classic elbow of capture and the incised character of the channel there, he found gravels along the abandoned course of the channel to the southeast. The capture added 16 % of the present drainage area of Sandy Creek (1,025 km^2). As part of a comprehensive study of the operation, evolution, and equilibria of Sandy Creek watershed (Shepherd, 1975), I investigated (a) the effects of type of bedrock on channel morphology and (b) the causes and effects of the stream capture event first described by Barnes.

GEOLOGIC SETTING

Within Texas the Llano region (fig. 1) is an exceptional geologic province for two reasons: (1) it contains the only exposures of Precambrian igneous and metamorphic rocks for hundreds of miles in any direction, and (2) the Llano domal structure has experienced two stages of uplift and denundation. The original dome, with its Precambrian core and arched Paleozoic strata, was covered by nearly flat Cretaceous strata; the entire structure is now partially exhumed.

Within the Llano region, Sandy Creek watershed is the largest watershed dominated by Precambrian igneous and matamorphic rocks.

PHYSIOGRAPHIC DEVELOPMENT OF THE LLANO UPLIFT

The regional elevation of the Edwards Plateau, beginning in Miocene time, initiated the present physiography (Paige, 1912). Drainage patterns that developed on the flat Cretaceous sediments covering the ancient Llano Uplift were not coincident with present patterns (Barnes et al., 1972, p. 46). As Pleistocene rivers incised through the Cretaceous strata, the ancient core of the domal structure was exposed. Its Precambrian igneous and metamorphic granite, gneiss, and schist, less resistant to erosion than the dominantly carbonate blanket, were exhumed. Thus, streams incised more rapidly into the Precambrian core, and the embryonic Llano topographic basin was formed.

The physiographic evolution of the underlying Llano Dome has been controlled by two distinct stages and types of structural deformation. The nearly flat, resistant limestone beds which form the marginal scarps do not reflect the domal structure beneath. Thus, as the scarps were eroded back from the center of the dome, they did not form strike valleys and hogbacks typical of the Black Hills Uplift in western South Dakota (Thornbury, 1966, p. 220-223). As erosion of the dome (domal exhumation) continued and the deeper, marginal strata with higher dips became exposed, the hogbacks became more pronounced, apparent dips increased and the dip slopes occupied

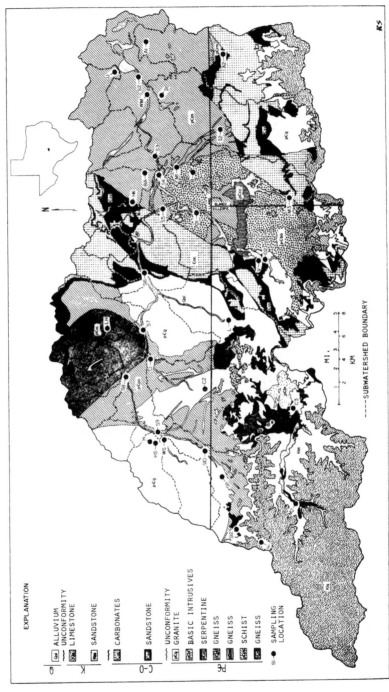

Figure 1. Map of lithic units providing distinctive sediments to the channels of Sandy Creek watershed, Llano County, central Texas.

EXPLANATION

ALLUVIUM
UNCONFORMITY
LIMESTONE

SANDSTONE

CARBONATES

SANDSTONE

UNCONFORMITY
GRANITE

BASIC INTRUSIVES

SERPENTINE

GNEISS

GNEISS

SCHIST

GNEISS

SI— SAMPLING
       LOCATION

----- SUBWATERSHED BOUNDARY

MI.

KM

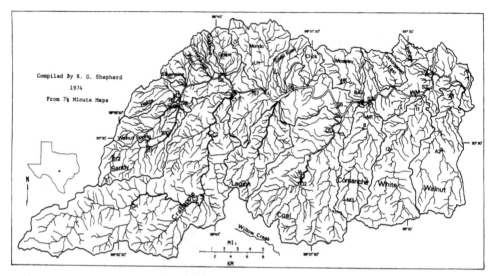

Figure 2. Drainage network of Sandy Creek watershed, Llano, Gillespie, and Blanco Counties, Texas.

less outcrop area. Finally, outcrop widths approached the thicknesses of the hogback units. Expansion of the area of surficial expression of a dome's core ceases when hogbacks become vertical. The structure thus places a definite spatial limit on the evolution of domal physiography.

In the Llano region, scarps receded until the stream capture of watersheds draining the Edwards Plateau around the topographic basin occurred. Following capture events, cuestas formed limestone buttes as a captured watershed was dissected. In fact, this stream capture process has eroded nearly all the Cretaceous limestone units and exposed the Paleozoic strata to the north and west of the Llano basin, where only a few outliers of Cretaceous rocks remain (fig. 1). The process of stream capture dominating the physiographic evolution of the Llano region is exemplified by the evolution of Sandy Creek watershed.

SANDY CREEK WATERSHED CHANNELS

Collection of Basic Data

Representative sediment samples, cross-sections, and gradients of channels were obtained in the field to investigate relationships within and between three types of environments: 1) along Sandy Creek between major tributaries, 2) at the mouth of major tributaries to Sandy Creek, and 3) at locations representative of the principal source rocks of the sediments. In addition, longitudinal profiles were taken from standard 7 1/2 minute U.S.G.S. topographic maps. The bedrock geology was compiled from existing maps and reports. Sampling methods and possible sources of sampling error may be found in Shepherd (1975), as well as a complete list of references for the bedrock map.

Sandy Creek Geometry and Sediments

The following important relationships were discovered from the data collected downstream along the channel of Sandy Creek:

1) Channel-bed width increases along Sandy Creek from its source to its confluence with Crabapple Creek. Downstream from there the width decreases (fig. 3).

2) The thickness of alluvial fill in Sandy Creek increases downstream until its confluence with Crabapple Creek. From there downstream the thickness decreases by a factor of more than four (fig. 4).

3) Downstream from the confluence with Crabapple Creek, mean grain size in Sandy Creek (fig. 5) decreases very slightly (all samples were very coarse sand) and sorting increases very slightly (all samples were poorly sorted).

4) Channel gradient fluctuates slightly along Sandy Creek downstream from Crabapple Creek, but the best fit regression line of slope on distance does not differ significantly from zero (fig. 6).

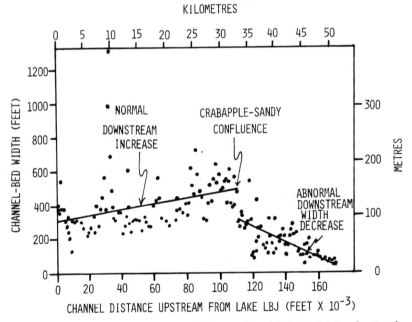

Figure 3. Relation of channel-bed width to channel length, Sandy Creek.

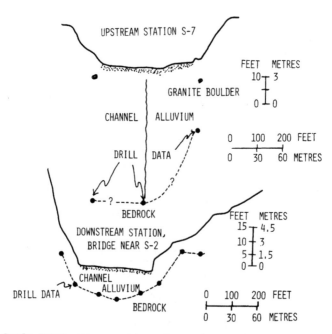

Figure 4. Sandy Creek cross sections from sites S-2 and S-7 showing the
downstream decrease in the thickness of alluvial fill.

Downstream from the Crabapple Creek confluence channel cross-sections
along Sandy Creek (fig. 7: sections S-7 to S-2) have a marked downstream
increase in depth of flow for floods of equal frequency.

Considered together, these data show that Sandy Creek downstream from
its confluence with Crabapple Creek does not conform to the usual changes
of channel geometry and sediment (Leopold, Wolman, and Miller, 1964; Leopold
and Maddock, 1953). Normally channel width, depth, thickness of fill, and
sediment sorting increase downstream, while sediment size and channel gradient decrease.

Southern Tributaries--Sediment, Channel Geometry, and Longitudinal Profiles

With few exceptions, the spatial distribution of source rocks and its consequent significance to channel morphology and watershed evolution are manifest in the six major southern tributaries of Sandy Creek (fig. 2). Most
significant is that the outcrop area of granite and gneiss generally increases
from east to west in the six southern watersheds, while the abundance of
schist and sedimentary source rocks decreases in the same direction (fig. 1
and table 1).

The basic source-rock characteristics that most influence stream channel characteristics are grain size and composition. The average grain size of the granite source material is greater than that of gneiss, which in turn is greater than that of schist. Thus, average source-rock grain size generally increases to the west through the six southern watersheds. In contrast, the maximum size of sediment in the channels decreases to the west (fig. 8). The granite forms grus with a relatively smaller mean size, compared to the wider range of grain-sizes that ultimately are concentrated in the limestone-dominated channels.

From east to west, sorting of channel sediments generally becomes better, with a marked decrease in the relative amounts of coarse particles.

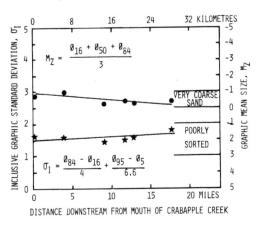

Figure 5. *Relations of mean bed-sediment size (M_Z) and sorting (σ_I) to distance along Sandy Creek downstream from its confluence with Crabapple Creek.*

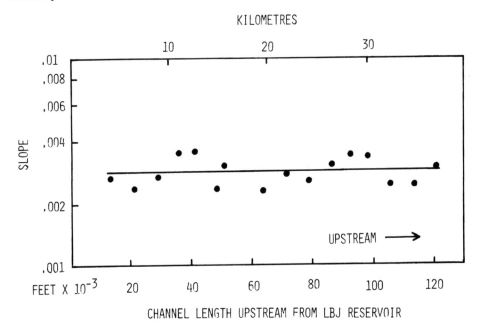

Figure 6. *Relation of channel gradient to channel length along Sandy Creek downstream from its confluence with Crabapple Creek.*

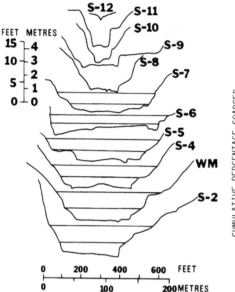

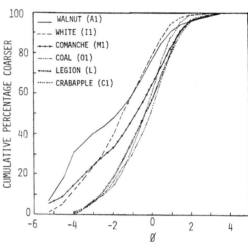

Figure 7. Channel cross sections along Sandy Creek. Sections are shown in downstream order, from section S-12 (upstream) to section S-2 (downstream). Refer to figure 2 for locations.

Figure 8. Size distributions of sediment samples from near the mouths of the six major southern tributaries of Sandy Creek. Legend lists streams from east (Walnut) to west (Crabapple).

The channel geometry of the eastern-most five southern tributaries directly reflects the prominent trends in the amount and distribution of the various source-rock types (fig. 9).

Walnut Creek is at the east end of the study area. The Walnut Creek drainage has the least amount of granite and gneiss of the five tributaries (fig. 1; table 1). The cross-section measured near its mouth has a low width-depth ratio (fig. 9). The channel-bed pattern is characteristically a pool and riffle sequence, prominently displayed in straight reaches (fig. 10). Bedrock outcrops are common, and they form the bottoms of most pools. Walnut Creek's overall channel pattern is ruggedly tortuous and controlled by structural patterns and local changes in lithologic resistance to erosion. Its sediment-source rocks of schist, limestone, and sandstone (fig. 1) all have similar qualities of sediment production. They are originally fine-grained; they provide dominantly fine detritus which is immediately

Table 1

Outcrop Areas of Source Rocks in the Six Southern Tributaries.
Streams are listed from east (Walnut) to west (Crabapple).

Source Rock	Outcrop Area (km^2)					
	Walnut	White	Comanche	Coal	Legion	Crabapple
Channel alluvium	0.3	0.3	3.4	4.9	1.0	3.1
Limestone & dolomite	43.5	31.6	12.4	37.0	3.6	146.6
Sandstone	6.0	10.9	2.1	16.8	4.7	1.6
Granites & intrusives	1.0	12.7	-	17.1	32.4	39.4
Serpentine	-	-	2.9	3.6	-	0
Schist	20.2	26.7	8.6	9.1	0.5	29.0
Gneiss	1.3	1.6	24.6	43.0	7.5	21.8

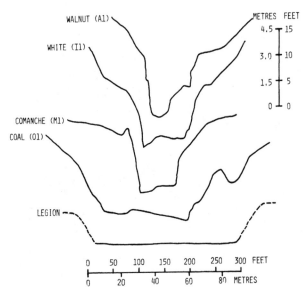

Figure 9. *Channel cross-sections from near the mouths of the eastern five southern tributaries of Sandy Creek. Streams are shown progressing from east (Walnut) to west (Legion).*

Figure 10. Aerial photograph of the channel of Walnut Creek near its mouth.

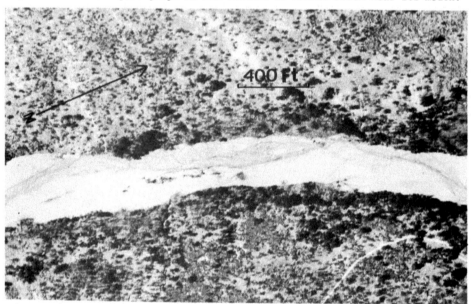

Figure 11. Aerial photograph of the channel of Legion Creek near its mouth.

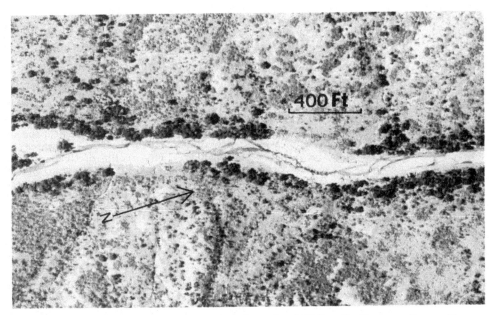

Figure 12. Aerial photograph of the channel of White Creek near its mouth.

available for suspended-load transport. This fine material is either transported rapidly out of the watershed or stored in flood-bank deposits. Hence, the channel contains dominantly coarse material which remains to be comminuted, or chemically disintegrated, or occasionally moved short distances by floods until it becomes protected from entraining forces and stored in the channel.

The drainage area of Legion Creek (the western-most stream except for Crabapple Creek) has the largest percentage of granite and gneiss of the six tributaries (table 1). This stream has the largest width-depth ratio and the straightest channel (fig. 11). Considering their similar drainage areas, Walnut Creek and Legion Creek are in marked contrast, each representing a different channel response to different types of sediment source rocks.

White Creek (fig. 12) and Comanche Creek are intermediate between Legion and Walnut Creeks in location and in channel type. They have proportionately less granite and gneiss in their drainage basins than Legion Creek, but more than Walnut Creek. Like Walnut Creek, their patterns are tortuous and controlled by channel-bedrock outcrops with variable small-scale structures and lithologic differences. However, there is much more sand in their beds, and width-depth ratios are characteristically larger than those of Walnut Creek, but smaller than those of Legion (fig. 9).

The longitudinal profiles of the six southern tributaries exhibit a variety of shapes and lengths, but several patterns are evident (figs. 13 and 14).

A simple quantitative method of comparing profiles was developed which distinguishes them on the basis of dimensionless concavity, C (fig. 15). The total length of the dimensionless stream profile is divided into tenths (or other division). The mean chord-profile distance, $\bar{h}$, is divided by the mean total chord axis distance, $\bar{H}$, to get concavity, C. As $\bar{h}$ increases relative to $\bar{H}$, C increases. This measure is actually an approximation of the area under each dimensionless profile, which could be used instead.

Proceeding from east (Walnut) to west (Crabapple), computed values of C for the six tributaries were 0.28, 0.35, 0.45, 0.44, 0.51, and 0.10, respectively. Thus profile concavity increases with distance westward in the watershed, until the Crabapple Creek profile is reached. The profile of Crabapple Creek is alternately concave, convex, and concave downstream, with a straight segment near the mouth. The profile reflects the capture of the headwaters of Willow Creek and the continuing erosional adjustment of the profile to the altered water and sediment discharges. Westward from Walnut Creek to Comanche Creek, the profiles become smoother, but from Comanche Creek to Crabapple Creek they are rugged and change abruptly in curvature. The profile of Sandy Creek is essentially straight downstream from its confluence with Crabapple Creek.

The type of channel bedrock exerts an influence on local variations in slope and profile concavity, but the influence is not consistent (fig. 14).

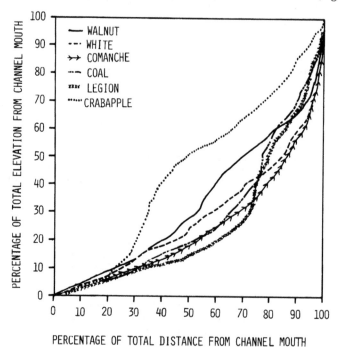

Figure 13. *Dimensionless longitudinal profiles of the six southern tributaries. Legend lists streams from east (Walnut) to west (Crabapple).*

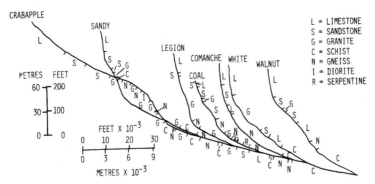

Figure 14. *Longitudinal profiles and channel bedrock of Sandy Creek and its six major southern tributaries, showing bedrock along channels. Stream locations proceed from east (Walnut) to west (Crabapple) except for Sandy Creek, the main stream.*

For example, the Paleozoic carbonate and sandstone units coincide with a mid-length profile convexity for Walnut Creek, but the profiles of Comanche and White Creeks exhibit fairly smooth concavities where these same rock types occur (fig. 14). The prominent, smoothly convex portion of Crabapple Creek coincides with eight lithologic contacts, and the Sandy Creek profile downstream from Crabapple Creek is nearly straight although ten changes in rock type occur. The most prominent concavity on Sandy Creek occurs where the bedrock is entirely granite. There is no consistent relation between channel slope and the type of channel bedrock.

For the six profiles shown in figure 13, the percentages of different source rocks in the basins were compared to dimensionless profile concavities (C) of the downstream 50 % of the profiles, because in that distance the channels and profiles exhibit pronounced responses to the type of source rocks in the watershed. The analysis showed that the percentage of granite and gneiss in each watershed is closely associated with the shape of the profiles (fig. 16).

SPECTRUM OF BEDROCK CHANNELS

Schumm (1963) classified alluvial river channels through distinctions of width-depth ratio, gradient, sinuosity, and relative amounts of sediment transported as bed load and suspended load. According to specific ranges of these factors, Schumm distinguished suspended-load, mixed-load, and bed-load alluvial channels, and characterized their dominant processes of erosion and deposition.

Two important factors limit the concise applicability of Schumm's classification in Sandy Creek watershed. First, precise data on the relative amounts of sediment transported as bed load and suspended load are not available for Sandy Creek watershed. Second, the channels of Sandy Creek watershed are not typically alluvial, because bedrock predominantly controls their patterns and shapes. Because channel patterns are rigidly maintained by bedrock banks, sinuosity is effaced as a variable. Only locally, such as in the down-stream portion of Coal Creek, are regular sinuosities established and both bed and banks composed of the same material as that transported by the stream. The type, amount, and distribution of channel-bed sediment are the major independent variables controlling the spectrum of bedrock streams in Sandy Creek watershed. The classification is, in many ways, analogous to that of Schumm (1963), however, and his terminology will be employed, with certain modifications.

Suspended-Load Bedrock Channels

Channels dominated by limestone or schist bedrock form one end of the spectrum of bedrock channel types in Sandy Creek watershed (fig. 17). With low width-depth ratios and pronounced pool-riffle thalweg patterns, they are analogous

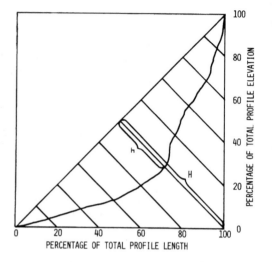

Figure 15.   Method of determining dimensionless concavity (C) of longitudinal stream profiles.

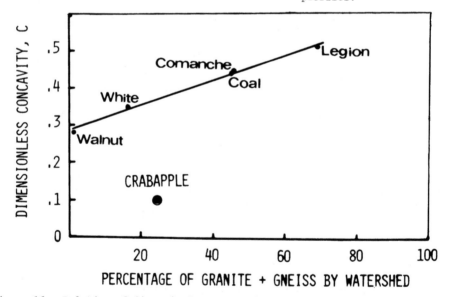

Figure 16.   Relation of dimensionless concavity of stream profiles to the percentage of granite plus gneiss in each of the six southern tributaries.

268

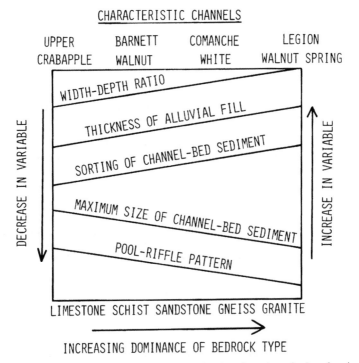

*Figure 17. Spectrum of channel types in Sandy Creek watershed, showing corres-
pondence between directions of change of important variables, the
types of source rocks, and specific examples of channels.*

to the suspended-load alluvial channels of Schumm (1963). The fine detritus
produced by the fine-grained schist and limestone source rocks principally provides suspended load. However, in contrast to the suspended-load channels
designated by Schumm, the bed material is dominantly gravel and well sorted,
and represents lag material entrained only during floods. Consequently, the
suspended-load bedrock channels in Sandy Creek watershed have the coarsest
bed material of all channel types. Schist-dominated bedrock channels have a
larger percentage of sand-sized bed material than do limestone-dominated channels; thus the latter occupy the suspended-load end of the spectrum (fig. 17).
An example of a suspended-load bedrock channel in Sandy Creek watershed is
Walnut Creek near its mouth (fig. 10).

Mixed-Load Bedrock Channels

Mixed-load bedrock channels are characterized by more granite and/or
gneiss in their drainages. White Creek (fig. 12) and Comanche Creek are representative examples. Mixed-load bedrock channels have more sand-sized
detritus than suspended-load channels, and such detritus is better sorted and
thicker. Pool-riffle sequences are present, but pools are commonly separated
by thick sand bars. Width-depth ratios are greater than in suspended-load

bedrock channels.

Bedload Bedrock Channels

Bedload bedrock channels occupy the end of the spectrum opposite from suspended-load bedrock channels (fig. 17). They are characterized by channels containing chiefly granitic detritus, such as Legion Creek (fig. 11). They have high width-depth ratios, thick bed deposits, the best-sorted detritus, and the smallest maximum size of bed material. They do not have pool-riffle sequences; rather, they have irregularly meandering courses at low flows, and braided to single-channel patterns during floods. Bed load is easily mobilized and produces moving bed forms during all discharges. Channels draining granite outcrop areas provide the most bedload sediment to the channel of Sandy Creek.

The spectrum of bedrock channel types is generally represented by the eastern five southern tributaries of Sandy Creek. From east to west, they change from suspended-load to bed-load bedrock channels. This is reflected most in the characteristics of their channels near their mouths (figs. 10, 11, and 12), where the cumulative effect of the east-to-west increase in the abundance of granite and gneiss is most pronounced.

Crabapple Creek, however, is a hybrid type of channel. In different reaches it conforms to different positions in the spectrum of channel types.

Classification of Sandy Creek

As the principal channel of the watershed, Sandy Creek downstream from its confluence with Crabapple Creek reflects both the evolution of the watershed system and the spectrum of channel types so pronounced in the southern tributaries. It is a bedload bedrock stream. However, it exhibits anomalous characteristics of channel geometry because it has aggraded from upstream (fig. 18). Nevertheless, it has adjusted to this aggradation to efficiently convey the amounts of water supplied to it. In doing so it has developed an essentially straight profile, in order to maintain the regular distribution of stream power along its course. The remarkably straight profile and remarkably uniform sediment characteristics while diverse types of channel and tributary sediments are encountered, support the statement of Leopold and Maddock (1953, p. 52):

> Slope is the one hydraulic factor which can be adjusted over a period of time by processes within the stream and may be considered the factor which makes the final adjustment, . . . after the interaction of all the other factors has resulted in establishment of mutually adjusted values of those factors.

Furthermore, in Sandy Creek, sediment size and sorting appear to respond equally as rapidly as slope.

CHANNEL RESPONSES TO STREAM CAPTURE

Sandy Creek watershed reacted to the stream capture of Willow Creek in a complex manner. Because the captured area was entirely underlain by

Figure 18. Aerial photograph showing changes of width along Sandy Creek due to aggradation, and elbow of capture on Crabapple Creek.

Cretaceous limestone and sandstone units and had suspended-load, bedrock streams, the discharges of Crabapple Creek diverted to Sandy Creek were deficient in bed-load sediment relative to those in the precapture network. The channel of Sandy Creek adjusted. In doing so it propagated adjustments in its own network, in the channel of Sandy Creek downstream from Crabapple Creek, and in the network of Sandy Creek upstream from Crabapple Creek. In addition, the downstream reaches of other tributaries of Sandy Creek were affected. The following interpretations of fluvial adjustment are based principally on relations developed by Schumm (1969, 1971, and 1973) and his associates.

Initial Adjustment

Following the capture, significant changes in several channel variables occurred. Lane   (1955) depicted the relationships between water discharge, Q_w, channel slope, S, sediment size, D, and bed-load discharge, Q_s, as

$$Q_s D \propto Q_w S \qquad (1)$$

Accordingly, if water discharge is abruptly increased relative to bed-load sediment discharge and if slope does not immediately change, a stream will react either by entraining more of the bed sediment to meet the increased capacity,

271

or by entraining larger sizes of bed materials to meet the increased competence. Usually both adjustments occur.

The initial reaction of Crabapple Creek to its abrupt increase of water discharge following stream capture logically was to flush the bed sediment from the channel in its previously aggraded reach near its mouth. Slope also was probably affected. Transposing slope to the denominator on the left side of equation 1, a slope decrease is indicated. As the bedrock floor of the channel was eroded, the slope of the channel was likely to have been reduced.

To include properties of channel geometry, Schumm (1969, 1971) extended Lane's qualitative relationship to the form of

$$Q_w \propto \frac{w\ d\ L}{S} \tag{2}$$

where w is channel width, d is channel depth, and L is meander wavelength. This relationship indicates that Crabapple Creek should have responded to the increased discharge by eroding its channel and thus increasing its width and depth. However, because the channel was bedrock and not alluvial, the width and depth increases occurred not by erosion, but by the increased discharges being conveyed in larger cross-sectional areas of the valley. Meander wavelength and sinuosity could not adjust because the channel walls were bedrock. Furthermore, the constricting effect of the bedrock bed and banks of the channel induced incision such that the increase of depth was greater than that of width (Shepherd and Schumm, 1974).

The channel of Sandy Creek downstream from its confluence with Crabapple Creek also degraded. The limestone drainage area which Crabapple Creek captured (approximately 165 km²) was larger than the drainage area of Sandy Creek upstream from Crabapple Creek (145 km²). Thus, the increased discharge conveyed through Crabapple Creek was effective further downstream. Consequently, the channel of Sandy Creek just downstream from Crabapple Creek adjusted in the same manner as Crabapple Creek, but with certain limits. Degradation probably was pronounced for only about 15 km downstream in Sandy Creek, because the resistant Paleozoic carbonates which the channel crosses provided a local base level of degradation (fig. 1). Furthermore, it is possible that degradation occurred for less than 5 km downstream from the confluence. Both the large amounts of granitic detritus from the Legion Creek granite body and the low relief in the vicinity of the channel where it crosses that body probably subdued and limited degradation. Moreover, a more important limit to the extent of degradation was the immediate secondary response of the system to the increased discharge of water.

Secondary Adjustment

The secondary response was caused by the reaction of the watershed to the initial cycle of degradation. As the channels of Crabapple and Sandy Creeks began to degrade, their entire drainage networks upstream likely began to degrade, and were rejuvenated. Consequently, the bedload in network channels upstream from their confluence began to enter Crabapple and Sandy Creeks at an increased rate. Thus, bed-load discharge increased. Schumm (1969, p. 260) depicted the relationships between the increase in bed-load and channel variables as

$$Q_s \propto \frac{w\ L\ S}{d\ P} \tag{3}$$

for alluvial streams, where P is sinuosity and the other variables are the same as in equations 1 and 2. However, the principal channels adjusted to increases of both water and bed load. Thus, the combined relationships are (Schumm, 1969, p. 261)

$$Q_w^+\ Q_s^+ \propto \frac{w^+\ L^+\ (w/d)^+}{P^-}\ S^{\pm}\ d^{\pm} \tag{4}$$

where + and - indicate the direction of change. Schumm suggested that although S and d could either decrease or increase, S will usually increase and d will usually remain constant or decrease, in alluvial channels. With some qualifications, equations 3 and 4 both apply to the response to initial degradation in Crabapple and Sandy Creeks.

As the network channels rejuvenated, primarily their bedload discharges increased (equation 3) as the gradients of the streams increased and nickpoints began to migrate headward. Previously, relatively adjusted regimes of the network channels were altered to degradational regimes. Consequently, alluvial terraces and valley-fill deposits began to be removed. This is currently evident in the upstream portions of the networks. For example, small headwater streams in the western part of the watershed are currently incising in solid granite. And larger channels, such as Sandy Creek near location S10 (fig. 1) are cutting laterally at bends into older valley-fill deposits.

The response of these degradational network tributary channels to the slope increase, can be depicted by rearranging equation 3:

$$S \propto \frac{d\ P\ Q_s}{w\ L} \tag{5}$$

Possible increases in depth and bed-load discharge and a decrease in width are indicated by equation 3 and observed in the field. However, the possible increase in sinuosity and decrease in meander wavelength indicated by equation 3 did not occur because of the bedrock control of pattern.

While equation 5 generally depicts the response of the network, a different response, generally indicated by equation 4, occurred first in the main channels and then was propagated upstream as the system adjusted to rejuvenation. In the channels of Crabapple and Sandy Creeks, both water and sediment discharges increased--the former from capture and the latter from network rejuvenation. Thus, width, slope, and width-depth ratio increased, and depth decreased. Sinuosity and meander wavelength changed only slightly in the bedrock channels. The result was channel aggradation in the previously degrading portions of the channels. Indeed, because of the large amounts of granitic detritus delivered from the drainage of Sandy Creek, its channel aggraded to the extent that at locations S-6 and S-7, the valley was filled with alluvium. Thus, the thickness of channel fill and the channel-bed width of Sandy Creek decrease significantly downstream. As noted previously, at locations S-7 the fill is over 15 m thick, while at S-2 it is only 3.5 m (fig. 4). The aggradational response has been most pronounced in Sandy Creek in the vicinity of its confluence with Legion, downstream from the Sandy-Crabapple confluence. At S-6 just below the Sand-Legion confluence, the channel-bed width is 160 m, and the depth of sand is 8 m. Aggradation has also been propagated upstream in tributary networks. In the channel of Legion Creek near its mouth this is manifest in the large width-depth ratio, the large volume of channel fill, and the straight profiles.

In summary, the response to the capture of Willow Creek by Crabapple Creek was complex and cyclic. Changes in stream regimen occurred rapidly, and different regimes were in effect at different places at the same time. The initial increase in water discharge from the captured limestone area caused degradation downstream. The degradation caused rejuvenation of tributary networks and larger sediment discharges, which in turn caused aggradation downstream.

Precisely this same response was obtained by Schumm and Parker (1973) in an experimental study of drainage evolution. But in their experiment, the response was caused by a change in base level, instead of stream capture, which was responsible in Sandy Creek watershed. However, changes in the relative discharges of water and sediment may result if any of several different types of thresholds are exceeded.

Schumm (1973) identified two fundamental types of geomorphic thresholds. Extrinsic thresholds are those which are controlled by some variable external to the system. An extrinsic threshold was crossed when Crabapple Creek captured Willow Creek and added a new drainage area to the system. But the rejuvenation of the tributary network and the aggradation downstream, caused by the initial degradation after capture, was an example of exceedance of an intrinsic threshold--one which existed in the system all along, but had not been triggered. Thus, both extrinsic and intrinsic thresholds were exceeded and balances between process and form were upset. In adjusting to these disturbances the Sandy Creek watershed system developed its current anomalous properties.

ACKNOWLEDGEMENTS

Financial support for the study came from the University of Texas at Austin and a Geological Society of America Penrose Grant. I thank Victor R. Baker, supervisor, and R.L. Folk, E.F. McBride, R.A. Morton, and M.G. Wolman for their help as committee members and for reviewing and improving the dissertation from which this paper is based. P.C. Patton, S. Hulke, S.M. Shepherd, and V.R. Barnes provided significant help during the field research. I thank Willard Owens Associates, Inc. for support in completing the final draft, and Pete Patton and Garnett Williams for reviewing it.

REFERENCES

Barnes, V.E., 1952, Crabapple Creek quadrangle, Gillespie and Llano counties, Texas: University of Texas, Bureau of Economic Geology Quadrangle Map.

Barnes, V.E., Bell, W.C., Clabaugh, S.E., Cloud, P.E., Jr., McGehee, R.V., Rodda, P.U., and Young, K., 1972, Geology of the Llano region and Austin area field excursion: University of Texas, Bureau of Economic Geology Guidebook 13, 77 p.

Gilbert, G.K., 1877, Report on the geology of the Henry Mountains: Washington, D.C., U.S. Government Printing Office, 160 p.

Lane, E.W., 1955, The importance of fluvial morphology in hydraulic engineering: American Society of Civil Engineers Proceedings, v. 81, paper 745, 17 p.

Leopold, L.B., and Maddock, T., Jr., 1953, The hydraulic geometry of stream channels and some physiographic implications: U.S. Geological Survey Professional Paper 252, 57 p.

Leopold, L.B., Wolman, M.G., and Miller, J.P., 1964, Fluvial processes in geomorphology: San Francisco, W.H. Freeman and Company, 522 p.

Paige, Sidney, 1912, Description of the Llano and Burnet quadrangles: U.S. Geological Survey Atlas 183, Llano-Burnet Folio, 16 p.

Schumm, S.A., 1963, A tentative classification of alluvial river channels: U.S. Geological Survey Circular 477, 10 p.

_____, 1969, River metamorphosis: American Society of Civil Engineers Proceedings, v. 95, p. 255-273.

_____, 1971, Fluvial geomorphology: channel adjustment and river metamorphosis; in Shen, H.W., ed., River mechanics; Fort Collins, Colorado, H.W. Shen, v. 1, p. 5-1 to 5-22.

_____, 1973, Geomorphic thresholds and complex response of drainage systems; in Morisawa, M., ed., Fluvial geomorphology: Binghamton, State University of New York, Publications in Geomorphology, p. 299-310.

Schumm, S.A., and Parker, R.S., 1973, Implications of complex response of drainage systems for Quaternary alluvial stratigraphy: Nature, v. 243, p. 99-100.

Shepherd, R.G., 1975, Geomorphic operation, evolution, and equilibria, Sandy Creek Drainage, Llano Region, Central Texas [Ph.D. dissertation]: Austin, University of Texas, 209 p.

Shepherd, R.G., and Schumm, S.A., 1974, Experimental study of river incision: Geological Society of America Bulletin, v. 85, p. 257-268.

Thornbury, W.D., 1966, Principles of geomorphology: New York, John Wiley and Sons, 618 p.

Woodruff, C.M., 1977, Stream piracy near the Balcones Fault Zone, Central Texas: Journal of Geology, v. 85, p. 483-490.

QUATERNARY FLUVIAL GEOMORPHIC ADJUSTMENTS

IN CHACO CANYON, NEW MEXICO*

DAVID W. LOVE

Department of Geology
University of New Mexico
Albuquerque, New Mexico

ABSTRACT

Quaternary fluvial geomorphic adjustments on three spatial and temporal scales have taken place in Chaco Canyon, in semiarid northwestern New Mexico. These adjustments are reflected in the morphology and sedimentology of Chaco Arroyo, and in the alluvial canyon fill and terrace deposits above the modern canyon floor. Facies preserved within the deposits, coupled with historical and archaeological data, aid in interpreting the processes of the fluvial adjustments in a temporal framework. Increases in stream power transport sediment beyond the canyon and/or modify the type of sediment deposited within the canyon. Decreases in stream power allow aggradation within the canyon.

The smallest fluvial adjustments include ephemeral changes within the modern arroyo and along the canyon margins. The development of Chaco Arroyo during the past 140 yrs is related to changes in precipitation, to changes in land management, and to inherent fluvial adjustments.

Intermediate adjustments include wide-spread alluviation of the canyon floor interrupted by cycles of cut-and-fill. The morphology and sedimentology of channel cuts and fills and the facies of the alluvial canyon floor indicate that there are (1) contrasting processes of channel formation, (2) different means of filling channels, and (3) that thin sheets of local- and headwater-derived sediments are intercalated when the canyon floor has no entrenched channels. The results of an evaluation of the influence of intrinsic and extrinsic variables on cycles of cut-and-fill are somewhat equivocal. Land use does not appear to have been a major influence on channels in Chaco Canyon, even during the extensive occupation of the canyon nearly 1,000 yrs ago.

The greatest adjustments in Chaco Canyon were induced by long-term cycles of erosion and deposition during which fluvial processes eroded alluvium from the canyon, cut the bedrock floor, and partially refilled the canyon with alluvium. Deposits indicating large, long-term fluctuations in discharge are preserved along the margins of the canyon at several topographic levels. Gravel deposits indicating stream power much larger than present form terraces and morphologically indistinct mounds. Deposits indicating a

---

*This paper is inventoried as contribution Number 26 of the Chaco Center, National Park Service and the University of New Mexico, for purposes of bibliographic control of research relating to Chaco Canyon.

hydrologic regime similar to the present are preserved by carbonate cementation in low-energy environments along the canyon margins. Alternation of sediment types suggest that there are geomorphic thresholds and complex responses even at this scale of adjustment. Long-term climatic fluctuations are responsible for alternations between these two types of fluvial regimes.

INTRODUCTION

The Chaco River is an ephemeral stream which drains 11,500 km^2 of semiarid northwestern New Mexico (fig. 1). The present stream undergoes large changes in morphology and sedimentology along its course. Alluvial deposits adjacent to the stream indicate that it has undergone similar changes in the past. Archaeological research in the upper part of the Chaco Canyon drainage basin provides an extensive historic and archaeologic context for studying the processes and products of morphologic and sedimentologic fluvial adjustments.

The purpose of this paper is to examine the different temporal and spatial scales of fluvial adjustment in Chaco Canyon. The scales of adjustment include (1) changes in channel morphology and sedimentology within Chaco Arroyo and similar adjustments outside the arroyo, (2) alluviation of the floor of Chaco Canyon and cycles of cut-and-fill, and (3) alternating episodes of erosion of the bedrock floor of the canyon and partial refilling of the canyon with alluvium. These adjustments are considered within the framework of a conceptual model of the geomorphic system in Chaco Canyon. The model considers the geomorphic and sedimentologic consequences of sediment transport and deposition in the Chaco drainage basin. The model can be used (1) to evaluate interpretations of the history of the canyon, (2) to assess (qualitatively) the influence of climatic conditions and land use on the behavior of the geomorphic system, and (3) to evaluate current concepts of fluvial geomorphology in semiarid areas (such as those in Cooke and Reeves, 1976; and Schumm, 1977) as applied to Chaco Canyon. This paper is organized to consider the geomorphology and sedimentology of each scale of adjustment.

Authors who have addressed the geomorphology, sedimentology, and/or geologic evolution of Chaco Canyon include Jackson (1878), Dodge (1920), Bryan (1925, 1926, 1954), Senter (1937), Schumm and Chorley (1964), Judd (1954, 1959, 1964), Tuan (1966), DeAngelis (1972), Siemers and King (1974), Nichols (1975), Hall (1975, 1977), Leopold (1976), and Ross (1978). Bryan's (1954) work, in particular, provides a geomorphic and historic

Figure 1. *Location of the Chaco River drainage basin and Chaco Canyon National Monument, northwestern New Mexico.*

model of the evolution of Chaco Canyon. This model and other interpretations of the history of Chaco Canyon are evaluated in the context of the fluvial adjustments and the historical/archaeological interpretation discussed below.

GENERAL GEOLOGIC AND GEOMORPHIC SETTING OF CHACO CANYON

The Chaco drainage basin above the mouth of Chaco Canyon may be divided into the upper 855 km^2 catchment basin and the lower 261 km^2 Chaco Canyon (fig. 2). The bedrock units of this portion of the Chaco drainage consists of Upper Cretaceous and Lower Tertiary sandstones and shales which dip gently (3°) northeastward. The shallow swales, discontinuous washes (relatively wide with low banks), and discontinuous arroyos (relatively narrow with high banks) of the upper catchment area generally cross the strike of the underlying bedrock. However, Chaco Canyon follows the strike of the Cliff House Sandstone. Tributary canyons commonly parallel regional fracture patterns which trend northeastward.

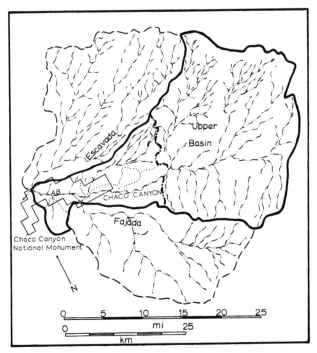

Figure 2. *Headwater drainage basins of the Chaco River. The upper Chaco drainage basin is enclosed by the heavy solid line. The boundary between the upper catchment basin and Chaco Canyon is marked by the heavy dot-dash line. Divides of major tributary drainages are marked with dashed lines, and drainages adjacent to Chaco Canyon are marked with dotted lines. Point A is Pueblo del Arroyo. Point B is Pueblo Bonito.*

Chaco Canyon is 32 km long, from 500 to 1,000 m wide, and is entrenched up to 180 m into the Cliff House Sandstone and underlying Menefee Formation. Because of the gentle northeastward dip of the bedrock, the canyon walls are higher and tributary canyons are more deeply dissected on the south side of the canyon. The walls of the canyon consist of two 30-m high cliffs separated by a 30-m high rise of slopes and benches (fig. 3). Below the cliffs are pediments and talus cones. Alluvial fans spread onto the alluvial floor of Chaco Canyon from the mouths of tributary canyons and reentrants. Sand dunes are common on the top of the cliffs and in some of the tributary canyons. The alluvial canyon floor is coursed by Chaco Arroyo and other tributary drainages.

Annual and seasonal weather in Chaco Canyon are extremely variable. Annual precipitation ranges from 85 mm to more than 471 mm and averages about 223 mm. The mean annual temperature is 9.9° C. The length of the summer season (the number of days between 0° C minimum temperatures in the spring and fall) has decreased from about 140 days to about 110 days since 1957. Large amounts of precipitation can occur at any time of the year, gene-

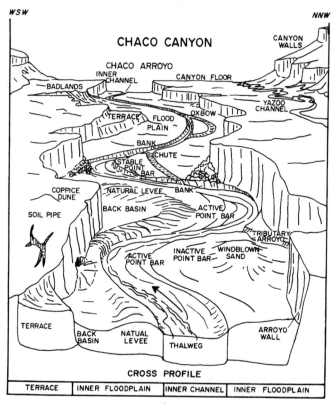

Figure 3. *Geomorphic features of Chaco Arroyo and the alluvial canyon floor in Chaco Canyon, New Mexico. The view is downstream.*

280

rating runoff to transport sediment. Although runoff events in winter and spring appear to be less flashy than those of the summer season, probably more geomorphic work is done during the winter and spring. Winter events appear to be important for making sediment available to channels in the headwaters. There is little evidence to support the idea that summer flows cut channels while winter flows fill channels (c.f. Schoenwetter and Eddy, 1964).

Measurements of water and sediment discharge in Chaco Arroyo have been initiated recently by the U.S. Geological Survey. Estimates of peak discharges range from 70 m^3/sec to 125 m^3/sec. The sediment load is commonly large. Discharge in Chaco Arroyo generally does not increase downstream through the canyon because Chaco Arroyo and tributary arroyos usually do not flow at the same time. Local runoff which reaches the main arroyo from the canyon margins tends to precede the floods from the headwaters.

Sediments which are transported to the canyon from the headwaters frequently can be distinguished from locally derived sediments by sedimentary structures, grain size, color, and clay   (Love, 1977).

SMALL SCALE FLUVIAL ADJUSTMENTS: CHACO ARROYO

Geomorphic Features of Chaco Arroyo and Adjacent Canyon Floor

Chaco Arroyo begins 15 km east of Chaco Canyon and continues for 37 km through the canyon to the point where Escavada Wash joins the drainage from the northeast (fig. 2). Below the junction of Escavada Wash, the Chaco River is a broad (200-450 m) braided sandy wash as it leaves the confines of the canyon. Chaco Arroyo meanders with a sinuosity of 1.26, and some meander loops point up canyon. The gradient of the canyon floor parallel to Chaco Arroyo ranges from 0.002 to 0.0048 m/m. The canyon floor adjacent to Chaco Arroyo has no oxbows, natural levees, or typical backbasin deposits or channels similar to those on the present inner channel and floodplain. Axial fans spreading down canyon are not cut by the modern arroyo.

Chaco Arroyo has walls 1 to 11 m high and averages about 65 m in width (fig. 3). The arroyo walls exhibit various stages of erosion, from freshly broken vertical walls to eroded, badland-like slopes of less than 30º. A terrace of alluvium 1.5 to 3 m below the canyon floor is present along the arroyo walls. An active meandering inner channel within the arroyo is from 3 to 10 m wide and is confined between banks 1 to 3 m high. Loose sandy sediment in the inner channel forms sand bars and ripples. Beneath the loose sand, the channel floor has typical pools and riffles. The local water table is reached in pools in the inner channel. The inner floodplain adjacent to the inner channel includes point bars, chutes, natural levees, oxbows, and backbasins (fig. 3). Aeolian deposits also occur on the inner floodplain. The inner channel and floodplain gradually disappear toward the head of Chaco Canyon. Above Chaco Canyon, the arroyo has a braided channel.

Most tributary arroyos appear to be geomorphically younger than the main arroyo. (One tributary appears to follow a prehistoric roadway or canal.) Some tributary arroyos and soil pipes join Chaco Arroyo at a level about 1.5 m below the level of the alluvial canyon floor. Others are entrenched to the level of the inner Chaco floodplain.

Sedimentary Facies of Chaco Arroyo

Many of the geomorphic features of Chaco Arroyo are sites for deposition of sediments. The dominant source of sediments for Chaco Arroyo is the head-water area, although gravel and some buff-colored sediment are locally derived. The sedimentary facies of the inner channel and inner floodplain in Chaco Arroyo are: (1) inner channel sand, (2) inner channel base, (3) inactive point bar and backbasin, (4) natural levee, (5) oxbow sandplug, (6) tributary channel sandplug, (7) oxbow, (8) aeolian dune, and (9) slope wash and blocks from arroyo walls.

The alluvial terrace preserved locally along the walls of the modern arroyo contains sediment facies similar to the modern inner channel and floodplain. An abandoned channel in this deposit retains some topographic expression below the adjacent stabilized point bar deposits. The channel was cut off by chutes across the point bar deposits. Some point bar deposits have gravel on their upper surfaces.

Historic Data

Table 1 summarizes historical observations pertinent to adjustments of Chaco Arroyo. Figures 4 and 5 show the climatic record for Chaco Canyon and the region. Figure 6 shows tree-ring indices (a complex indicator of climatic conditions; Fritts, 1971) of the Chaco area for the past 200 yrs. Climatically sensitive tree-rings do not exist for Chaco Canyon after 1848.

Table 1

Summary of Historic Data Pertinent to the

Development of Chaco Arroyo

Date	Event
post-1630	Navajo families tended small herds (4 to 6 animals) of sheep and cattle in region surrounding Chaco Canyon (Van Valkenburg, 1938).
ca. 1770's	There was a slight increase in habitation of the Chaco Canyon area as descendants of survivors of the Pueblo Revolt moved away from the Rio Grande valley (Brugge, 1976, personal commun.).
ca. 1840	An old Navajo recalled moving to the lower end of Chaco Canyon at this time and said that there was more rain and no deep channel in the canyon before the Anglos came (reported by Pepper, 1920, and Judd, 1959).
1849	Simpson was the first Anglo to describe the ruins of Chaco Canyon. He described other deep arroyos in the region and mentioned that there was water 0.3 m deep and 2.4 m wide in

Table 1 (continued)

Date	Event
1849 (contd.)	a channel in Chaco Canyon. Simpson's guide named one of the ruins adjacent to the present arroyo "Pueblo del Arroyo" (Simpson, 1852).
1850's	Bryan (1954) reported on old Navajos who recalled only shallow channels in their youth and placed their recollections at about this time. Judd (1959) stated that the same Navajos reported that the recalled time period was during the 1880's.
1877	Jackson (1878) described Chaco Canyon and its ruins in detail, but none of his photographs developed. He described Chaco Arroyo as being continuous from 15 km above Chaco Canyon through the canyon. His map shows few tributary arroyos to Chaco Arroyo. The arroyo was 4.9 m deep and 12 to 18 m wide south of Pueblo del Arroyo. His map appears to show an inner channel, and Jackson described young cottonwoods and willows along the margins of the arroyo. He described a shallow "old arroyo" 2.4 m deep between Pueblo del Arroyo and the deeper modern arroyo. Jackson also discovered a buried arroyo channel 4.3 m deep exposed in the walls of the modern arroyo. Anglo cattle companies did not bring large herds into the area until after 1877. They moved to Utah in less than 10 yrs. Vegetation in Chaco Canyon was devastated by local herds of sheep and cattle prior to the 1920's when Judd photographed the canyon.
1880's	Navajos and a freight driver reported that Chaco Arroyo was a shallow channel with trees along it.
1896	Pepper and others photographed Chaco Arroyo. The arroyo of Gallo Wash was not entrenched (Malde, 1979, personal commun.).
1900	Holsinger (1901) photographed Chaco Arroyo. The arroyo was much narrower than in later photographs. The alluvial terrace within the arroyo looked uneroded. Jackson's "old arroyo" near Pueblo del Arroyo was eroded extensively compared to its representation on Jackson's map.
1901, 1902	Dodge (1920) studied the alluvial geology and decided that the buried channel described by Jackson was a modern deposit and that the artifacts found in the buried channel had come from a modern tributary arroyo which eroded the Pueblo Bonito trash mound.

Table 1 (continued)

Date	Event
1920's	Photographs taken during the 1920's and published by Bryan (1954) and Judd (1954, 1959, 1964) show that the arroyo had widened considerably by 1922. Bryan called the remanent alluvial terrace a "false bank" and noted that it contained sheep dung. Judd described cottonwood trees which fell with collapsing arroyo walls and reestablished themselves on the margins of the modern arroyo. Some tributary arroyos had not yet become entrenched.
1929	Charles Lindbergh took pictures of Chaco Canyon and its ruins from the air. Chaco Arroyo was a braided stream. Vegetation on canyon floor was sparse. Lizard House Arroyo was entrenched only 1.5 m.
1934	The Soil Conservation Service took aerial photographs of the region. Chauvenet (1935) and others took ground-based photographs. The weather record for Chaco Canyon began. Lizard House Arroyo was entrenched 4 m. The Soil Conservation Service, National Park Service, and Civilian Conservation Corps initiated an erosion control program in Chaco Canyon National Monument. During the following decade, the Monument was fenced, 700,000 trees and shrubs were planted (most in Chaco Arroyo), and numerous wood, stone, and wire erosion control devices were built in Chaco Arroyo. Tributary arroyos were modified by erosion control structures. Levees were built on the canyon floor and native grasses were planted in plowed areas in the canyon.
1939	Judson took photographs of Chaco Arroyo which show the aggradation of the inner floodplain. Some aggradation was directly related to erosion control structures.
1941	Record year for precipitation (481 mm). Threatening Rock fell (Schumm and Chorley, 1964).
1940's	A Navajo sheep reduction program took place.
1950's–1960's	Mathews photographed Chaco Arroyo with inner banks apparent.
1961	Leopold (1976) surveyed a profile across Chaco Arroyo.
1963	Tuan (1966) photographed the inner floodplain and channel of Chaco Arroyo. He ascribed the features to a period of aggradation across the entire arroyo floor, followed by a period of entrenchment of the inner channel.

Table 1 (continued)

Date	Event
1965	The U.S. Geological Survey took aerial photographs of Chaco Canyon and the surrounding region.
1972, 1973	The Chaco Center took detailed black-and-white and color transparency aerial photographs. The inner Chaco channel formed chute cutoffs and oxbows in several reaches and some meanders shifted slightly downstream since 1965.
1973	Northwestern New Mexico experienced an extremely wet spring. The inner Chaco channel aggraded several centimeters.
1975	A large flash flood swept sand and gravel out of the inner channel and deposited them on point bars. The base of the inner channel was scoured.

Historic Adjustments

Bryan (1954) suggested that the modern Chaco Arroyo began to cut about 1860. This conclusion was based on his interpretation of historical accounts and on his ideas about arroyo behavior. Bryan believed that arroyos form by headward cutting of previously unchanneled alluvial canyon floors during periods following extended droughts. In addition, he regarded overgrazing as a direct cause of modern arroyos. He suggested that Simpson in 1849 and Jackson in 1877 had seen stretches of trenched and untrenched alluvium in a series of discontinuous arroyos. Bryan (1954) used this idea to explain why some stretches of the arroyo were shallow during the 1880's (table 1) even though Jackson (1878) had described a deep arroyo in the 1870's. Bryan suggested that the "false banks" (described above as the inset alluvial terrace) had formed during the arroyo cutting episode.

The morphology and sedimentology of the arroyo and canyon floor do not corroborate Bryan's interpretation. The longitudinal profile of the canyon floor is not broken by any steep reaches which might have initiated cutting of discontinuous arroyos. Furthermore, no axial fans spread down canyon as would be expected if there had been discontinuous arroyos along the canyon floor (such as in the model proposed by Schumm and Hadley, 1957). Arroyo meanders, with some loops pointing up canyon, indicate that a channel was established on the canyon floor before the arroyo was cut, and that the channel was not entrenched by progressive headcutting up the canyon floor. Chaco Arroyo formed along a meandering channel which had not previously existed on the canyon floor. This conclusion is based on the lack of evidence of levees, oxbows, and backwater deposits adjacent to the margins of the modern arroyo. The meander wavelengths of Chaco Arroyo are different from those of other smaller channels which already existed on the canyon floor.

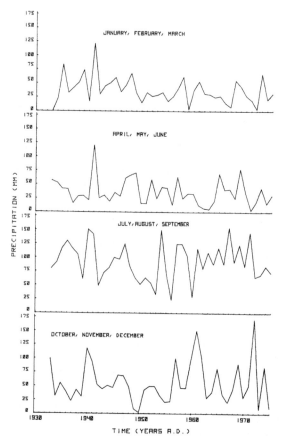

Figure 4. *Variation in seasonal precipitation for Chaco Canyon National Monument, New Mexico, from 1933 to 1975. The three-month segments nearly correspond to natural breaks in the annual cycle (e.g. July, August, and September are the months of summer thunderstorms). Note the periodicity from the late 1950's to the early 1970's in each season. During this period of time, the growing season declined from about 140 days to 110 days between 0°C temperatures.*

In the western part of the canyon, two or more channels appear to have competed to form the modern arroyo. Jackson's (1878) unfilled "old arroyo" (table 1) may have been such a channel. Either the arroyo was entrenched by a single headcut migrating 52 km up the channel in the canyon, or it entrenched by progressive erosion of the channel base. The morphology of the narrow, meandering, incised channel suggests that the channel was gradually lowered by erosion of pools and riffles along the base of the channel rather than by a headcut. Leighly (1936) described such a process in other arroyos in western New Mexico.

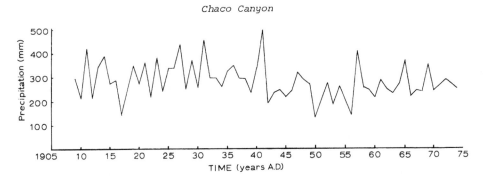

Figure 5. *Average annual precipitation of the northwestern plateaus section of New Mexico from 1909 to 1974.*

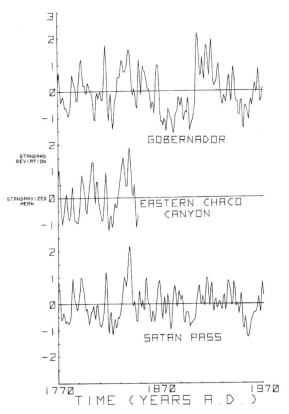

Figure 6. *Three-year weighted average standardized tree-ring indices for Gobernador (80 km north of Chaco Canyon), eastern Chaco Canyon, and Satan Pass (56 km south of Chaco Canyon) for the period from 1770 to 1970. Positive values indicate more than average moisture and negative values indicate less than average moisture. (Data from the Tree-ring Laboratory of the University of Arizona).*

287

After the arroyo formed, the alluvial terrace formed as an inner flood-plain and channel similar to the modern inner floodplain and channel. The abandoned floodplain and channel aggraded 3 to 4 m. When the floodplain was within 1.5 m of overtopping the arroyo walls, chute cutoffs formed, floods swept gravel out of the channel, and the arroyo became reentrenched. This reentrenchment took place prior to 1896, because early photographs show the inner terrace and entrenched arroyo.

After reentrenchment, the arroyo widened and eroded most of the alluvial terrace. Tributary arroyos became entrenched. Chaco Arroyo had a braided channel from wall to wall. By 1934, the braided channel was no longer stable. By 1939, the arroyo channel had changed from braided to a narrow inner chan-nel between partially stabilized point bars and the inner floodplain. The sedi-mentary facies of the floodplain indicate that aggradation occurred above the former braided channel. The amount of sediment stored in the inner flood-plain within Chaco Canyon is approximately 4×10^6 m^3. If this fill accumulated between 1935 and 1950, the rate of deposition would have been about 268,000 m^3/yr.

The inner channel and floodplain continue to be stable. Minor accretion on the floodplain and minor shifts in the channel have taken place since the early 1960's (Leopold, 1976). Since 1965, chute cutoffs have formed oxbows. A flood in 1975 deposited gravel on the stabilized point bars and swept most of the loose sand from the inner channel.

The two major questions about development of Chaco Arroyo concern (1) the timing of the early development of Chaco Arroyo and (2) the reasons for the fluvial adjustments in Chaco Arroyo. By examining the reasons for the rela-tively well documented adjustments in Chaco Arroyo since 1934, reasons for the timing of these adjustments may become apparent and the principles may be used to establish the timing of past adjustments in the arroyo.

Causes of Changes

The adjustment of Chaco Arroyo from a braided channel to an inner chan-nel and floodplain could have several explanations. Changes within the drain-age basin (intrinsic changes) which could affect the adjustments include (1) changes in channel gradient and width, and (2) changes in type or amount of sediment. Extrinsic changes (changes which take place independent of the drainage basin) include changes in (1) land use and management and (2) weather, particularly precipitation.

Comparisons of photographs taken near the turn of the century with later photography reveal that the arroyo widened extensively (perhaps doubled in width) prior to 1934. Extension of meander loops may have lengthened the channel and thereby decreased the gradient. These changes in width and gradient may have caused the development of a single channel between aggrad-ing point bars. One explanation for this change is that more infiltration may occur in wide braided channels with decreased gradient so that excess sedi-ments are stored in point bars and channel margins.

The kind and amount of sediment entering Chaco Arroyo from the head-waters also may have changed. Perhaps a pulse of sediment from the head-

288

waters was flushed into the canyon, or perhaps an increase in fine-grained sediment caused the change in channel morphology. Erosion of the arroyo walls may have furnished excess sediment as well. Further investigations of the hydrology of the headwater drainages are necessary to test these suggestions.

Climatic conditions in the region changed after the early 1930's. The regional average precipitation and tree-ring indices both show a decrease in moisture in northwestern New Mexico (figs. 5 and 6). Thomas (1963) documented a period of less than normal rainfall from the mid-1930's to the mid-1950's throughout the southwestern United States. Burkham (1966) demonstrated that less than the expected runoff occurred for the Rio Puerco (the drainage basin adjacent to the Chaco drainage east of the Continental Divide) during this period.

The precipitation records at Chaco Canyon National Monument show extreme variability with time (fig. 4). The major exception to less-than-normal precipitation occurred from July 1940 to December 1941, a period during which more than 750 mm fell. According to National Park Service records, Chaco Arroyo flowed during most of the winter and spring of 1941, and Threatening Rock collapsed onto Pueblo Bonito (Schumm and Chorley, 1964). During the period of less than average precipitation there was no increase in precipitation intensity. Thus, there were fewer large storms capable of producing runoff and floods to transport sediment out of the canyon.

Several changes in land use which occurred could have affected channel morphology. The National Park Service, Soil Conservation Service, and Civilian Conservation Corps conducted an extensive erosion control program in the monument area from 1934 to 1952 (table 1). During the 1940's, a program to reduce Navajo livestock to prevent further overgrazing in the region was also underway.

Without more quantitive data, it is impossible to evaluate the extent to which each of the above factors influenced the change in channel morphology. Natural widening of the arroyo, the erosion control measures, decreased precipitation, and the decrease in large runoff events were all probably important factors leading to aggradation of the inner floodplain within the National Monument.

Upstream from Chaco Canyon National Monument, and in other arroyos in the area unaffected by the erosion control program, inner floodplains also formed along some reaches. Therefore, man-made changes probably were not the fundamental causes of the adjustments in the monument. Some of these floodplains do not have extensive vegetation, and a slight decrease in grazing pressure is unlikely to have caused such rapid changes in channel morphology. Moreover, the grazing control program started after the inner floodplain had begun to stabilize. Probably Chaco Arroyo had already begun to form point bars and backwater areas, and decreased stream power coupled with artificial impediments aided the deposition. The decrease in large floods during the following decade could have helped stabilize the new morphology.

Timing of Stages of Development

The possible correlation between the development of the inner channel and floodplain and the decrease in precipitation may be used to interpret the early development of Chaco Arroyo.

The initial cut, partial fill, and reentrenchment could have taken place during any one of three possible intervals of time. First, the initial cut and partial fill, may precede historical records so that the inner floodplain formed before 1849, when Simpson visited the canyon. Second, the initial cut, partial fill and reentrenchment could have taken place between 1849 and 1877. Third, the initial cut could have taken place prior to 1877 and the alluvial terrace could have formed between 1877 and the late 1880's.

This third alternative coordinates the adjustments of the arroyo with tree-rings and recollections of early visitors. If the present fill is associated with a decrease in moisture, as reflected by tree-rings, a period of greater than normal precipitation could have initiated the modern arroyo. Tree-ring indices indicate a wetter than normal period around 1840. Old Navajos have also reported that there was increased precipitation during this period. If the meandering channel progressively entrenched, Simpson (in 1849) may have seen the channel when it was still relatively shallow. Perhaps Jackson (in 1877) saw the arroyo when it had entrenched but was beginning to fill during a period of decreased moisture (as shown by tree-rings). The Navajos and freight driver who recalled a shallow channel with trees in the 1880's (table 1) may have seen the floodplain which is now the alluvial terrace.

The arroyo may have formed chute cutoffs and reentrenched during the early 1890's, leaving the former floodplain as a terrace. Dodge (1920) could have studied this terrace in 1901 and 1902 and concluded that the artifact-bearing gravel at its base was a modern deposit. During the moist period from 1905 to the 1930's the arroyo was widened, most of the alluvial terrace was removed, and a number of tributary arroyos were created. Bryan (1925) studied the arroyo after much of the alluvial terrace had been eroded, so that Jackson's (1878) buried channel was reexposed.

This interpretation attributes the channel adjustments to climatic variability. It uses the somewhat tenuous correlation between modern inner floodplain accretion and decreased precipitation and extrapolates this relationship to past conditions. Changes in land use (particularly the arrival of large herds of cattle) do not appear to have been factors in the formation and modification of Chaco Arroyo. However, one tributary does seem to follow a prehistoric road or canal, and other tributary arroyos do appear to have been cut after overgrazing had taken place in the area. Complex response in the headwaters, furnishing water and sediments to the canyon, is possible but undocumented. The sediment volume and the facies of the fill in the alluvial terrace and the modern floodplain appear to preclude complex response as the sole cause of these features. Variations in stream power and sediment supply in this drainage basin are as likely to be due to variations in precipitation and runoff as to complex response.

INTERMEDIATE SCALE FLUVIAL ADJUSTMENTS IN
CHACO CANYON: AGGRADATION OF CANYON FILL
AND CYCLES OF CHANNEL CUT-AND-FILL

Geomorphic Data

Two sets of geomorphic conditions occur episodically on the alluvial floor of Chaco Canyon. The canyon floor may be relatively flat (perhaps with no channel at all), or the floor may be dissected by entrenched channels. The observed cycles of cut-and-fill imply (1) an initial condition of no entrenched channel, (2) a phase of channel entrenchment, and (3) a phase of channel fill to recreate a flat canyon floor. The geomorphic features on the alluvial floor of Chaco Canyon indicate that when no entrenched channels are present, there are: small meandering or anastomosing channels; fans and distributary channels from the canyon margins; sand and clay-rich aeolian dunes; and areas of soil development. When entrenched channels are present, canyon-margin fans may continue to exist, or they may be trenched by tributary channels. In addition, the alluvial flats between the canyon margins and the arroyos experience very little deposition or erosion.

Channel cuts in the alluvial canyon fill have several geometries and a large range in size. Not all present and past channels are arroyos. The planforms of modern channels range from nearly straight to extremely sinuous. In cross channels vary in shape from vertical or even overhanging walls on either side of a flat channel base to wide, shallow parabolas. The shape of channel cuts appears to depend on the characteristics of the surrounding sediment and on the amount of time the margins of the channels have been exposed to erosion. Tributary and yazoo channels are commonly smaller and shallower (less than 4 m wide and less than 1 m deep) than Chaco Arroyo. The small channels appear to be more ephemeral than the large arroyo channels. Most buried channels are smaller than the modern Chaco Arroyo.

Sediment Movement and Facies

The two possible sets of geomorphic conditions of the canyon floor affect the transport and distribution of sediment in Chaco Canyon. As the modern arroyo shows, when channels are entrenched in the canyon, the sediments and water within the channels are derived predominantly from the headwaters. Headwater-derived sediments are confined to the arroyo, whereas locally-derived sediments from the canyon margins may spread out onto the canyon floor, or they may be confined in tributary arroyos.

The path of water and sediment from the headwaters changes when there is no confining channel to convey them through the canyon. If the canyon had a relatively flat floor, and if flood discharges as large as the present typical flood (125 m^3/sec) were to enter Chaco Canyon, the narrow parts of the canyon could be flooded from margin to margin. Sediments from the headwaters (containing montmorillonite and mixed-layer clay) would be spread across the canyon floor.

The exposed 12 m of alluvial fill in Chaco Canyon (fig. 7) contains three general types of facies: (1) fan-like deposits shed from the canyon margins

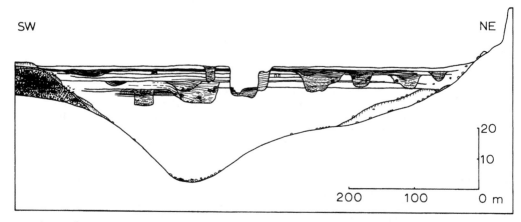

SW

NE

20

10

200 100 0 m

Figure 7. *Generalized cross-section of the alluvial fill of Chaco Canyon to a depth of 12 m. Maximum depth of fill may be 38 m (Ross, 1978). Remnents of earlier fill are perched on the margins of the canyon but are partially buried by modern alluvium. Horizontal lines indicate widespread inundation of the canyon floor by sediments from the headwaters and are connected with filled-in entrenched channels and laminated fills of swales and yazoo channels. The blank intervals between the horizontal lines represent deposition of locally derived or reworked sediments.*

which spread toward the middle of the canyon; (2) sediments from the headwaters which form widespread thin layers across the canyon floor; and (3) local- and headwater-derived sediments which fill channels. Carbon-14 dates associated with buried canyon floor surfaces indicate that the alluvial floor of Chaco Canyon accumulates about one meter of sediment per 1000 yrs.

Channel fills in Chaco Canyon vary in size and character. The first type of fill (fig. 8A) is similar to that produced by the aggradation of the modern inner channel and floodplain. This type of fill is seen in once buried-channels, which date from the period of pueblo occupation and from an arroyo dated about 3,700± 250 radiocarbon yrs ago.

The second type of channel (arroyo) fill (fig. 8B) has a coarse-grained braided channel sand at the base, but the sediment becomes finer-grained higher in the fill. The upper parts of the fill are graded beds of sandy silt and clay, and ripple-bedded silt and clay. One of the channels containing this type of fill post-dates the occupation of the pueblos. The other channel appears to be the oldest example of cut-and-fill exposed in the modern arroyo walls. It may be several thousand years old.

The third type of channel fill (fig. 8C) consists of nearly horizontal units of cross-bedded coarse sand, similar to braided channel deposits. There are no fine-grained backwater deposits along the margins of the channel. The flat-lying layers of sand fill the buried arroyo to the top of the former cut

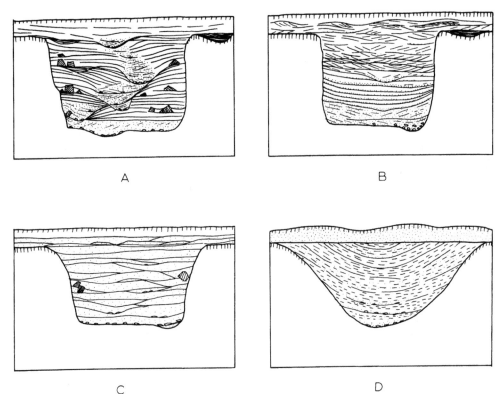

Figure 8. *Ways of filling former channels generalized from exposures in the
arroyo walls of Chaco Arroyo, Chaco Canyon National Monument, New
Mexico. All channels may not be the same size. A. Fill by an
aggrading channel and vertical accretion on the adjacent inner
floodplain. B. Braided channel base with graded beds and cross-
laminated silt derived from the headwaters in the upper part of the
fill. C. Fill by an aggrading braided sandy channel. D. Poorly
sorted sand, silt, and clay in crescentic layers. The layers are
truncated at the top by an aeolian deposit. All these sediments
are locally derived.*

and spread out across the former canyon floor. An example of this type of
channel fill apparently predates ceramic-producing occupants in Chaco Canyon.

The fourth type of fill observed in Chaco Canyon results from the aggradation of small channels that have sloping walls (fig. 8D). The fill is made up
of crescentic layers of either fine sand, silt, and clay of headwater origin,
or poorly sorted, locally-derived sediment. In the latter case, clay
platelets are not disaggregated, but occur as distinct clasts. At least one

293

channel of this type dates from the period of pueblo occupation.

As described above, buried tributary channels and swales are commonly partially filled with laminated backwater deposits similar to modern oxbow deposits. Some charcos (small natural depressions) may have been filled by minor floods to create oxbow-like deposits.

Present exposures along the walls of the modern arroyos and in trenches at archaeological sites do not provide enough control to show the courses of buried channels in plan view. Bryan's (1954) map of his so-called "Post-Bonito channel" appears to incorporate more than one buried channel. Channels which are aggraded to the present surface of the alluvial canyon floor do not directly influence the soil or vegetation on them. The subtle changes in soil and vegetation seen on the canyon floor do not appear to be directly related to buried channels.

When there are no entrenched channels, the facies on the canyon floor are very different. As fans from the canyon margin continued to aggrade, thin layers of gray, clay-rich sediments were spread in nearly continuous bands across the canyon floor. Tributary channels, yazoo channels, and other low areas were filled with laminated backwater sediments similar to modern oxbow deposits. Commonly these deposits are thin (less than several centimeters), implying either: (1) the canyon commonly has a channel which confines headwater sediments, or (2) more or less steady-state transport of sediment through the canyon is achieved when there is no confining channel and therefore, little sediment accumulates, or (3) large flood events in Chaco Canyon are rare, particularly when there is no confining channel.

Summary of Data Concerning the Past Ecosystem in Chaco Canyon

Climatic conditions, the water table, vegetation, and land use by prehistoric occupants are the major factors which could have affected geomorphic and sedimentologic conditions on the canyon floor.

Evidence of past climatic conditions includes tree-ring indices, pollen profiles, and macrovegetal remains in packrat middens. Problems exist with each of these indicators of climatic conditions. Yet, these indicators show no major changes in climatic regime or vegetation for thousands of years (Hall, 1977; Neller, 1976a and 1976b). However, the tree-ring indices do show numerous minor episodes of less-than-normal and greater-than-normal moisture (Dean and Robinson, 1977).

Data from drill holes in the alluvial fill indicate that the present water table is about 13 m below the surface of the canyon floor in areas where the base of the inner channel of Chaco Arroyo is at about the same level (Ross, 1978). Within the exposed sediments of the canyon fill, little evidence exists of a previously higher ground water table. The water table level in Chaco Canyon appears to be related to the elevation of the junction of Chaco Arroyo and Escavada Wash, near the mouth of Chaco Canyon. The Chaco River below the junction has a steeper gradient and a shallow water table. There are groundwater seeps in the Chaco Arroyo channel just above the junction and evaporite salts (bloedite) along the channel margins. Salts fill cavities in the arroyo walls (only 2 m high in this part of the canyon), but are not found more than 1 m above the modern channel upstream from this area. This could indicate that the water table was never high in the upper part of Chaco Canyon.

Human activity in Chaco Canyon during the development of the pueblos spans the period from A.D. 500 to 1350 (T. Windes, 1979, personal communication). During the major period of pueblo occupation, the inhabitants probably were not dependent on food or wood resources from the canyon. Some of the large pueblos may have been occupied seasonally, with the population dispersed around the adjacent region at other times of the year. The estimated 100,000 large pine trees used to build the major pueblos appear to have been brought into the canyon along prehistoric roadways.

Evidence of land use during the pueblo periods includes (1) quarried rock used to build pueblos, small irrigation structures, and for other cultural activities, (2) small canals which directed local runoff, (3) small agricultural (?) plots on the margin of the canyon floor, (4) the use of arroyo channels for disposal of refuse, (5) the use of locally available small trees and willows, and (6) a network of specially prepared roadways.

Discussion

Widespread alluviation of the floor of Chaco Canyon and cycles of cut-and-fill raise several interrelated questions: (1) What are the possible intrinsic and extrinsic factors affecting the cycles of cut-and-fill? (2) What are the processes of cutting channels? (3) What controls the geometry of the cuts? (4) What are the processes of channel filling? (5) How did the geomorphic system operate during the time when Anasazi culture developed in Chaco Canyon over 1,000 yrs ago?

Variables

Possible intrinsic causes for crossing thresholds and inducing channel cutting in the alluvial floor of Chaco Canyon include: (1) localized steep gradients, (2) localized concentration of discharge, (3) lateral shifts of channels, (4) changes in sediment and soil type on the canyon floor, and (5) changes in vegetation related to changes in sediment and soil. Intrinsic causes for filling channels include: (1) decreased gradients, (2) dispersal of discharge, (3) lateral shifts of channels, (4) increases in sediment load, and (5) changes in vegetation. The data concerning climatic conditions and land use summarized above are not adequate to demonstrate causal relationships between these extrinsic factors and adjustments of the geomorphic system in Chaco Canyon.

Channel Erosion

The modern channels exhibit at least four modes of incision: (1) incision along a meandering course by eroding pools and riffles (the modern Chaco Arroyo), (2) headcuts or coalescence of discontinuous arroyos (tributary arroyos), (3) extension and collapse of soil pipes, and (4) progressive incision of the channel further and further upstream without a headwall. Some swales may be eroded depressions, or charcos, formed during large floods along the alluvial floor of Chaco Canyon.

Incision of some channels in the alluvial floor of Chaco Canyon may be due to intrinsic causes which finally exceed threshold conditions. Other channels, particularly those similar to the modern arroyo, may form because of increased stream power generated by unusual amounts of precipitation.

Geometry and Depth of Cuts

The geometry of the small channels may be explained in terms of the configuration of the underlying sediments, the formation of discontinuous arroyos (Schumm and Hadley, 1957; Leopold and Miller, 1956), and the stabilization of hydraulic geometry after the channel was formed. The large arroyos may adjust their geometry by similar means, but it appears that the depth of the present incision is related to the groundwater level in the canyon. The relationship between the modern arroyo and groundwater table has not been investigated thoroughly. It is not clear whether erosion of the channel controls the water table or whether the water table controls the depth of incision. Because the sediments beneath the water table are plastic, it is possible that these sediments, squeezed by the weight of the alluvial fill, would flow into and fill cuts made below the water table. Further entrenchment would then cease.

Why Chaco Arroyo cut through 12 m of the nearly 40 m of fill in Chaco Canyon is not known. It is not clear what hydraulic constraints were met at the 12 m depth which could not be met at 5 m or 25 m instead. Several of the buried channels were incised to nearly the same level.

Processes of Channel Fill

The observed types of channel fill are very similar to Schumm's (1960) hypothetical channel fills except that there are more truncations and cross-cutting relationships within the fills in Chaco Canyon.

Fining-upward cycles occur within fill types 1, 2, and 4 (figs. 8A, 8B, 8D respectively), particularly in the upper portions of the fill where graded beds and ripple-laminated units with clay drapes occur at the top. The fill of the fourth type does not always contain a large amount of fine sediment.

Fill types 1 and 2 imply a decrease in stream power and therefore, a consistent change in sediment type as the channels fill. These changes could be due to either intrinsic changes in the headwaters or to a consistent shift in climatic conditions.

The texture of the braided sandy fill (type 3) (fig. 8C) suggests that either suspended sediment was not present (extremely unlikely) or that it was transported out of the canyon. The coarse fill is evidence that a consistent supply of sand-size sediment was available during aggradation. Perhaps the channel gradient was steep and the channel was connected only with sandy washes in the headwaters. These tributaries would have delivered to the main channel a load composed mostly of coarse sand, or perhaps water loss to the stream bed during aggradation resulted in deposition of bedload without deposition of finer suspended sediment. In either case, this type of fill provides no evidence of a decrease in stream power during aggradation. Therefore, the hydrologic controls during deposition may be totally intrinsic. The initial deposition may have been caused by either intrinsic or extrinsic controls, but there is no evidence to indicate which.

Small type 4 fills appear to be a result of the normal adjustment of discontinuous arroyos located along the margins and in tributaries of Chaco Canyon. Deposition may be related to a decrease in gradient or an increase in local sediment supply. Thus, such cycles of cut-and-fill may be caused by the adjustment of intrinsic controls.

The presence of montmorillonite-rich layers which are spread across the canyon floor suggests one possible intrinsic cause for cycles of cut-and-fill. The montmorillonite-rich layers are high in sodium, have low permeability, swell when wet, are sticky, and support little vegetation. An alluvial flat covered with such a deposit would favor increased runoff, leading to a period of channel formation. Once the channel was formed, the clays on the former alluvial flat could be eroded, blown into clay dunes when dry, leached of sodium, and mixed with sand and kaolinite-rich sediments from the canyon margins. After the channel filled again, clay-rich sediments would cover the alluvial flats and the process could begin anew. This simple mechanism does not explain (1) why channels and fills vary in size, (2) their frequency within the alluvium, or (3) the variable facies. The clay layers do not have similar characteristics at all levels in the canyon fill, possibly indicating that other processes also influenced the configuration of the canyon floor.

Past Geomorphic Processes

Knowledge of conditions on the alluvial floor of Chaco Canyon during the development of pueblo culture, from A.D. 500 to 1350, is critical for some interpretations of cultural evolution. Bryan (1954) suggested that the canyon floor had not been entrenched during most of the occupation. He thought the water table was high at this time and that the inhabitants of the canyon raised their crops by floodwater farming. Bryan suggested that the inhabitants had cut down trees in the canyon and thus had ruined the vegetation on the canyon floor. He reasoned that runoff after a drought had initiated an arroyo which formed a migrating headcut up the canyon, destroying the farm land and forcing the people to abandon the canyon.

Since Bryan's work, archaeological research in Chaco Canyon has produced the following interpretations: (1) Channels up to 4 m deep were present during the development of the pueblos (Judd, 1964). (2) A major entrenched channel filled during extensive occupation of the canyon, between A.D. 1130 and 1220. (3) Deposits overlying the buried channel(s) indicate repeated flooding of the canyon floor after the channel(s) filled. (4) The large pine trees used to build the pueblos came from outside the canyon (T. Windes, 1979, personal communication). (5) Evidence of extensive agriculture on the alluvial canyon floor is poorly documented at present. The development of irrigated plots using water from the headwaters may have occurred late in the occupation (Loose and Lyons, 1976). (6) Tree-ring indices show numerous periods of less than normal moisture during the period of pueblo development (Dean and Robinson, 1977).

The limited data presented above do not support Bryan's (1954) scenario of ecological disaster resulting from tree cutting, extensive devegetation, in-

creasing aridity, and consequent arroyo cutting as a cause for pueblo abandonment. It appears that the inhabitants had already learned to deal with
arid periods, arroyos up to 4 m deep, and the lack of trees long before the
large pueblos were completed. It could be argued that the decline in population coincided with the filling of a major channel. There were definite advantages to having entrenched channels during the time of pueblo development. An entrenched channel would control flooding of the canyon floor,
confine alkali-bearing clay to the channel, provide water in low parts of the
channel, and keep subterranean structures such as pithouses, kivas, and
storage cists from being flooded. The sandy parts of an inner floodplain
could have been used for agriculture. No evidence discovered so far indicates that the occupants caused the arroyos or that the cutting of the arroyos
caused the occupants to abandon the canyon.

To summarize the fluvial adjustments which take place during cycles of
cut-and-fill, morphologic and sedimentologic evidence in Chaco Canyon indicates that different intrinsic and possibly extrinsic factors influence different
phases of the cycle. Some cuts-and-fills apparently reflect minor intrinsic
adjustments. The large channels which presently are filled by different sedimentary facies may indicate the influence of variable climatic conditions and
the response of headwater channels in the drainage net. There is no strong
evidence concerning the relationship between entrenched channels and prehistoric people in Chaco Canyon. The meager data available suggest that land
use was not a major factor in arroyo behavior in the past, and that arroyo
behavior did not influence the major cultural phases nor the abandonment of
the canyon.

LARGE SCALE FLUVIAL ADJUSTMENTS IN CHACO CANYON: ALTERNATING EPISODES OF CANYON EROSION AND AGGRADATION

Geomorphic and Sedimentological Data

Deposits which are not related to the present canyon-floor aggradation are
found at several topographic levels within the canyon and beyond the canyon
margins. There are basically two types of deposits: (1) fine-grained alluvial
deposits and (2) coarse-grained (gravel) alluvial deposits. These deposits
represent (1) a canyon fill similar to that of the present semiarid regime, and
(2) gravel deposited during episodes when stream power was much greater
than it is presently. Some deposits have characteristics of both types of
sedimentation and may be mixtures formed during later semiarid episodes.

The facies of deposits of the modern semiarid regime were described above.
Deposits from earlier episodes of semiarid alluvial aggradation in the canyon are
preserved locally along the southern margin of the canyon. The major deposits cap bedrock spurs jutting into the canyon at levels 3 to 10 m above the
modern alluvium. The deposits are well-cemented with calcite microspar. The
cementation appears to be related to groundwater movement rather than to
pedogenic processes. Calcium carbonate probably leached by groundwater
from the northward-dipping Cliff House Sandstone was deposited in the sediments on the south side of the canyon. Presumably the groundwater on the

north side of the canyon would have moved down the dip of the sandstone and not cemented the deposits on the north side of the canyon.

In some places, these old deposits have a general morphology similar to their modern analogs. The cemented talus forms steep slopes but is separated from the modern cliffs by as much as 50 m of erosion. Up-canyon from the talus, the slope becomes more gradual and finer-grained sediments and soil-like horizons are common. The deposits near the center of the canyon contain coarse-grained, gravelly, cross-bedded sandstone similar to the channel sand of the modern arroyo.

The range in elevation of the deposits and lack of continuity between exposures make it impossible to determine whether there was more than one period of formation of these cemented deposits. There are indications that the canyon was partially filled with semiarid deposits more than once during its history.

Gravel of type 2 deposits is found on at least five topographic levels within the canyon. Some deposits form distinct terraces along the canyon margins, but most are morphologically indistinct. In some localities, the gravel is cross-bedded with sets up to 1 m thick. The presence of gravel with clasts ranging to more than 30 cm in diameter is the major characteristic of this type of deposit. The clasts are predominantly subangular sandstone pebbles and cobbles which could have been derived locally. Most of the deposits also contain pebbles and cobbles of jasper, agate, plutonic and volcanic rocks, multicolored quartzite, and foliated metamorphic rocks all of which appear to have been derived from the Tertiary formations exposed in the headwaters.

An extensively exposed gravel unit contains nearly horizontal layers of sand intercalated with the gravel (fig. 9). Blocks of alluvium similar to alluvial blocks which collapse into the modern channel are incorporated in the gravel deposit.

Some gravel deposits are uncemented while others are well-cemented with sparry calcite. The clasts in some deposits are coated with manganese oxides.

Age of the Deposits

The ages for both kinds of deposits are not established. Hall (1977) assigned a late Pleistocene age to the deposits shown in figure 9. A red paleosol with caliche beneath it is present in the aeolian and fluvial deposits overlying the gravel. Laterally, this paleosol bifurcates into three or four less well-developed paleosols on the flanks of the deposits. The number of paleosols suggests that the underlying gravel deposit is older than late Pleistocene. The extensive cementation and geomorphic position of other gravel deposits and semiarid facies in the canyon suggest that they are much older than this gravel.

Discussion

The two kinds of deposits indicate changes in base level and stream power of such a magnitude that there must have been a climatic control. Alternating types of deposits suggest episodic changes in both the magnitude of the

Figure 9. *Exposure of cross-bedded gravel with lenses of sand.* *Large blocks of fine-grained alluvium are included in the gravel.* *The top third of the exposure is colluvium and windblown sand with more than one red paleosol and caliche horizon developed on it.*

fluvial thresholds and the drainage basin processes. Changes in stream power affected the rates of erosion and sedimentation as well as the types of sediment.

The blocks of alluvium caught in the gravel deposit described above suggest that an uncemented alluvial fill existed in Chaco Canyon prior to a rather sudden increase in stream power. The small amount of gravel available to the modern channel suggests that a large amount of erosion would be necessary to provide the quantity of gravel in this old deposit. The deposit could be interpreted as having formed early in a change from semiarid to much wetter conditions. As such, erosion in the headwaters would have provided gravel to the main stream even though the previously deposited alluvial fill had not been flushed from the canyon. Thus, this remanent gravel deposit appears to have been left behind as erosion of the alluvial fill continued and as the stream established a course deeper in Chaco Canyon. This older gravel could have been deposited as a result of a complex response in channel behavior after a major change in stream power.

At this third and largest scale of adjustment, the history of Chaco Canyon is a series of alternating episodes of two types: (1) periods of greater stream power, extensive canyon erosion, and canyon entrenchment in bedrock, and (2) semiarid episodes during which the canyon partially filled. Complex response of the fluvial system appears to take place at this scale of adjustment.

CONCLUSIONS

Similar processes operate at three scales of fluvial adjustment to produce geomorphologically and sedimentologically distinct features in Chaco Canyon. The scales are (1) adjustments within the modern Chaco Arroyo and adjacent area, (2) cycles of cut-and-fill and wide-spread alluviation of the canyon floor, and (3) alternating episodes of erosion and deposition in which Chaco Canyon was eroded and partially refilled. Each of these scales of adjustment was affected by geomorphic thresholds and the complex response of the system. At each scale, greater stream power tended to transport sediment through the canyon while decreased in stream power caused sediment storage within the canyon. At each scale of adjustment, increases in stream power could have been caused by greater discharge resulting from increased precipitation. The response of the two smaller scales of adjustment could have been complicated by other extrinsic variables, such as changes in land management, and by intrinsic fluvial adjustments. The scales and factors of adjustment are summarized in table 2. The geomorphic and sedimentologic history of the canyon consists of a unique series of events within a framework of episodic change. The history of Chaco Canyon is summarized in figure 10.

Land use does not appear to be a major variable in this drainage system at present nor in the past. In the future, changes in land use which affect the discharge and sediment supply in Chaco Arroyo could modify the arroyo. If abundant sediment was produced but not transported through the canyon, Chaco Canyon National Monument could be flooded with excess sediment. If, at another extreme, excess discharge were to flow into the canyon, the arroyo might be eroded further. If both sediment and water from the head-waters did not reach the canyon, local processes would slowly modify the arroyo and canyon floor. If it is desirable to "manage" the behavior of Chaco Arroyo, some balance of stream power and sediment supply must be achieved. To protect the ruins, erosion of arroyo channels and soil pipes must be minimized, but sediment must not accumulate in the canyon to cause flooding of the canyon floor.

ACKNOWLEDGEMENTS

I am grateful for the support and guidance of S.G. Wells and C.T. Siemers. I appreciate J. Hawley's and A. Gutierrez's helpful criticisms of the manuscript. G.P. Williams' and D.D. Rhodes' advice and patience are also appreciated. I thank the Chaco Center (National Park Service), the New Mexico Geological Society, and the Student Research Allocations Committee of the University of New Mexico for financial assistance.

Table 2

Summary of Scales of Adjustment, Associated Intrinsic and Extrinsic Variables,
and Resulting Geomorphic and Sedimentologic Features

Scale			Variables			
Type	range of vertical magnitude at a location	common time span for geomorphic adjustments (years)	Intrinsic			
			biologic	sedimentologic	morpho-logic	hydrologic
1. changes within and adjacent to Chaco Arroyo	less than or equal to depth of channel incision $(10^{-2}$ to $10^{1.1}$m)	10^{-2} to 10^2	ecotones, natural variations in species and number of individuals	changes in erodability and availability of sediment, types of sediment, pulses of sediment, available storage along channel, facies in equilibrium with channel conditions	local changes in gradient, hydraulic geometry, physiographic expression of facies	variations in discharge, flashiness, seasonality, groundwater level
2. alluviation of the canyon floor and cycles of cut-and-fill	less than or equal to depth of alluvial canyon fill $(10^0$ to $10^{1.6}$ m)	10^0 to 10^4	ecotones, natural variations in species and number of individuals	adjustment of canyon floor facies of local- and headwater-derived sediments, facies of channel fills (some may not be intrinsic)	local changes in gradient, hydraulic geometry, lateral channel shifts, morphology within cycle of cut-and-fill	variations in discharge, flashiness, seasonality, groundwater level
3. alternations between flushing sediments from the canyon, eroding the bedrock floor, and partially filling the canyon with alluvium	greater than depth of alluvial fill $(10^1$ to $10^{2.5}$ m)	10^3 to 10^5	variation and evolution of species	intrinsic adjustments depend on climatic conditions complex response to changing discharge conditions		(large discharge conditions) flow of carbonate-bearing groundwater (small discharge conditions)

Table 2

Summary of Scales of Adjustment, Associated Intrinsic and Extrinsic Variables,
and Resulting Geomorphic and Sedimentologic Features

		Results	
Extrinsic			
land use	climatic variations	morphologic	sedimentologic
changes in morphology, sediments, soils, vegetation	changes in precipitation i.e. seasonality intensity amounts changes in temperature i.e. seasonality severity length of growing season wind, albedo and other climatic factors	ephemeral changes in drainages and canyon floor, development of Chaco Arroyo	changes in erosion and deposition, development of sedimentary facies within arroyo, minor facies on canyon floor
changes in morphology, sediments, soils, vegetation extent of influence appears to be minimal	changes in precipitation i.e. seasonality intensity amounts changes in temperature wind, and other climatic factors	three phases of cycles of cut-and-fill	canyon floor facies, transport of sediment out of canyon during cut phase, downstream relocation of sediment in all phases
no evidence of association	large variations in climate definitely influence rest of system	geomorphic levels of stability and incision, terraces	erosion of fine-grained facies, deposition of gravel, erosion of bedrock floor of canyon and redeposition of fine-grained facies
	greater precipitation	Chaco Canyon is incised to deeper levels, gradient high large channels(?)	gravel available in large quantities at least briefly
	less precipitation (semiarid conditions)	floor aggrades, gradient low, small or no channels	canyon fills with fine-grained alluvium, semiarid facies

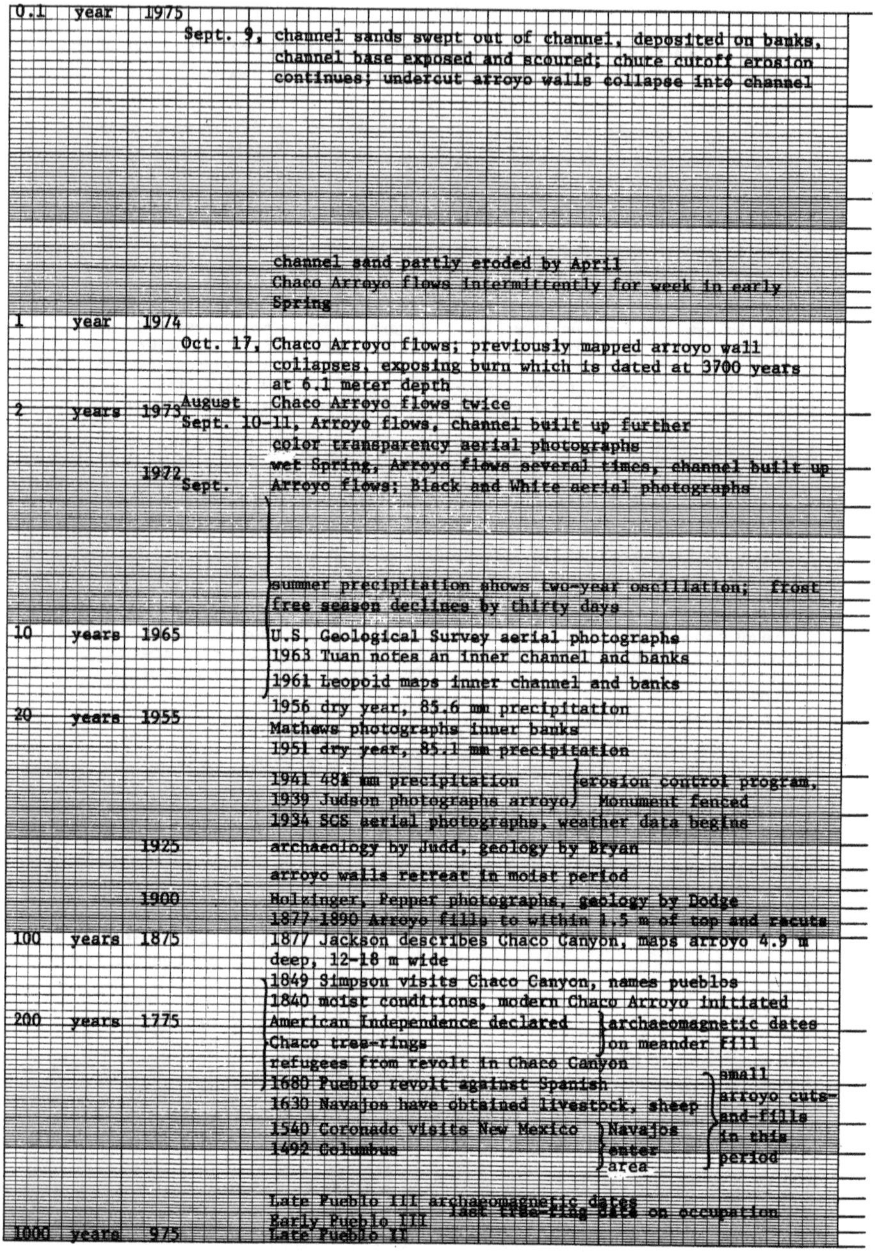

Figure 10. *General outline of geomorphic history of Chaco Canyon. The scale at the left is logarithmic and backward in time, starting in 1975. Asterisks indicate radiometric dates.*

1000 years 975
A.D.
* Early Pueblo II
Pueblo I
Chaco tree-rings
Basketmaker
pithouses built on floor of canyon

2000 years B.C.
? arroyo cut-and-fill, undated
* arroyo cut-and-fill (inferred by Hall, 1975)
* arroyo cut-and-fill

* Hall's oldest C^{14} date, 5.5 m depth in colluvial sediments
? arroyo cut-and-fill only top 1.5 m exposed
7 m below present surface
End of Pleistocene

10,000 years

beginning of "Wisconsin" glacial stades. If discharge increased during this time, Chaco River may have eroded to bedrock floor of the canyon, and alluviated the canyon floor in late Pleistocene

100,000 years
? several red soils form on sandy colluvial and windblown deposits

? gravel deposits in gravel quarry may have formed at this time of high discharge. Chaco River on bedrock part of this time
? canyon fills during arid interglacial
? loose and cemented gravel benches below first Cliff House cliff
? cemented talus, soils and arroyo-like deposits formed under semiarid conditions
? cemented gravel on bench above first Cliff House cliff

? cemented sand dune above junction of Escavada
? loose gravels on mesa margins

1,000,000 years

BEGINNING OF PLEISTOCENE
? initial cutting of Chaco Canyon began, at least two gently rolling geomorphic surfaces with 1 m of caliche developed

PLIOCENE

10,000,000 years

Archaic & Basketmaker

Figure 10 continued.

REFERENCES

Bryan, K., 1925, Date of channel trenching (arroyo cutting) in the arid Southwest: Science, v. 62, p. 338-344.

_____, 1926, Recent deposits of Chaco Canyon, New Mexico, in relation to the life of the prehistoric peoples of Pueblo Bonito [abs.]: Journal of the Washington Academy of Science, v. 16, p. 75-76.

_____, 1954, The geology of Chaco Canyon, New Mexico, in relation to the life and remains of the prehistoric peoples of Pueblo Bonito: Smithsonian Miscellaneous Collections, v. 122, pub. 4141, p. 1-65.

Burkham, D.E., 1966, Hydrology of Cornfield Wash area and effects of land-treatment practices, Sandoval County, New Mexico 1951-1960: U.S. Geological Survey Water Supply Paper 1831, 87 p.

Chauvenet, W., 1935, Erosion control in Chaco Canyon, New Mexico, for the preservation of archaeological sites [M.A. thesis]: Albuquerque, University of New Mexico, 60 p.

Cooke, R.U., and Reeves, R.W., 1976, Arroyos and environmental change in the American Southwest: Oxford, Oxford University Press, 213 p.

Dean, J.S., and Robinson, W.J., 1977, Dendroclimatic variability in the American Southwest A.D. 680 to 1970: Tucson, Laboratory Tree Ring Research, University of Arizona, Final Report to National Park Service, Department of the Interior Project: Southwest Paleoclimate, Contract CX-1595-5-0241, 9 p. + 2 appendices.

DeAngelis, J.M., 1972, Physical geography of the Chaco Canyon country: unpublished report to the National Park Service, Chaco Center, University of New Mexico, 113 p.

Dodge, R.E., 1920, Report of field studies; in Pepper, G.H., ed., Pueblo Bonito: American Museum of Natural History Anthropological Papers, v. 27, p. 23-25.

Fritts, H.C., 1971, Dendroclimatology and dendroecology: Quaternary Research, v. 1, p. 419-449.

Hall, S.A., 1975, Stratigraphy and palynology of Quaternary alluvium at Chaco Canyon, New Mexico [Ph.D. dissertation]: Ann Arbor, University of Michigan, 66 p.

_____, 1977, Late Quaternary sedimentation and paleoecologic history of Chaco Canyon, New Mexico: Geological Society of America Bulletin, v. 88, p. 1593-1618.

Holsinger, S.J., 1901, Report on prehistoric ruins of Chaco Canyon National Monument: Washington, D.C., General Land Office Monuments, National Archives, Part I, 85 p., Part II, 110 p.

Jackson, W.H., 1878, Report on the ancient ruins examined in 1875 and 1877: 10th Annual Report, U.S. Geological and Geographical Surveys (for the year 1876), p. 431-450.

Judd, N.M., 1954, The material culture of Pueblo Bonito: Smithsonian Miscellaneous Collection, v. 124, 398 p.

Judd, N.M., 1959, Pueblo del Arroyo, Chaco Canyon, New Mexico: Smithsonian Miscellaneous Collections, v. 138, 222 p.

_____, 1964, The architecture of Pueblo Bonito: Smithsonian Miscellaneous Collections, v. 147, 349 p.

Leighly, J., 1936, Meandering arroyos of the dry Southwest: Geographical Review, v. 26, p. 270-282.

Leopold, L.B., 1976, Reversal of erosion cycle and climatic change: Quaternary Research, v. 6, p. 557-562.

Leopold, L.B., and Miller, J.P., 1956, Ephemeral streams--hydraulic factors and their relation to the drainage net: U.S. Geological Survey Professional Paper 282A, 37 p.

Loose, R.W., and Lyons, T.R., 1976, The Chetro Ketl field: a planned water control system in Chaco Canyon: in Lyons, T.R., ed., Remote sensing experiments in cultural resource studies: non-destructive methods of archaeological exploration, survey, and analysis: Albuquerque, Chaco Center, National Park Service, Department of the Interior and University of New Mexico, p. 133-156.

Love, D.W., 1977, Dynamics of sedimentation and geomorphic history of Chaco Canyon National Monument, New Mexico: New Mexico Geological Society Guidebook, 28th Field Conference, San Juan Basin III, p. 291-300.

Neller, E., 1976 a, Sleeping dune: unpublished report to the Chaco Center, National Park Service and the University of New Mexico, 58 p.

_____, 1976 b, Botanical analysis in Atlatl Cave, preliminary report: unpublished report to the Chaco Center, National Park Service and the University of New Mexico, 34 p.

Nichols, R., 1975, Archaeomagnetic study of Anasazi-related sediments of Chaco Canyon, New Mexico [M.S. thesis] : Norman, University of Oklahoma, 111 p.

Pepper, G.H. ed., 1922, Pueblo Bonito: American Museum of Natural History Anthropological Papers, v. 27, 398 p.

Ross, J.R., 1978, A three dimensional facies analysis of the canyon alluvial fill sequence Chaco Canyon, New Mexico [M.S. thesis]: Albuquerque, University of New Mexico, 87 p.

Schoenwetter, J., and Eddy, F.W., 1964, Alluvial and palynological reconstruction, Navajo Reservoir District: Museum of New Mexico Papers on Anthropology, no. 13, 155 p.

Schumm, S.A., 1960, The effect of sediment type on the shape and stratification of some modern fluvial deposits: American Journal of Science, v. 258, p. 177-184.

_____, 1977, The fluvial system: New York, Wiley-Interscience, 338 p.

Schumm, S.A., and Chorley, R.J., 1964, The fall of Threatening Rock: American Journal of Science, v. 262, p. 1041-1054.

Schumm, S.A., and Hadley, R.F., 1957, Arroyos and the semi-arid cycle of erosion: American Journal of Science, v. 255, p. 161-174.

Senter, D., 1937, Tree rings, valley floor deposition, and erosion in Chaco Canyon, New Mexico: American Antiquity, v. 3, p. 68-75.

Siemers, C.T., and King, N.R., 1974, Macroinvertebrate paleoecology of a transgressive marine sandstone, Cliff House Sandstone (Upper Cretaceous), Chaco Canyon, northwestern New Mexico: New Mexico Geological Society Guidebook, 25th Field Conference, Ghost Ranch (Central-Northern New Mexico), p. 267-277.

Simpson, J.H., 1852, Journal of a military reconnaissance from Santa Fe, New Mexico, to the Navajo country: report by the Secretary of War, 31st Congress, 1st Session, Senate Executive Document, no. 64, 140 p.

Thomas, H.E., 1963, The meteorological phenomenon of drought in the Southwest: U.S. Geological Survey Professional Paper 372A, p. 1-43.

Tuan, Yi-Fu, 1966, New Mexican gullies: a critical review and some recent observations: Association of American Geographers Annals, v. 56, p. 573-597.

Van Valkenburgh, R., 1938, A short history of the Navajo People (unpublished monograph): Window Rock, Arizona, U.S. Department of Interior, Navajo Service, 68 p.

CHANNEL ADJUSTMENT TO SEDIMENT POLLUTION

BY THE CHINA CLAY INDUSTRY

IN CORNWALL, ENGLAND

K. S. RICHARDS

Department of Geography
University of Hull
Hull, United Kingdom

ABSTRACT

Extraction of the reserves of kaolin in the St. Austell granite outcrop of Cornwall, to supply the china and paper industries since the late eighteenth century, has imposed an abnormal load of up to 3 % by weight of fine suspended sediment on the streams draining the area. This pollution has caused adjustment of channel form, the width-depth ratio being substantially reduced, but channel capacity relatively unaltered. Analysis of the suspended sediment load of one stream by X-ray diffraction and Coulter Counter methods reveals downstream coarsening and mineralogical adjustment that jointly imply selective deposition of fines. These deposits are probably related to flocculation of kaolin in water of increasing downstream ionic concentration. Bank material of polluted and unpolluted streams differs in terms of plasticity, clay and silt content, and clay mineralogy. Bank clay content and mineralogical indices are used in multiple regressions to predict plasticity index and width-depth ratio, which themselves appear to be inversely related. The best description of width-depth ratio variation is provided by a regression model incorporating discharge or catchment area as a 'scale' effect, and indices of clay mineralogy. This satisfactorily predicts the different channel shapes of polluted and unpolluted streams as a result of the altered mineralogical balance of the fine fraction of the bank material of the former.

INTRODUCTION

Established relationships between form and process in equilibrium alluvial channels are now being used by fluvial geomorphologists to explain adjustments of river channels, and to predict the effects of changes in catchment hydrology and sediment yield on river geometry. Ultimately, these applications may permit suggestions of the means of forestalling or controlling those channel changes which could be economically or socially undesirable. Particular emphasis has been placed on analysis of the channel erosion and enlargement associated with increased flood magnitude in urbanizing catchments (Hammer, 1972; Gregory and Park, 1976). A number of difficulties are encountered in these applications, however.

First, there is a lack of established process-response models. Schumm (1969) outlined a qualitative model of river metamorphosis which is partially calibrated by empirical relations between discharge and sediment properties

309

and channel morphometric indices. Rango (1970) used similar methods to predict adjustments to precipitation and sediment yield changes. However, many
investigations have been indirect <u>post factum</u> studies of adjustment. They have
used space-time substitution by comparing upstream channel properties with
altered downstream characteristics, or by using adjacent control catchments.
While these illustrate the magnitude of change, they do not always yield insight
into the mechanisms of adjustment, and do not contribute to a general quantitative process-response model.

Secondly, sedimentological influences have received less attention than discharge changes. This has resulted in sometimes contradictory results in studies of streams adjusting to urban development, because of the non-coincident
cycles of change in flood hydrology and sediment yield (Wolman, 1967; Leopold,
1973; Hollis and Luckett, 1976). Most channel changes reflect complex environmental influences in which several independent variables alter simultaneously or
in sequence, making explanation of their separate effects difficult and complex
responses the rule (Schumm, 1973).

Thirdly, there remains room for debate over the relevant independent variables, especially in terms of sediment influences. Channel shape (width-depth
ratio) is related to the channel perimeter percentage of silt and clay (Schumm,
1960), but this is an empirically convenient sediment index rather than a theoretically sound one. Soil mechanical measures can be used to define threshold
tractive forces leading to channel instability (Flaxman, 1963); however, difficulties are often encountered in recreating field conditions in laboratory tests
(McQueen, 1961) and in satisfactorily quantifying the spatial variability of bank
material which particularly influences the channel geometry of small streams.
Whether bank sediment is an inherited independent variable reflecting valley
fill by palaeochannels, or a direct adjustment to contemporary rates and types
of sediment transport (Schumm, 1971), is also not always clear. The status
of sediment type as a variable may change from stream to stream.

This study is an investigation of fluvial adjustment in small streams draining the area of the Hensbarrow Downs granite outcrop in Cornwall, which has
been a centre of intensive extraction of kaolin (china clay) since the late 18th
century (fig. 1). These streams have been subject to extremely heavy inputs
of suspended sediment, as well as possible changes in flood hydrology. An
attempt is made here to define the adjustments resulting from each, with the
above problems providing central themes in the analysis.

STUDY AREA

The Hensbarrow, or St. Austell, granite boss is one of five major outcrops
in the South West peninsula of England, all of which may be connected at depth
to form a continuous batholith from Dartmoor to the Isles of Scilly (Edmonds
et al., 1969). Evidence of hydrothermal alteration of the granite is widespread,
and includes tourmalinization by boron-rich gases and greisening by fluorine.
Kaolinization, which involves hydrothermal alteration of plagioclase feldspar, is
evident on S.W. Dartmoor, S. Bodmin Moor, and on Tregonning Hill, a small
outcrop of granite near Helston in West Cornwall. However, the St. Austell
granite is most extensively affected. The attitude of the various hydrothermal
alterations suggests here that kaolinization was a late process in the granite
cooling.

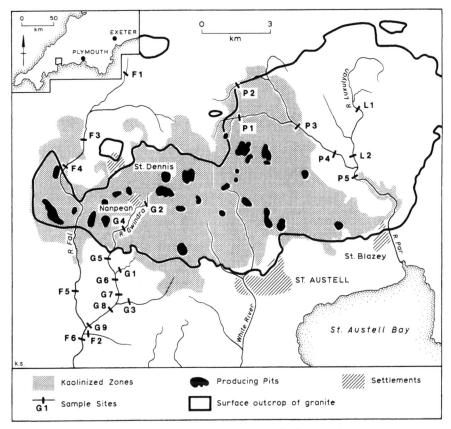

Figure 1. *The study area showing: the outline of the Hensbarrow Downs granite outcrop, the area designated as having potential for china clay extraction, and the location of cross-sections surveyed in the analysis of channel adjustment.*

Keller and Hanson (1975) demonstrated that scanning electron micrographs of kaolinites reveal dissimilar fabrics: hydrothermal kaolin is associated with low inter-grain porosity and elongated crystals, and sedimentary kaolin is characterized by relatively large (5 - 10 µm) expanded books of high porosity. Samples from china clay pits in the St. Austell area (Keller, 1976) display the latter fabric, and suggest that hydrothermal alteration may have been followed by extensive weathering. The fabric of kaolin may be significant in relation to its plasticity, with the hydrothermal kaolins having limited water-retention capacity as a result of low inter-grain porosity, and hence lower plasticity (Bain, 1971).

The kaolinized zone of the St. Austell granite is outlined on figure 1. After initial exploitation of kaolin (china clay) at Tregonning Hill in ca. 1750,

311

one William Cookworthy opened a pit south of Nanpean in 1770 to supply his porcelain factory in Plymouth. The industry developed rapidly when Josiah Wedgwood began to supply his factories in the Potteries. Many small, competing pits opened up, and overproduction rapidly became a major problem. Although today the industry is centralized, mechanized, and efficient, the mining methods are essentially similar. Hydraulic processes are widely used. The clay is washed from the sides of the pits by high-pressure hoses and separated by differential settlement of the quartz sand and mica waste from the kaolin clay. Not surprisingly, by 1799 a traveller in Cornwall described a stream as "white as milk, and very turbid" (quoted by Barton, 1966).

Control of pollution was impossible before the small competing firms were consolidated ca. 1950. The Cornwall River Authority was faced with a historic legacy of streams, such as the Par and St. Austell (White) Rivers, carrying up to 5.3 % by weight of suspended sediment momentarily, or 2.8 % over 24 hours. Pollution was unpredictable; although runoff from waste tips still containing large quantities of clay increased the loads during rainfall, industrial outfalls were uncontrollable. For example, wash water from a pit was often transferred to "mica drags" owned by another company where settlement of mica was undertaken in lagoons. Table 1 lists some data sampled on one day in January, 1956, at the intake and outfall from a mica lagoon. Wind conditions evidently influence the efficiency of settlement considerably. Only since the early 1970's has it become possible to attempt recovery of some of the affected water courses.

Some of the abnormal influences of the industrial history of this area on channel processes can be gaged by comparing discharge and sediment transport in the streams of the St. Austell area with others in Devon and Cornwall. In Figure 2A, mean suspended sediment concentrations (mg/l) are related to catchment area (A, km^2) and rainfall (P, mm) by the regression shown. Although both partial correlations are significant, the physical basis of this relationship is weak. The negative rainfall exponent reflects the tendency of rainfall to decline towards the northeast. The rainfall exponent is also associated with agricultural differences which result in more arable farming and higher sediment yield in areas of S.E. Devon underlain by "soft" rocks. For example, the

Table 1

Suspended Concentrations at the Intake and Outfall from a Mica

Lagoon at Various Times during one Day in January, 1956

Time	Intake concentration (ppm)	Outfall concentration (ppm)	Wind conditions
11.00	14000	50	Dry, moderate S.W. wind
13.00	32000	225	Dry, strong S.W. wind
15.00	1200	975	Dry, very strong S.W. wind

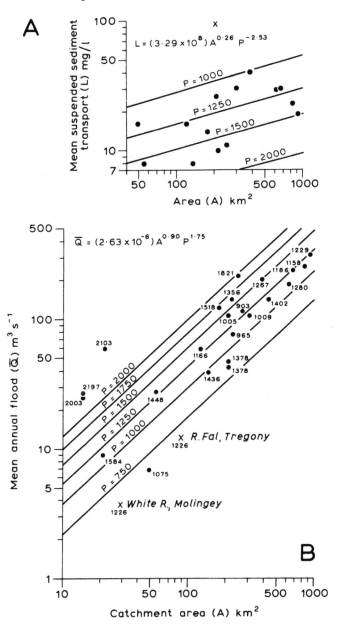

Figure 2.   A: Average suspended sediment load in West Country streams.   The point defined by a cross is the R. Otter in South east Devon.   B: Mean annual flood of West Country streams.   The two streams heading in the study area were not included in the multiple regression analysis.

cross in the diagram is the R. Otter at Dotton, draining an area of Permo-Triassic sandstones and marls. Mean suspended sediment loads of the Par, White, Crinnis, and Fal Rivers are even today at least ten times greater than those predicted by the regression, and clearly would have been even more extreme in the past. It is more difficult to examine the effect on streamflow. Present day excessive sediment loads reflect runoff from tips, scour of clay from bed and banks, leakage from old workings and accidental outfalls. It might be expected that higher runoff rates per unit area occur in the mining area, given the frequent waste tips of largely unvegetated sand, gravel, and clay at the angle of rest. However, this may be compensated for by drainage into (and deliberate storage of water in) abandoned pits, and by extensive modern land drainage schemes to control runoff. Thus, in figure 2B the White and Fal Rivers are seen to have mean annual floods ($\bar{Q}$, m^3/sec) of about 1/3 the amount expected for their size and rainfall. In the multiple regression, the model is more physically reasonable, and the overall $R^2 = 0.77$, with partial correlations both significant at $p = 0.01$.

CHANNEL CHARACTERISTICS

The high pollution in streams of this area has made stream gaging on a regular basis virtually impossible until recently. The White River at Molingey, for example, still exhibits a shifting stage-discharge relationship. A provisional rating curve defined in 1975 tended to slightly underestimate current meter gaging checks from 1975-76, and overestimate them from 1977-78. However, these current meter data permit establishment of the hydraulic geometry of the section. The width (b), depth (f), and velocity (m) exponents are listed in table 2, with those for the River Fal at Tregony (a relatively unpolluted stream) for comparison. Both sets of exponents were derived from data obtained from 1975-1978. The smallest correlation coefficient is $r = 0.89$, and there is no systematic change in any exponent if the data sets are subdivided into temporal blocks. The exponent j is the predicted exponent in the load-discharge relationship based on Leopold and Maddock's (1953, fig. 18, p. 25) diagram relating j to b, f, and m. From these it can be seen that much more rapid increases of sediment concentration with increasing discharge are to be anticipated today in the White River than in the River Fal.

Table 2

At-a-station Hydraulic Geometry Exponents for two Gaging Stations.

The 'j' exponent is the predicted load-discharge exponent,

based on the method of Leopold and Maddock (1953).

River	Gaging Station	b	f	m	m/f	j
White	Molingey	0.11	0.45	0.46	1.02	3.0
Fal	Tregony	0.22	0.54	0.24	0.44	1.8

Measurements of bankfull channel geometry were made at the survey sections defined in figure 1 in order to identify adjustments in the total cross-
section form. The sections were on polluted and unpolluted streams; the former including the Gwindra and Par, the latter the upper Fal, Luxulyan, and
upper Par. Classification is, however, rather arbitrary in some cases, and
some of the measured sections were not obviously members of one group. The
polluted streams have large proportions of their catchment areas affected by
china clay extraction, and are generally recognizable by their continued milky,
turbid nature.

Channel capacity increases with catchment area (fig. 3A). Because there
was no significant difference between the regression equations for polluted and
unpolluted streams, a single combined relationship was employed. In the polluted catchments, however, the complexities of abstraction, outfall, and leats
render catchment area an unsatisfactory discharge index. This is reflected in
correlations of r = 0.91 for unpolluted streams and r = 0.63 for polluted
streams. The similarity of the regressions, in spite of this, tends to suggest
that discharge adjustments have been relatively minor in terms of their effects
on channel geometry. However, it is possible that joint modification of flow
magnitudes and bank material might change the magnitude-frequency relationship between process and form (Pickup and Warner, 1976).

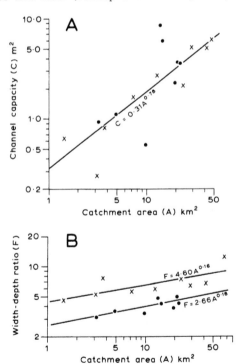

Figure 3. A: *Channel capacity as a function of catchment area.* B: *Width-
depth ratio related to catchment area. Polluted streams shown
by dots, unpolluted streams by crosses.*

Because width generally increases downstream faster than depth (Leopold and Maddock, 1953), it is necessary to compare width-depth ratios (F) for polluted and unpolluted streams in relation to their sizes. The relationships of F with catchment area are well-defined (fig. 3B). Polluted streams have a width-depth ratio averaging 0.58 that of unpolluted streams of comparable catchment area, and are therefore narrower and deeper. This result is confirmed if F is plotted against channel capacity to eliminate doubt concerning the validity of catchment area as a discharge index. Thus it would seem that the main effect of the china clay extraction industry has been to cause a change in channel shape, as defined by the width-depth ratio. The altered shape probably reflects the high suspended sediment loads of polluted streams. Figure 3B summarizes the net effect of changes in sediment load and discharge (if the latter have been significant). Had flood discharges been augmented by the runoff from waste tips, there would have been more discharge per unit catchment area, which should result in increased width/depth ratios. The gross sediment influence would have had to be more than that required to reduce F by 42 % with no discharge change. If the true effect is the reduction of mean annual flood to 1/3 of the value in an unpolluted catchment (fig. 2B), this implies an 'effective' discharge-producing area of 0.3 that for unpolluted catchments of the same discharge (since the discharge-area exponent is 0.9). Because F varies with area to the 0.17 power, such a reduction of area implies the reduction of F by only 18 %, whereas the average reduction indicated by figure 3B is 42 %. Thus, the sediment load effects would appear to be dominant, however the discharge changes are interpreted.

SUSPENDED SEDIMENT CHARACTERISTICS

Variations in Concentration

Samples of suspended sediment load collected at the study sites (fig. 1) during a summer discharge of about 60 % duration indicate that natural sediment loads vary with basin area, as in the general analysis of West Country streams (fig. 2A). The suspended sediment data from sites on polluted streams display an inverse relationship with basin area, declining downstream from a maximum of 950 mg/1 at the head of the River Gwindra. However, the decline is erratic, reflecting the irregular distribution of point source sediment inputs and diluent, clean tributaries. These spatial patterns are summarized in figure 4, where discharge is plotted in place of basin area.

Clay Mineralogy

The mineralogy of the suspended sediment load of the River Gwindra was examined by X-ray diffraction methods. Six of the mainstream sites (i.e., excluding G9) produced sufficient sediment to enable this analysis to proceed, although it was necessary to prepare the slides directly from the residue of sediment left on Millipore 0.22 μm filter papers. A Phillips PW 1050 diffractometer was used for the analysis. The diffractograms (fig. 5) were obtained using nickel-filtered CuKα radiation at 20 mA and 40 kV, at a chart speed of 10 mm/min and range of 2×10^3 cps. A 1 second time constant was used at a scanning rate of $1°$ 2θ per minute. The divergent and receiving slits were respectively $1°$ and 0.1 mm. Although each slide was prepared from 0.05 g of sediment, this weight control is not sufficiently accurate to merit direct quan-

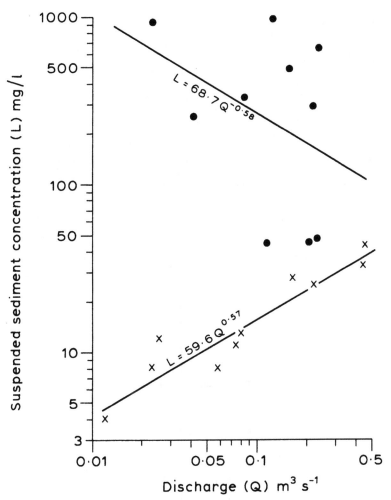

Figure 4.   Trends in low flow suspended sediment concentration in polluted
(dots) and unpolluted (crosses) streams with increasing discharge
downstream.

titative comparison of peak heights between diffractograms. For comparison,
ratios of various peak heights above the background count are used.

The diffractograms are relatively simple to interpret, consisting of kaolin
peaks at 7.14 and 3.58 Å, muscovite-illite at 10.05, 5.01, and 3.32 Å, some
possible goethite at 4.18 Å, gibbsite at 4.35 Å, and quartz at 4.26 Å. Some
confusion may arise with subsidiary peaks for disordered kaolin in the 4.4 to

3.6 Å range (Noble, 1971), but the quartz peak remains diagnostic in these diffractograms. Although the 3.34 Å quartz peak is stronger, it is usually obscured by the 3.32 Å illite peak. The tendency for non-parallel alignment of micaceous minerals broadens this peak such that, even at a scanning rate of 1/4 ° 2θ per minute and 25 mm/min chart speed, resolution of two peaks was impossible. However, the 4.26 Å quartz peak is usually an acceptable guide even in quantitative petrology (Cosgrave and Sulaiman, 1973), so it has been used in this study. Schultz (1964) accepted the validity of such a measurement as long as the peak is 5 - 10 cps above the background. This criterion is satisfied. All slides were air dried, but some were subsequently heated for 1 hour at 550ºC to check for disappearance of the kaolin peaks on collapse of the lattices. Three of the samples (from sites G2, G7, and G8) were subsequently sedimented, after ultrasonic dispersion, for 20 hours. The suspension was withdrawn to a depth of 25 cm in order to isolate the fraction less than 2 μm.

Mean peak ratios are listed in table 3 for the six samples of polluted sediment load, the three fine fractions, and for a single combined sample of unpolluted suspended sediment. The mineralogy of the combined sample is essentially similar to that of the other samples in terms of presence and absence of specific minerals, but the ratio measures differ. This use of peak ratios in

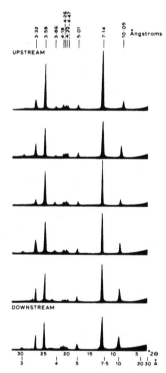

Figure 5. X-ray diffractograms for the suspended sediment of the six upstream sites on the mainstream of the River Gwindra.

Table 3

Mean Ratios of Diffractogram Peak Heights;

K = kaolin, M = muscovite-illite, Q_z = quartz

Type of sample	Mean ratios		
	K/Q_z	K/M	M/Q_z
Suspended sediment, polluted sites	35.4	4.4	8.3
Suspended sediment, <2μm, polluted sites	62.9	11.5	5.4
Suspended sediment, unpolluted sites	29.2	7.2	4.1

environmental discrimination is exemplified by the study of estuarine clays by Sawhney and Frink (1978). Downstream variation in the strength of the kaolin, quartz, and muscovite-illite peaks is evident, with kaolin decreasing and the others increasing. This trend, interrupted by the diffractogram for site G4, is evident in figure 5 and is summarized by the strong trend of reducing kaolin/quartz ratio in figure 6C.

The ultrasonically dispersed suspended sediment samples were diluted to a known concentration in particle-free salt solution. Using a Coulter counter with a preset threshold, a count was made of the number of particles greater than 20μm nominal diameter. The count, expressed as number of particles per μg of sediment, increases consistently with basin area tributary to the sediment sampling cross-section (fig. 6B). Thus the amount of kaolin relative to quartz in suspended sediment decreases downstream, and the number of coarse grains (>20μm) per unit weight of sediment increases. The second site downstream (G4) is anomalous, because it has a lower K/Q_z ratio and higher coarse particle count than its position in the catchment suggests. This section is unlike all the other measured sections in that it has recently experienced considerable bed aggradation. Whereas all the other sections have pebble/cobble beds, the streambed at site G4 is composed of sand and fine gravel delivered from an upstream construction site and depot. For this reason it was omitted from the cross-section shape analysis. However, the supply of sand is entirely consistent with its low K/Q_z ratio and high particle count. The inverse relationship between these two suspended sediment characteristics is confirmed by the very high correlation evident in figure 6A.

These results confirm that kaolin dominates the finer fraction of suspended sediment, and that quartz and mica predominate in the coarser fraction. The enhancement of K/Q_z and K/M ratios in the diffractograms for the <2μm suspended sediment fraction (table 3) emphasizes this point. It would seem, therefore, that analysis of the changing mineralogy of suspended sediment of the

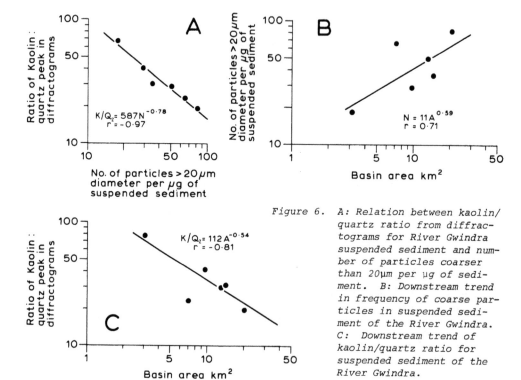

Figure 6. A: Relation between kaolin/ quartz ratio from diffractograms for River Gwindra suspended sediment and number of particles coarser than 20μm per μg of sediment. B: Downstream trend in frequency of coarse particles in suspended sediment of the River Gwindra. C: Downstream trend of kaolin/quartz ratio for suspended sediment of the River Gwindra.

River Gwindra from its headwaters demonstrates a mechanism by which channel adjustment can occur. The suspended sediment generally coarsens downstream as the relative importance of the kaolin component declines and the fine fraction is selectively removed. That this is the result of flocculation of the kaolin clay mineral is suggested strongly by the slimy white deposit on the bed material of the stream. Supporting this conclusion is the analysis by Edzwald and O'Melia (1975) of estuarine clays. As a result of laboratory flocculation tests and analysis of clay mineral distribution in estuarine sediments, they concluded that kaolin is less stable in suspension than illite or montmorillonite and flocculates out of suspension to become the dominant clay mineral at the head of an estuary. In the River Gwindra, although rapid and unpredictable fluctuations in water quality can result from spillages of acids and alkalis used in the bleaching process associated with china clay production, the general downstream trend parallels that for West Country rivers in general. Conductivity (solute load) and pH increase, particularly as headwater runoff becomes less important quantitatively in comparison with throughflow from agricultural areas, and as granite bedrock is replaced by metamorphic and sedimentary rocks. For example, along the River Fowey, an unpolluted stream draining the Bodmin Moor granite close to the study area, Ca concentration increases from 3 to 9 mg/l and Mg concentration increases from 1 to 3 mg/l over a down-stream distance of 33 km. The resulting higher concentrations of bivalent Ca^{++} and Mg^{++} cations permit increased cation bridging between clay particles, and the floccules are deposited. This mechanism provides a reasonable explanation of sedimentation of fines whenever streams are subjected to increased suspended sediment load dominated by clay minerals.

BANK MATERIAL CHARACTERISTICS

Samples of approximately 150 g of bank material, collected at each cross-section (fig. 1), were analyzed in order to assess the differences between polluted and unpolluted sites and to relate bank material characteristics to channel shape (width-depth ratio). About 1/3 of each sample was sieved down to 63μm, and the silt and clay fractions suspended in sedimentation tubes. Withdrawals were made at 10 cm after 4 min 48 sec (<20μm) and 8 hr (<2μm). For each sample, median diameter (D_{50}), percentage of silt and clay (SC), and percentage of clay (Cl) were estimated. Another third of each sample was used to obtain the plastic (PL) and liquid (LL) limits and the plasticity index (PI) for the fraction passing a BS 36 (420μm) sieve (British Standards Institute, 1968). Finally, the remaining sediment was treated to prepare air-dried slides for X-ray diffraction. This involved the removal of organic material from clay obtained by sedimentation after ultrasonic dispersal. The clay (<2μm) was suspended in 50 ml N NaOH/HOAc buffer (pH 5.0), and 10 ml of 6 % H_2O_2 was added when the water-bath-heated suspension reached 70ºC. This method of organic removal prevents degradation of the clay minerals (Douglas and Fiessinger, 1971). X-ray diffraction analysis proceeded as above, except that the range employed was 4×10^3 cps.

A number of indices describing the bank material characteristics, and averages for polluted and unpolluted stream sites, are listed in table 4. Although the data in this table indicate that the banks of polluted streams are composed of finer sediment, with enhancement of the kaolin content with respect to quartz and muscovite-illite, the Atterberg limits are very similar. The plasticity index averages 4 - 5 %, with plastic and liquid limits being slightly higher for unpolluted bank material. The highest plasticity index is at site F2, where the bank material has the lowest kaolin/quartz ratio and high chlorite content. These values are a result of the location of this site, well off the granite outcrop in an area underlain by slates and shales.

The nine indices inevitably repeat information, and a principle components analysis (PCA) reflects this. Three principal components have eigenvalues greater than 1.0, and these account for 87% of the variance of the initial 22-sites by 8-indices matrix. The intercorrelation of the Atterberg indices is illustrated in figure 7, which is a Casagrande Plasticity Chart. The plotting positions of the data indicate that most of the samples can be classed as inorganic silts of medium compressibility (Casagrande, 1947). The principal component loadings matrix (table 5) indicates that the first component measures the "plasticity" of the sediments, and the second measures the "grain size". The clay mineralogy ratios, of which only two were used because adding the third closes the number system, load on the third component, but also on the first "plasticity" component. The inverse loading of the Atterberg limits and mineralogy ratios on this first component attest to the complex relation of the

321

Table 4

Mean Values of Bank Material Indices for Polluted

and Unpolluted Cross-Sections

Index	Polluted sites	Unpolluted sites
Median diameter (D_{50})	66μm	104μm
Percent silt/clay (SC)	50.1 %	40.5 %
Percent clay (Cl)	17.4 %	13.3 %
Plastic Limit (PL)	31.9 %	34.5 %
Liquid Limit (LL)	36.5 %	39.2 %
Plasticity Index (PI)	4.6 %	4.7 %
Kaolin/quartz ratio (K/Q_z)	213.5	48.2
Kaolin/mica ratio (K/M	13.8	6.1
Mica/quartz ratio (M/Q_z)	16.2	8.3

latter to the former. Intuitively, a high kaolin/quartz ratio should give high plasticity, but this is complicated if other minerals also vary in importance. For example, Dumbleton and West (1966) have shown that addition of muscovite to kaolin tends to reduce the Plasticity Index while maintaining the Liquid Limit, whereas quartz silt and sand added to kaolin reduces both. In addition to this, the diffractogram peaks at 10.05, 5.01, and 3.32 Å may include muscovite and illite in varying proportions, and each has a rather different influence on plasticity. However, it is evident that the principal components analysis identifies three major sources of information in the data matrix, relating to "plasticity", "grain size", and "mineralogy". It is necessary to determine whether these relate to channel characteristics.

The width-depth values plot around the general trend of Schumm's (1960) relation of F to the weighted mean silt-clay index (fig. 8). The silt-clay measure in this study refers only to bank samples.) The correlation of F with SC in the present sample is, however, non-significant, which is a not uncommon finding in studies of small upland British streams. (There is no significant relation in the upper River Fowey on nearby Bodmin Moor, described by Richards, 1976). The data from polluted and unpolluted sections are separated by Schumm's regression equation (fig. 8). This suggests that, although the higher silt-clay percentage in polluted streams is associated with lower width-depth ratios, this effect is augmented by that of another independent variable. Because the principal components analysis suggested a "plasticity" component, the plasticity index (PI) is a suitable measure of this, and produces the best multiple regression predictor of F in association with SC this relation is:

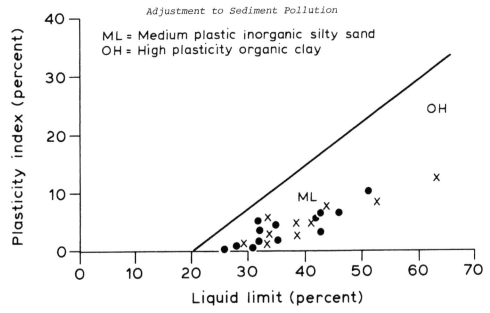

Figure 7. *Casagrande Plasticity Chart for the bank material samples from polluted (dots) and unpolluted (crosses) streams.*

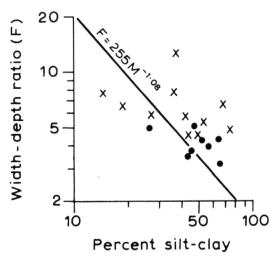

Figure 8. *Plot of channel width-depth ratio against percent silt and clay in channel banks for polluted (dots) and unpolluted (crosses) streams. Schumm's (1960) relation shown for comparison.*

323

Table 5

Principal Component Loadings Matrix and Cumulative Percentage

of Variance Accounted for by the First Three Components

Variable	Component 1	Component 2	Component 3
Median diameter	-0.128	-0.733	-0.539
Percent silt clay	-0.082	0.941	0.159
Percent clay	-0.363	0.810	0.075
Plastic Limit	-0.885	0.001	-0.311
Liquid Limit	-0.934	0.105	-0.298
Plasticity Index	-0.732	0.235	-0.228
Kaolin/quartz ratio	0.586	0.449	-0.628
Kaolin/mica ratio	0.674	0.424	-0.564
Cumulative percent variance	39.3	70.9	86.8

$$F = 12.3 \, SC^{-0.20} \, PI^{-0.11} \tag{1}$$

However, this equation explains only 20 % of the variance of F, although it seems physically reasonable, with streams being narrower and deeper for higher plasticity indices. A high plasticity index is associated with high liquid limit (fig. 7), and the bank material in question remains stable at higher moisture contents, thereby enabling the channel to remain narrow. However, if plasticity is an important control of channel shape, the clay content of bank material is more critical than the silt and clay content, because only the clay affects plasticity. In addition, the clay mineralogy is important as the PCA results suggest, because of the loadings of mineralogy indices on the "plasticity" component (table 5).

The plasticity index was correlated with other sediment characteristics, and the best empirical prediction equation, which explains 50 % of the variance of the dependent variable, and has partial correlations significant at $p = 0.10$, is

$$PI = 0.36 Cl^{0.80} (K/M)^{-0.95} (K/Q_z)^{0.48} \tag{2}$$

Clearly, therefore, the plasticity index is related to the clay content (Cl) and clay mineralogy. All else being equal, both clay content and kaolin/quartz ratio are directly related to the plasticity index. Although K/Q_z and K/M are strongly positively correlated, both appearing to be inversely loaded to the plasticity indices on the first principal component, the partial regression of K/M on PI is a negative relationship while that of K/Q_z is positive. The explanation of this remains obscure, but may reflect a tendency of the muscovite-illite diffractogram peak to imply more of the latter clay mineral than fine-grained mica. Illite would tend to enhance plasticity, whereas mica would dilute the plastic effect of clay minerals in bank sediment. A comparable prediction model for width-depth ratio, with $R^2 = 0.57$, is

$$F = 23.3(K/Q_z)^{-0.43}(K/M)^{0.41}Cl^{-0.17} \qquad (3)$$

With more clay, and more kaolin relative to quartz in the clay fraction, streams are narrower and deeper. This influence of the sediment character appears to act indirectly through the plasticity, because the same set of independent variables predicts both F and PI, but with opposite signs. The clay content and its mineralogy appear to exert a significant control over bank stability in streams, and therefore influence the equilibrium width-depth ratio. However, more detailed analysis of the clay mineralogy is required before the full physical significance of the correlations of the K/M ratio with both the plasticity index and width-depth ratio can be ascertained.

Bank sediment alone cannot explain channel shape, because as the trends in figure 3B demonstrate, shape adjusts to variations in discharge. Therefore, a complete prediction of shape variation requires a discharge index. If catchment area is used as the index, the partial correlation for Cl becomes non-significant, and the best multiple regression is

$$F = 10.9(K/Q_z)^{-0.77}A^{0.75}(K/M)^{0.5} \qquad (4)$$

with $R^2 = 0.79$, all partials significant at $p = 0.05$, and predictors ranked according to the strength of the partials. For a fixed catchment area, and mean values of K/Q_z and K/M for polluted and unpolluted streams as listed in table 4, the predicted width-depth ratios of polluted streams are slightly less than half those of unpolluted streams.

An alternative model incorporates the reduction in discharge implicit in figure 2B, by adding an estimate of mean annual flood $(\overline{Q})$, based on the assumption that this discharge in polluted catchments is one-third of the value in unpolluted catchments of the same area. This multiple regression also, of course, has $R^2 = 0.79$ and takes the form

$$F = 9.8\,\overline{Q}^{\,0.17}(K/Q_z)^{-0.30}(K/M)^{0.22} \qquad (5)$$

again with predictors in order of the strength of the partials. The polluted streams have width-depth ratios of 0.58 those of unpolluted streams (fig. 3B). A stream in the study area of 10 km^2 catchment area is associated with a mean annual flood of 5.31 m^3/sec if unpolluted, and 1.77 m^3/sec if polluted (fig. 2B). The above model predicts $F = 6.76$ in the former and $F = 4.30$ in the latter, for the average mineralogical ratios of table 4. This implies that, after accounting for their different discharges, a polluted stream has a width-depth ratio of 0.64 that of an unpolluted stream of the same catchment area. This is a reflection of enhanced clay content and of adjustment of the mineralogical composition of the clay fraction, resulting in more stable bank material and narrower, deeper channels in streams polluted by china clay waste. The relationship emphasizes the dominant role played by the suspended sediment compared to that of altered flood discharge regime.

ADDITIONAL ADJUSTMENTS

The main sedimentological change resulting from pollution by kaolin and micaceous waste has occurred in the bank material. However, some evidence exists of slight changes in the bed material, in addition to the superficial slime of flocculated kaolin. This can only be indicated indirectly by comparing polluted and unpolluted sites. Hack (1957) demonstrated that stream gradient varies directly with bed material calibre and inversely with basin area (as a discharge surrogate). From this, it would appear that bed material size should relate to the product of gradient and area, which provides a stream power index. Polluted streams show a tendency to have median pebbles with slightly longer intermediate axis lengths than unpolluted streams of the same area-slope product (fig. 9). Three possible explanations for this may be considered. First, area is a discharge surrogate, and variations in runoff per unit area may invalidate its use as such. If polluted stream catchments experienced greater runoff rates, the stream power for a given catchment size and stream slope would be larger, and the bed material size greater than expected. However, there is no consistent evidence to support the suggestion of increased runoff. In fact, the opposite is the case. A second explanation could be that the increased silt and clay content of bank material has been associated with increased sinuosity (Schumm, 1963), thereby reducing slope and area-slope product without affecting bed material size. The result would be that the median diameter appears to have been increased. There is no evidence of adjustment of sinuosity, which would in any case appear contradictory in association with stabilization of bank material. These streams are of low sinuosity with coarse pebble-cobble bed material, and are unlikely to alter their sinuosity significantly.

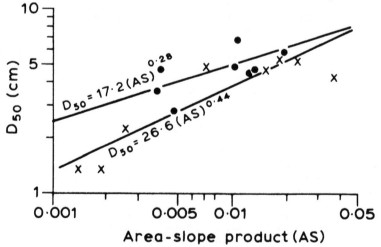

Figure 9. *Relationships between median intermediate axis length and the product of catchment area and stream gradient for polluted (dots) and unpolluted (crosses) streams.*

The third possibility relates to the adjustments of sub-bankfull flow geometry within the narrower cross-section which results from bank stabilization
by clay deposition. These adjustments are illustrated by plotting width, depth,
and velocity of flow as functions of downstream increase in discharge at the
time of suspended sediment sampling (fig. 10). The polluted streams are
associated with narrowed flow widths, as is to be expected given the smaller
channel widths at given catchment sizes. The downstream trend in this low
flow discharge is very strongly correlated with catchment area and is statistically indistinguishable for polluted and unpolluted catchments. At low flow,
the narrower width of flow is compensated by a larger velocity and depth. The
higher flow velocity of polluted streams may reflect the turbulent-damping ef-

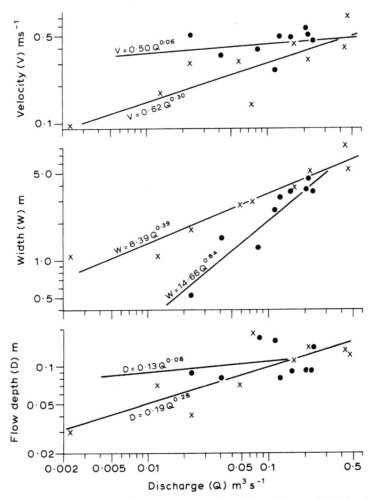

Figure 10. *Downstream hydraulic geometry at low flow (approximately 60 % duration) for polluted (dots) and unpolluted (crosses) streams.*

fects of their higher sediment concentrations. In sand-bed streams this effect is often offset by the development of form roughness (dunes), but it may be dominant in these streams because of the coarse nature of bed material, and the absence of mobile bedforms. Thus, although the slightly larger bed-material size may increase bed roughness, the more obvious differences in sediment concentration, which declines erratically downstream in polluted streams (fig. 4), encourages higher velocities (Vanoni, 1946; Vanoni and Nomicos, 1960). In the longer term, therefore, it is possible that the coarsening of bed-material reflects the influence of a more dense fluid-sediment mixture capable of increased bed shear. Increased bed shear would result partly because of the high suspended sediment concentration (Gerson, 1977). In addition, the normal effect of accelerated flow due to damped turbulence, which results in lower depth and bed shear, is offset by adjustment of flow depth in the narrower section. Progressive removal of the finer bed-material (sand and fine gravel) would increase the median intermediate axis. This possibility is strongly suggested by the highly imbricated beds of the polluted streams, which are extremely difficult to sample because of the interlocking of coarse pebbles and cobbles with little intervening fine matrix.

CONCLUSIONS

River channels in the St. Austell area have made significant adjustments to the heavy loads of fine micaceous and kaolinite silt and clay imposed upon them since the late eighteenth century by a china clay industry. Mines have always relied on water for washing clay from the pits, separating clay from the waste products, and transporting undesirable residues. However, the results of this investigation are of more general interest for four main reasons.

First, even though the study has perforce measured channel form after the channel has adjusted, it has demonstrated a mechanism of selective sedimentation of clay minerals to explain how the adjustment has occurred. Secondly, the results imply that adjustment to the nature and quantity of sediment load presently carried by the stream does occur. This reinforces Schumm's (1971) suggestion, based on very limited data, that bank material reflects sediment load. Sediment load may thus be treated as an independent variable on the recent time scale, and bank material characteristics do not necessarily represent an inherited control of present channel geometry. Thirdly, clay content and clay mineralogy may influence bank material plasticity and the channel cross-section shape. It may be possible to develop satisfactory multiple regression predictions of both these variables, with clay content and mineralogy as predictors. The silt-clay index may summarize the influence of bank materials on channel form, but in a very generalized way. A fruitful avenue for future research could involve a more detailed exploration of the sedimentological properties controlling bank stability, particularly in the context of the link between clay mineralogy and plasticity, which is clearly complex. Finally, streams subjected to human interference provide laboratories for the analysis of process-response behavior. It may be as important to exploit these laboratories to gain new insights as to apply existing knowledge to explain the observed adjustments.

ACKNOWLEDGEMENTS

The assistance of the following people is gratefully acknowledged; Professor R.J. Chorley for his comments on a draft of the manuscript, Dr. S. Ellis and Dr. A.G. Fraser for advice on X-ray diffraction, Mr. N. Sutherland for laboratory assistance, the South West Water Authority for providing gaging data and the data on West Country streams in general, and Miss H. Baslington for typing the text.

REFERENCES

Bain, J.A., 1971, A plasticity chart as an aid to the identification and assessment of industrial clays: Clay Minerals, v. 9, p. 1-17.

Barton, R.M., 1966, A history of the Cornish china-clay industry: Truro, Bradford Barton Ltd., 212 p.

British Standards Institute, 1968, Methods of testing soils for civil engineering purposes: London, British Standards Institute BS-1377, 234 p.

Casagrande, A., 1947, Classification and identification of soils: American Society of Civil Engineers Proceedings, v. 73, p. 783-810.

Cosgrave, M.E., and Sulaiman, A.M.A., 1973, A rapid method for the determination of quartz in sedimentary rocks by X-ray diffraction incorporating mass absorption correction: Clay Minerals, v. 10, p. 51-55.

Douglas, L.A., and Fiessinger, F., 1971, Degradation of clay minerals by H_2O_2 treatments to oxidize organic matter: Clays and Clay Minerals, v. 19, p. 67-68.

Dumbleton, M.J., and West, G., 1966, Some factors affecting the relation between the clay minerals in soils and their plasticity: Clay Minerals, v. 6, p. 179-193.

Edmonds, E.A., McKeown, M.C., and Williams, M., 1969, British regional geology - south west England: London, H.M. Stationery Office, 130 p.

Edzwald, J.K., and O'Melia, C.R., 1975, Clay distributions in recent estuarine sediments: Clays and Clay Minerals, v. 23, p. 39-44.

Flaxman, E.M., 1963, Channel stability in undisturbed cohesive soils: American Society of Civil Engineers Proceedings, Journal of Hydraulics Division, v. 89, p. 87-96.

Gerson, R., 1977, Sediment transport for desert watersheds in erodible materials: Earth Surface Processes, v. 2, p. 343-361.

Gregory, K.J., and Park, C.C., 1976, Stream channel morphology in north west Yorkshire: Review de Geomorphologie Dynamique, v. 25, p. 63-72.

Hack, J.T., 1957, Studies of longitudinal stream profiles in Virginia and Maryland: U.S. Geological Survey Professional Paper 294B, 10 p.

Hammer, T.R., 1972, Stream channel enlargement due to urbanization: Water Resources Research, v. 8, p. 1530-1540.

Hollis, G.E., and Luckett, J.K., 1976, The response of natural river channels to urbanization: two case studies from south east England: Journal of Hydrology, v. 30, p. 351-363.

Keller, W.D., 1976, Scan electron micrographs of kaolins collected from diverse environments of origin - II: Clays and Clay Minerals, v. 24, p. 114-117.

Keller, W.D., and Hanson, R.F., 1975, Dissimilar fabrics by scan electron microscopy of sedimentary versus hydrothermal kaolins in Mexico: Clays and Clay Minerals, v. 23, p. 201-204.

Leopold, L.B., 1973, River channel changes with time - an example: Geological Society of America Bulletin, v. 84, p. 1845-1860.

Leopold, L.B., and Maddock, T., 1953, The hydraulic geometry of stream channels, and some physiographic implications: U.S. Geological Survey Professional Paper 252, 56 p.

McQueen, I.S., 1961, Some factors influencing streambank erodibility: U.S. Geological Survey Professional Paper 424B, p. 28-29.

Noble, F.R., 1971, A study of disorder in kaolinite: Clay Minerals, v. 9, p. 71-81.

Pickup, G., and Warner, R.F., 1976, Effects of hydrologic regime on magnitude and frequency of dominant discharge: Journal of Hydrology, v. 29, p. 51-75.

Rango, A., 1970, Possible effects of precipitation modification on stream channel geometry and sediment yield: Water Resources Research, v. 6, p. 1765-1770.

Richards, K.S., 1976, Channel width and the riffle-pool sequence: Geological Society of America Bulletin, v. 87, p. 883-890.

Sawhney, B.L., and Frink, C.R., 1978, Clay minerals as indicators of sediment source in tidal estuaries of Long Island Sound: Clays and Clay Minerals, v. 26, p. 227-230.

Schultz, L.G., 1964, Quantitative interpretation of mineralogical composition from X-ray and chemical data for the Pierre shale: U.S. Geological Survey Professional Paper 391C, 31 p.

Schumm, S.A., 1960, The shape of alluvial channels in relation to sediment type: U.S. Geological Survey Professional Paper 352B, p. 17-30.

_____ , 1963, Sinuosity of alluvial rivers on the Great Plains: Geological Society of America Bulletin, v. 74, p. 1089-1100.

_____ , 1969, River metamorphosis: American Society of Civil Engineers Proceedings, Journal of the Hydraulics Division, v. 95, p. 255-273.

_____ , 1971, Fluvial geomorphology - the historical perspective; in Shen, H.W., ed. River mechanics: Fort Collins, Colorado, H.W. Shen, v. 1, p. 4.1-4.30.

_____ , 1973, Geomorphic thresholds and complex response of drainage systems; in Morisawa, M., ed., Fluvial geomorphology: Binghamton, State University of New York, Publications in Geomorphology, p. 299-310.

Vanoni, V.A., 1946, Transportation of suspended sediment by water: American Society of Civil Engineers Transactions, v. 111, p. 67-133.

Vanoni, V.A., and Nomicos, G.N., 1960, Resistance properties of sediment-laden streams: American Society of Civil Engineers Transactions, v. 125, p. 1140-1175.

Wolman, M.G., 1967, A cycle of sedimentation and erosion in urban river channels: Geografiska Annaler, v. 49A, p. 385-395.

HYDRAULIC GEOMETRY, STREAM EQUILIBRIUM

AND URBANIZATION

MARIE MORISAWA

ERNEST LAFLURE

Department of Geological Sciences and Environmental Studies
State University of New York
Binghamton, New York

ABSTRACT

Hydraulic geometry variables are not only interrelated to each other but also depend upon climate, geology, soils, land use, and vegetation. These are first-order variables that determine the hydrologic and sedimentologic regimen of the stream and, thus, the morphology. Whenever any of these factors change, effects are felt in the dynamics and morphology of the system. Man's activities may cause significant changes in fluvial systems. Many authors have documented the hydrologic and sedimentologic alterations in streams caused by urbanization.

Studies in urbanizing watersheds near Pittsburgh, PA, and Binghamton, NY, indicate changes in hydraulic geometry brought about by urbanization. The initial effect of increased discharge resulting from development in a basin is proposed to be an increase in velocity of flow. Channel enlargement is a secondary result of the changed hydrologic regimen. There is a tendency for an accelerated rate of increase in channel cross-sectional area with increased basin area of urbanized rivers. An inflection point indicates a threshold where peak velocity is capable of increasing the channel size. Thus, there is a lag between changes in land use and final morphologic adjustment of the channel. The time lag is determined by the relation of velocity increase to the ability of the stream to move bed and/or bank material.

Readjustment of a stream to changes in its regimen are governed by natural laws of dynamics and morphologic response.

INTRODUCTION

The morphology and hydraulic geometry characteristics of a stream are not only interrelated to each other, but are also dependent upon variables such as climate, lithology, topography, soil, land use, and vegetation (fig. 1). These first-order independent elements determine the discharge and sediment load, which are second-order factors. All of these variables and subvariables interact through a complicated feedback mechanism that determines the morphologic configuration of a river and the processes and rates of their operation in any individual watershed (fig. 2). Because the hydrology and other environmental conditions may vary in time and space over a river basin, the whole river system is dynamic, changing both temporally and spatially.

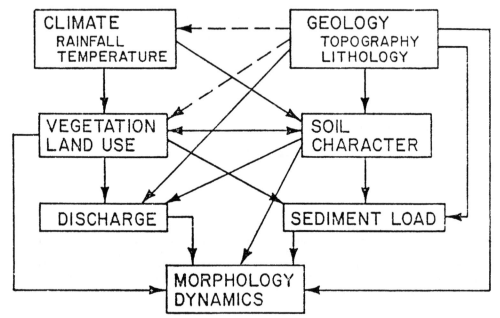

Figure 1. Factors influencing stream morphology and dynamics (from Morisawa and Vemuri, 1976).

Over a long period of time, each river seeks to achieve a natural adjustment of its hydraulics and morphology to the prevailing discharge and sediment load as determined by the general environment of its basin. This principle of adjustment means the river reaches an equilibrium between energy and load under the specific environmental conditions of the watershed. Whenever any of the conditions change, there is a response in the system that affects the morphology and dynamics.

Man's activities cause important disturbances in watersheds. In changing land use during urbanization, most of the first order, independent variables of figure 1 are altered. In turn, the second-order factors of hydrology and load are changed. Numerous studies have shown the effects of urbanization on the hydrology of watersheds. Specifically, land use changes and the disruption of the environment result in larger and more frequent floods (Carter, 1960; Wilson, 1967; Leopold, 1968; and Anderson, 1970), more total surface runoff (Harris and Rantz, 1964; Seaburn, 1969; and Miller and Viessman, 1972), and decreased time lag in runoff response (Brater and Suresh, 1969). The effects on sediment load have been studied by Guy (1970), Wolman (1967) and Knott (1973). A corresponding adjustment in the morphology of the stream channel is to be expected. Hammer (1972), Leopold (1973), Robinson (1976), and LaFlure (1978) are among those who have documented changes in stream channel morphology that resulted from urbanization.

Several basic principles must be considered in analyzing morphologic response in river systems. First, the system is integrated so that an environ-

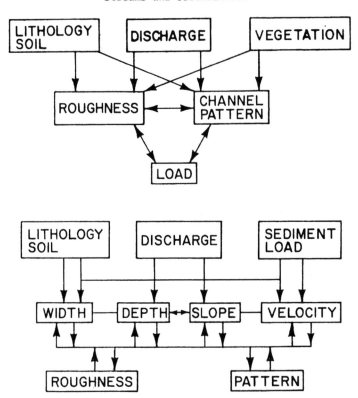

Figure 2. Factors affecting channel morphology (from Morisawa and Vermuri, 1976).

mental change in one part of a watershed is felt elsewhere in the system. For example, Smith (1973) gave examples of both upstream and downstream controls on aggradation of two rivers in Canada. Second, the system in balance is self-regulatory, so that a change necessitates response. The most important concept illustrated by figures 1 and 2 is that "everything affects everything else." And because response is generally negative (unless change continues), a new equilibrium will be reached in which a changed morphology is the outcome of the new conditions. Third, the response to change of the hydrologic and sedimentologic regimen may not necessarily be immediate (Allen, 1974). Schumm (1973) has shown that adjustment may be delayed until a threshold value is reached.

This paper presents examples of changes in hydraulic geometry and stream morphology caused by urbanization in watersheds near Pittsburgh, PA, and Binghamton, NY.

M. Morisawa and E. LaFlure

THE STUDY WATERSHEDS

Three areas--two near Pittsburgh, PA, and one in the Binghamton, NY region--are under study. The first area consists of four watersheds (Squaw Run, Guyasuta Run, Seitz Run, and Powers Run) in the townships of O'Hara and Fox Chapel near Pittsburgh (fig. 3). The second area consists of three basins (Mosside, Dirty Camp, and Abers Creeks) in the town of Monroeville, east of Pittsburgh (fig. 3). Two small basins (Brixius Creek and Fuller Hollow) comprise the New York study area (fig. 4).

Both Pennsylvania areas are on the unglaciated Allegheny Plateau. The areas are underlain by essentially flat-lying Pennsylvanian cyclothem sequences of sandstone, shale, limestone, claystone, and coal units. The broad flat summits of the plateau are dissected by streams that have cut valleys 60-120 m deep. Valley side slopes are generally steep. Although smaller valleys are narrow with limited flat areas, larger tributaries (i.e., Squaw Run and Abers Creek) have wide floodplains.

Climate is humid, with a mean annual precipitation of 1,016 mm and an average annual runoff of 406 mm (Geraghty et al., 1973). The highest discharge generally occurs in March, the lowest in September. Forests are mixed hardwoods, usually covering the valley side slopes. Some farms and open fields are located on the plateau surface. Urbanization has occurred on both the lower floodplains and the broad summits.

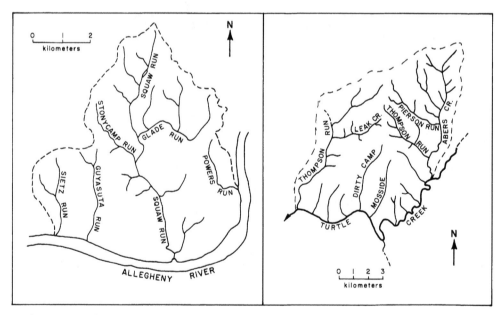

Figure 3. Pittsburgh area watersheds.

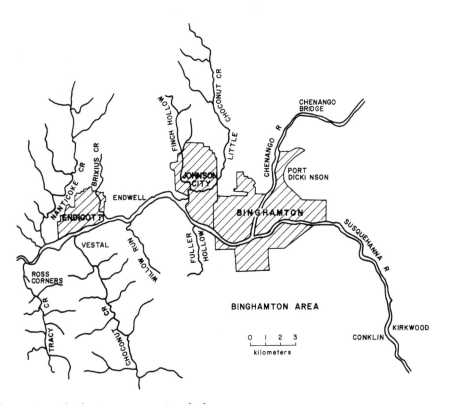

Figure 4. Binghamton area watersheds.

Two nearby unurbanized watersheds were chosen for control purposes. Rawlins and Shafer Runs are just north of the Fox Chapel area. These watersheds are natural (2 % urbanized) and served as a basis for comparison of channel enlargement.

The two stream basins under study in the Binghamton area are highly urbanized. Brixius Creek is on the north side of the Susquehanna River in the town of Endicott. Fuller Hollow is located on the south side of the river in the town of Vestal. Both basins are within the glaciated Allegheny Plateau. The bedrock is essentially flat-lying sandstone, siltstone, and shale of Devonian age.

The region has been glaciated, resulting in a topography characterized by smoothed and rounded hills. Relief is of the order of 90-210 m. The major valleys were broadened and deepened by glaciation, and many small tributaries flow in valleys rather large for the present-sized streams. The upper headwaters of both streams drain the till-covered open rolling summits. Middle reaches narrow as the streams cut through bedrock. Lower stretches are wider and terraced as the streams flow over glacio-fluvial deposits.

The Binghamton climate is humid, with 890 mm of rainfall yearly and an annual runoff of 510 mm (Geraghty et al., 1973). High discharges occur in April and low flow in September. Vegetation is mixed hardwood forest with stands of evergreens. Farms and open fields occupy part of the flat divide surface. There are numerous glacial or man-made ponds. Urbanization, which started in the lower watershed, has expanded upward on to the hill-slopes and broad divides.

Table 1 gives some of the physical characteristics of the study basins.

Table 1

Characteristics of Basins Studied

Watershed	Drainage Area km^2	Max. Relief m	Gradient m/m	% Urban [a]
Pittsburgh, PA area				
Area 1 [b]				
Squaw Run	21.9	135	.015	48
Seitz Run	3.73	105	.038	75
Guyasuta Run	3.21	115	.034	64
Powers Run	2.32	90	.036	91
Area 2				
Mosside Creek	3.47	105	.032	37
Dirty Camp Creek	5.57	115	.023	29
Abers Creek	27.5	120	.020	26
Control Basins				
Rawlins Run	4.01	90	.025	2
Shafers Run	4.01	90	.025	2
Binghamton, NY area				
Area 3				
Brixius Creek	7.93	115	.023	80
Fuller Hollow	9.89	215	.038	25

(a) Percent of area greater than 5 % impermeable

(b) Data for Squaw Run, Guyasuta Run, Seitz Run, and Powers Run from LaFlure (1978) and Nelson (1979).

338

All streams have gravelly pebble beds with some silt and cobbles. Generally the stream banks are of stratified gravel, sand, and silt. Upstream channel banks in the Pittsburgh area watersheds are composed of weathered rock and soil. Binghamton watersheds on till have channel banks of mixed clay, silt, and pebbles or cobbles. All streams studied have stretches where bed and/or banks are cut in bedrock.

The O'Hara-Fox Chapel watersheds (study area one) have been studied over two summers (1977 and 1978) (LaFlure, 1978 and Nelson, 1979). Hydrologic data were obtained from recording gages installed on Squaw and Powers Runs. Measurements of instantaneous discharge and bankfull morphology were made periodically on all O'Hara-Fox Chapel streams during both summers. Bankfull morphology of Rawlins and Shafers Runs was measured by Nelson (1979). Measurements of the Monroeville streams (study area two) were made by Lorraine O'Day and M. Morisawa in the summer of 1978. The Binghamton streams (area three) have been monitored periodically since 1972-73 by Morisawa and her students.

CHANGES IN HYDRAULIC GEOMETRY AND CHANNEL SIZE

Documentation of changes in the hydrologic regimen of the watersheds under study has been done elsewhere. LaFlure (1978) collected hydrologic data which indicated that urbanization of the river basins in O'Hara-Fox Chapel boroughs has resulted in higher peak flows and increased frequency of overbank flooding. Morisawa and Vemuri (1976) showed that urban development had greatly increased direct storm runoff from Brixius Creek watershed and Fuller Hollow.

The following is a discussion of the hydraulic geometry and morphologic data, comparing adjustments made by the streams of the three areas in response to the augmented discharge.

Downstream Hydraulic Geometry

In terms of bankfull hydraulic geometry (table 2) each stream responds to the downstream increase in discharge in a distinctive way. In the Pittsburgh area, Guyasuta Run, Squaw Run, Mosside Creek, and Abers Creek accommodate increasing discharge downstream primarily by adjusting the flow velocity, as indicated by the relatively large m-exponent. Seitz and Powers Runs increase in depth (exponent f) downstream more acutely than in either width (b) or velocity (m). Dirty Camp increases its width at a slightly greater rate than its depth in a downstream direction. Powers Run was the earliest developed of these watersheds, has the greatest percentage of its surface greater than 5 % impermeable, and is the smallest. The hydraulic geometry exponents show its channel has enlarged so that velocity change downstream is small in comparison with the other streams.

In the Binghamton area, a number of years of record show how Brixius Creek and Fuller Hollow have changed their modes of adjustment over time. The lower basin of Fuller Hollow was developed in the mid-1960's, and concomitantly, the lower reaches were rerouted and channelized in part. A subdivision was started in the headwaters about 1970. As of 1973 the stream

Table 2

Downstream Hydraulic Geometry

Watershed	Hydraulic Exponents [a]		
	b	f	m [b]
Squaw Run [c]	0.17	0.34	0.49
Guyasuta Run [c]	0.04	0.49	0.47
Seitz Run [c]	0.29	0.45	0.26
Powers Run [c]	0.37	0.55	0.08
Mosside Creek	0.33	0.23	0.44
Dirty Camp Creek	0.44	0.38	0.18
Abers Creek	0.38	0.03	0.59
Fuller Hollow			
1973	0.05	0.32	0.63
1975	0.20	0.30	0.50
1977	0.21	0.25	0.54
1978	0.54	0.05	0.41
1979	0.53	0.06	0.41
Brixius Creek			
1973	0.30	0.56	0.14
1975	0.03	0.65	0.32
1979	0.05	0.49	0.46

(a) $W \propto Q^b$, $D \propto Q^f$ and $V \propto Q^m$.

(b) $m = 1 - (b+f)$

(c) data from LaFlure, 1978

showed a very high rate of increase in velocity downstream. Both f and m values have decreased from 1973 to 1979, as urbanization has progressed. On the other hand, b has increased over this time. All the exponents were fairly stable during 1978-79. The present mode of adjustment is by a high rate of increase in width downstream, a somewhat lower (but still high) rate of increase in velocity downstream, and very little change in depth.

The lower part of Brixius Creek watershed is highly impermeable, being covered by industrial and commercial enterprises and high-density housing. After 1973 development slowly extended along the western edge of the basin on the upper slopes and divide. As of 1973 adjustment in a downstream direction was made primarily by a high increase in the depth exponent. With the subsequent development of the middle and upper watershed area, the m exponent has increased, b decreased greatly, while f increased and then decreased (but still remains high).

The downstream hydraulic geometry effects are highly complex, as the manner of adjustment may vary from station to station. This is partly because of individual station environment and partly because the effects of increased runoff may be felt at one station and not at another, depending on where in the watershed urbanization is taking place. At-a-station hydraulic geometry can be used to show adjustments at particular cross sections.

At-A-Station Hydraulic Geometry

Although there are definite downstream trends in the hydraulic geometry (fig. 5), adjustment varies from one transect to another on these small streams because station response to change in discharge is a result of the complex interaction of hydrology, hydraulics, and the comparative resistance of bed and bank material. In the Pittsburgh area this is exemplified by the 1977 Squaw Run at-a-station hydraulic geometry (table 3). Three stations (10, 4, and 6) adjust to increasing discharge primarily by a relatively high rate of velocity increase. Basins above these stations have undergone recent development. Four stations (1, 3, 5, and 7) divide adjustment to increasing discharge approximately equally among the channel size parameters (width and depth) and velocity, i.e., two-thirds of the change is in channel size and one-third in velocity. Station 2, below an impoundment, has a high f-value; station 8, at a river bend, shows major adjustments in width and depth. Station 9, on the floodplain above the mouth, has an extremely high rate of change in width with increasing flow, a high change in depth and a decreasing rate of flow at high discharges with backwater effects.

Figure 6 shows the difference in channel behavior before and after a storm at transect 8 on Squaw Run. The hydraulic geometry trends are quite different after the storm of July 7, 1977, than before it. Although the storm itself was a minor rainfall event, the stream responded to the storm runoff by eroding its bed and deepening its channel (fig. 7). With the new channel configuration the hydraulic geometry variables changed.

That this event is common as the stream tries to adjust during watershed development is further substantiated by data from Fuller Hollow (table 4). This transect is located just below the College Park subdivision in the headwaters, where development began in 1970. The hydraulic geometry variables indicate a high m-value in 1972. The rate of increase of velocity with discharge (m) decreased from 1972 to 1975, and the stream was tending towards a more equable distribution of effect among the hydraulic variables by 1975. Then another spurt of

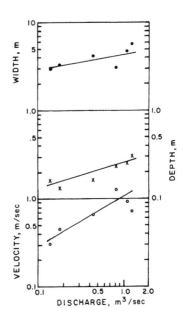

Figure 5. *Downstream hydraulic geometry, Fuller Hollow, 1975.*

Table 3

At-A-Station Hydraulic Geometry[a], Squaw Run, 1977

| Station | Hydraulic Exponents | | | Comments |
	b	f	m [b]	
1	.16	.49	.35	Below a tributary, older low density housing
2	.25	.63	.12	Below a small reservoir
10	.11	.23	.66	Below a tributary, recent housing, floodplain narrow
3	.11	.49	.40	Below a major tributary, recent urbanization
4	.06	.42	.52	Below a major tributary, floodplain narrows
5	.20	.43	.37	Small tributary, old development
6	.23	.26	.51	Floodplain, recent development
7	.32	.35	.33	Floodplain
8	.45	.45	.10	Floodplain, meander bend
9	.84	.41	-.25	Above mouth

(a) data from LaFlure (1978)

(b) $m = 1 - (b+f)$

development in 1976-77 resulted in a rise in the rate of velocity increase. Concomitantly, there was an increase in b and a drastic decrease in f.

These two detailed at-a-station examples suggest morphologic adjustments are still in a state of flux. The at-a-station effect from increased discharge with urbanization apparently begins with an increase in velocity. This seems a viable proposition, especially if the bed and banks are highly resistant. Augmented velocity raises the flow competency at the station and this, in turn, results in an increase in channel size (fig. 8). Enlargement can take place as deepening or widening or both, depending on local bed and bank conditions. Enlargement causes a decrease in velocity so that in time the stream achieves a new equilibrium at the transect, whereby velocity is just that required to maintain a stable channel configuration. However, under continuous change (i.e., continuous development and augmentation of discharge) the stream response continues to change, and this is reflected in changing hydraulic geometry variables.

Channel Enlargement

    An effort was made to determine
the magnitude of the changes in channel morphology in these watersheds.
Because only a limited field season
was available for the Pittsburgh watersheds, better accuracy was attempted
by using basin area to the transect in
place of a calculated bankfull discharge. The regression exponent,
e, in the power relation between bankfull channel cross-sectional area (the
dependent variable) and basin area to
the transect (independent variable),
was calculated for all watersheds (table
5). In addition, the channel enlargement ratio, R_e, the ratio of actual increase in cross-section size to the enlargement under natural conditions,
was calculated. Because all the watersheds under study were already urbanized, two nearby streams, Rawlins and
Shafers Runs were used as controls in
the Pittsburgh area. These two watersheds are contiguous to the Fox Chapel –
O'Hara basins and similar to them en-
vironmentally, except that they are un-
developed. The 1975 cross-sections of
both streams were used as controls for
Fuller Hollow and Brixius Creek basins. Although the 1975 cross-sections
may have been enlarged by urbanization, they can still serve as a basis
for comparison of further enlargement.

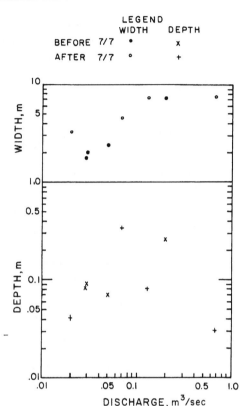

Figure 6. *Hydraulic geometry of
transect 8, Squaw Run, before and after the storm
of July 7, 1977 (data from
LaFlure, 1978).*

Table 4

At-A-Station Hydraulic Geometry, Fuller Hollow, Binghamton Area

Year	b	f	m[(a)]
1972	0.04	0.14	0.82
1973	0.22	0.41	0.37
1975	0.24	0.42	0.34
1977	0.38	0.05	0.57

(a) $m = 1 - (b+f)$

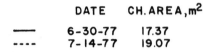

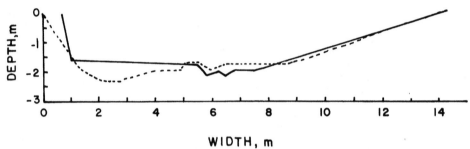

WIDTH, m

Figure 7.   Cross-sections at transect 8, Squaw Run, before and after the storm of July 7, 1977 (after LaFlure, 1978).

Figures 9 and 10 indicate the relationship of bankfull cross-section area to drainage basin area at the transect for all the watersheds studied. All regressions are significant at the 0.05 level except that for Guyasuta Run where scatter is quite large. Five of the nine watersheds show an increasing rate of channel enlargement in the lower part of the streams.

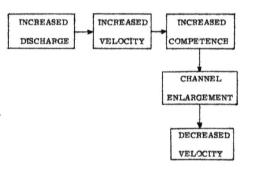

Figure 8.   Proposed response of a stream to increased discharge.

Measurements on the lower reaches of Brixius Creek could not be made as it is piped underground and only reappears occasionally, in a channelized course. The upper quarter of Fuller Hollow basin is also piped underground beneath College Park. Both of these streams increased substantially in channel size between 1975 and 1979 (fig. 9). Both had a higher rate of enlargement in a downstream direction in 1979 than in 1975. Relations for the upper part of Brixius Creek, shown on figure 9, are linear; relations for the mid and lower stretches of Fuller Hollow indicate an increasing rate of enlargement downstream.

For a given drainage area, e.g. 1.3 km^2, the channel cross-section area of Powers Run is larger than that of Squaw Run (fig. 10). Powers Run is much more highly urbanized than Squaw Run. Both, however, have an increasing rate of channel erosion downstream. The control streams, Rawlins and Shafers Runs increase in channel size linearly with increased basin area. The slope of the regression is 0.52. Both Guyasuta and Seitz Runs increase channel size linearly downstream, but at a higher rate than do Rawlins and Shafers Runs.

Table 5

Channel Area Changes with Basin Area and Urbanization

Watershed	e	R_e [(a)]	Percent of area > 5 % impermeable
Squaw Run [(b)]	0.88	1.3	48
Powers Run [(b)]	0.48	6.0	91
Seitz Run [(b)]	0.87	3.4	75
Guyasuta Run [(b)]	0.67	1.95	64
Rawlins-Shafer Runs [(b)]	0.52	1.0	2
Mosside Creek	1.59	1.31	37
Dirty Camp Creek	1.50	1.13	29
Abers Creek	0.58	1.14	26
Fuller Hollow	1.56	1.3	25
Brixius Creek	0.86	3.30	80

(a) Reduced to a basin of 2.2 km^2

(b) Data from Nelson (1979)

Both Mosside and Abers Creeks have an increasing rate of enlargement downstream. Mosside Creek may be somewhat aberrant because a large part of the middle stretches was altered during road building. The relationship in Dirty Camp Creek, although delineated by a straight line, represents only the upper stretches of the watershed lying in the town of Monroeville. Overall, these graphs indicate an increased rate of channel enlargement as compared to the natural streams, or in the case of Brixius and Fuller Hollow, to enlargement during some previous year.

Channel enlargement ratio plotted against the percent of the watershed that is greater than 5 % impervious indicates a delay in response to development and increased runoff until about 25 % of the area of the watershed has more than 5 % impermeability (fig. 11). Then the rate of downstream channel enlargement gradually increases until approximately 30-40 % of the watershed is greater than 5 % impermeable. After this level of urbanization is achieved the rate of increase in channel size is accelerated. There seems to be a lag in morphologic response of the stream to development until a threshold-value of urbanization is reached. After the threshold is crossed, channel enlargement increases greatly. The threshold of channel enlargement depends upon the percent of the basin that is greater than 5 % impermeable. Nelson (1979) found that 5 % impermeable area best represents both imperviousness and sewering. The lag in morphologic response may be due to the effects proposed previously, i.e., that the initial impact of increased runoff from urbani-

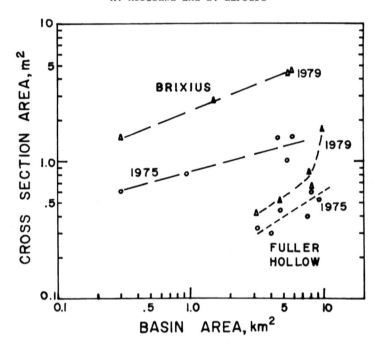

Figure 9. Channel size (cross-sectional flow area) as a function of upstream drainage area, Brixius Creek and Fuller Hollow, NY.

zation is increased velocity. When velocity is increased enough for the stream power to overcome resistance of bed and/or banks, enlargement takes place.

SUMMARY

Studies in urbanizing watersheds near Pittsburgh, PA, and Binghamton, NY, illustrate the changes in hydraulic variables caused by increased impermeability of basin surface and altered hydrologic regimen. The primary effect of urbanization is increased runoff. The initial hydraulic result appears to be an increase in velocity of flow. This causes increased flooding on the one hand (Leopold, 1968) and channel enlargement on the other (fig. 12). Channel enlargement should result in decreased frequency of over-bank flow and decreased velocity of flow in the channel. This response continues until a new equilibrium is reached, as evidenced by stable channel size and velocity. If change continues in the watershed, the state of equilibrium cannot be attained until urbanization ceases.

There is a tendency for an accelerated increase in channel size with increase of basin area in urban watersheds. The inflection point, at which the rate of enlargement starts to increase, indicates a threshold where flow velocity associated with increased flow magnitude is capable of changing the channel

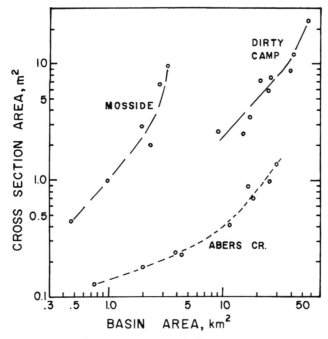

(A) Area 2 watersheds

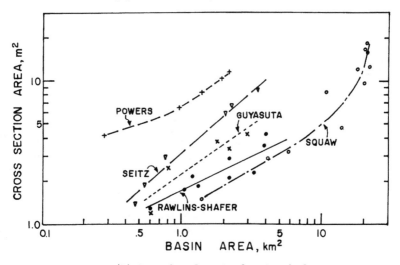

(B) Area 1 and control watersheds.

Figure 10.   Channel size (cross-sectional flow area)
as a function of upstream drainage area,
Pittsburgh area streams;

347

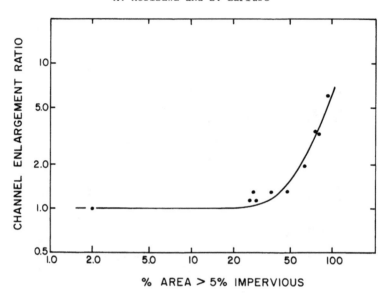

Figure 11. *Channel enlargement ratio as a function of percentage of area greater than 5 % impervious.*

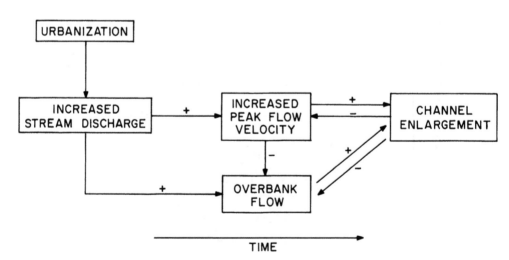

Figure 12. *Effects of urbanization on channel enlargement (after LaFlure, 1978).*

size. There is, thus, a time lag between changed hydrologic regimen and channel enlargement. This time lag is determined by the relation of the velocity increase to the power required to remove bed or bank material.

Readjustments of a stream to alterations in its regimen by man are governed by natural laws of dynamics and morphologic response. These laws and their applications can only be completely understood by further research.

ACKNOWLEDGEMENTS

The writers wish to acknowledge the help in data collection of Bruce Nelson and Lorraine O'Day and in computer analysis by Robert Gillespie. We thank Donald Coates and Gar Williams for critically reading the manuscript and offering valuable comments. We also owe our appreciation to the townships of O'Hara, Fox Chapel, and Monroeville for financial support for these studies.

REFERENCES

Allen, J.R.L., 1974, Reaction, relaxation and lag in natural sedimentary systems: general principles, examples and lessons: Earth Science Reviews, v. 10, p. 263-342.

Anderson, D.G., 1970, Effects of urban development on floods in northern Virginia: U.S. Geological Survey Water Supply Paper 2001C, 22 p.

Brater, W., and Suresh, S., 1969, Effects of urbanization on peak flow; in Moore, W.L., and Morgan, C.W., eds., Effects of watershed changes on stream flow: Austin, University of Texas, Water Resources Symposium 2, p. 201-213.

Carter, R.W., 1960, Magnitude and frequency of floods in suburban areas: U.S. Geological Survey Professional Paper 424B, p. 9-11.

Geraghty, J.J., Miller, D.W., Van Der Leeden, F., and Troise, F.L., 1973, Water atlas of the United States: Port Washington, N.Y., Water Information Center, Inc., p. 2, 21, 24-25.

Guy, H.P., 1970, Sediment problems in urban areas: U.S. Geological Survey Circular 601E, 8 p.

Hammer, T.R., 1972, Stream channel enlargement due to urbanization: Water Resources Research, v. 8, p. 1530-1540.

Harris, E.E., and Rantz, S.E., 1964, Effect of urban growth on streamflow regimen of Permanente Creek, Santa Clara County, California: U.S. Geological Survey Water Supply Paper 1591B, 18 p.

Knott, J.M., 1973, Effects of urbanization on sedimentation and floodflows in Colma Creek basin, California: U.S. Geological Survey Open File Report, 54 p.

LaFlure, E., 1978, The hydrologic and morphologic response to urbanization of four small watersheds in the Pittsburgh, PA, region [M.A. thesis]: Binghamton, State University of New York, 110 p.

Leopold, L.B., 1968, Hydrology for urban land planning: U.S. Geological Survey Circular 554, 18 p.

_____ , 1973, River channel change with time: Geological Society of America Bulletin, v. 84, p. 1845-1860.

Miller, C.R., and Viessman, W., 1972, Runoff volumes from small urban watersheds: Water Resource Research, v. 8, p. 429-434.

Morisawa, M.E., and Vemuri, R., 1976, Multi-objective planning and environmental evaluation of water resource systems: Final Report, Project C-6065, U.S. Department of the Interior, OWRT, p. 1-1 to 1-99.

Nelson, B, 1979 (in preparation), The morphologic and sedimentologic response of four small watersheds to urbanization, Pittsburgh, PA [M.A. thesis]: Binghamton, State University of New York.

Robinson, A.M., 1976, The effects of urbanization on stream channel morphology; in National symposium on urban hydrology, hydraulics and sediment control University of Kentucky, Lexington, p. 115-127.

Schumm, S.A., 1973, Geomorphic thresholds and complex response of drainage systems; in Morisawa, M., ed., Fluvial geomorphology symposium: Binghamton, State University of New York, Publications in Geomorphology, p. 299-310.

Seaburn, C.E., 1969, Effects of urban development on direct runoff to East Meadow Brook, Nassau County, Long Island, New York: U.S. Geological Survey Professional Paper 627B, 14 p.

Smith, D., 1973, Aggradation of the Alexandra-North Saskatchewan River, Banff Park, Alberta; in Morisawa, M., ed., Fluvial geomorphology: Binghamton, State University of New York, Publications in Geomorphology, p. 201-219.

Wilson, K.V., 1967, A preliminary study of the effect of urbanization on floods in Jackson, Mississippi: U.S. Geological Survey Professional Paper 575D, p. 259-261.

Wolman, M.G., 1967, A cycle of sedimentation and erosion in urban river channels: Geografiska Annaler, v. 49A, p. 385-395.

SOME CANADIAN EXAMPLES OF THE RESPONSE

OF RIVERS TO MAN-MADE CHANGES

DALE I. BRAY

Department of Civil Engineering
University of New Brunswick
Fredericton, New Brunswick, Canada

ROLF KELLERHALS

Kellerhals Engineering Services
Heriot Bay, British Columbia, Canada

ABSTRACT

Hydraulic engineering projects are capable of changing the hydrological and sediment regimes of a river system by creating artificial storages, by diverting flows from one basin to another, by changing the sediment input into a river, etc. Because many of the responses of a river system to these changes cannot be predicted from theory alone, it is important to document case histories which will aid in the solution of problems as they arise in the future.

This paper documents a few Canadian cases of known changes in fluvial systems which can be attributed to the works of man. These cases include: the downstream effects of a large storage reservoir on a major river; the effects of major interbasin diversions along the diversion routes; the effects of channel straightening; the effects of the closure of a tidal estuary; and the effects of road construction along a northern river in permafrost.

The documentation of changes in natural rivers due to human interference often points to the difficulty of directly transferring the results from analytical or laboratory studies to the field case. The writers believe that the best estimates of the change of a fluvial system due to the works of man must be made by a combination of the qualitative geomorphological approach with the more quantitative approach which is generally established from two dimensional steady uniform flow concepts.

INTRODUCTION

Fluvial systems continually undergo slow change due to natural causes. For the past several thousands of years the works of man have also induced minor local changes to rivers. In the last few decades, however, some major civil engineering projects have been undertaken which resulted in drastic changes to river systems. Lane (1955), Kuiper (1965), Kerr (1973), Gregory and Walling (1973), Kellerhals et al. (1979) and many others have presented case histories which illustrate various consequences of major interferences with rivers and river systems.

351

Much has been written concerning methods for evaluating the response of rivers to man-made changes, yet accurate prediction remains elusive in all but the simplest physical settings and no major improvement of the predictive capabilities seems to be in sight. The two main causes for this unsatisfactory state of the art are:

1. The three-dimensional nature of most river problems contrasts sharply with the purely two-dimensional formulation of present sediment transport theory. The few existing three-dimensional formulas, such as Lacey's (1958) regime formulas, are empirical and cannot be used beyond the narrow range of data on which they are based (e.g., straight sand-bed canals with cohesive banks which carry a certain type of water-sediment mixture).

2. Data requirements for most prediction techniques are prohibitive. If the expected changes involve any degradation, bank erosion, or lateral channel shifting, accurate prediction invariably requires detailed knowledge of the materials which will be exposed to the flow, and this is almost always unobtainable. Alternatively, even if the data are obtainable, the materials are likely to be so complex that quantitative methods are again precluded.

Simplified qualitative and quantitative methods for estimating the type and relative magnitude of changes resulting from human interference with rivers are useful, provided they are based on well-established physical principles and are well supported by past experience. The geomorphological and hydrological settings of any river system are prime considerations in evaluating past experience and in prediction. Neill and Galay (1967) present useful guidelines for the evaluation of river regime. Kellerhals et al. (1976) present a practical method of evaluating the geomorphological setting of a river reach under study.

Against this background, a brief documentation of a few Canadian cases illustrating river response to man-made changes may be useful.

The literature abounds with cases concerning field and laboratory studies of sand-bed channels. In Canada, bedrock-controlled and gravel-bed channels predominate, even among the largest rivers. Permafrost introduces a further variable in the northern areas of the country.

Although it may never be possible to have enough information to predict all changes that will take place in a natural river system at the design phase of a particular civil engineering project, the case studies presented herein should help to point out some of the potential problems. From another point of view, the geomorphologist can make inferences concerning the natural responses of rivers in the past based on a knowledge of the accelerated responses of river systems due to the works of man.

SOME CASE STUDIES

Five types of river response due to the works of man have been selected for presentation. Others could have been added, but the following five classes are typical of major problems associated with physical changes to river systems in Canada due to human interference:

1. Downstream effects of a large storage reservoir on a major river,
2. The effects of major interbasin diversions along the diversion routes,
3. The effects of channel straightening,
4. The effects of the closure of a tidal estuary, and
5. The effects of road construction along a northern river in permafrost.

Each of the five classes is illustrated with an example taken from the Canadian experience. We trust that no significant points are omitted due to the necessarily short summary of the cases presented herein. References for further detailed information are given wherever possible. Finally when evaluating past events, one seems to have "20/20" vision.

The Downstream Effects of a Large Storage Reservoir

The cases considered are associated with the W.A.C. Bennett Dam on the Peace River in British Columbia. Two aspects of downstream change in response to the modified hydrological regime in the Peace River are presented. Figure 1 shows the location of the river system.

First, the effects of the dam at a site far downstream of the structure are considered. For smaller projects, in the past, it may have been adequate to consider downstream effects over a "reasonable" reach of a few kilometers.

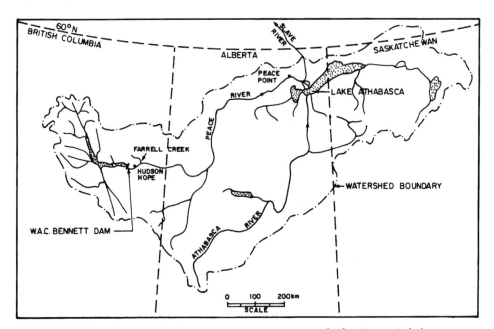

Figure 1.   *Location map of the W.A.C. Bennett Dam and the Peace-Athabasca Delta.*

This case demonstrates the importance of looking at the entire river system when evaluating human interference on a large river with relatively small natural storage.

The large storage reservoir associated with the W.A.C. Bennett Dam significantly modifies the downstream hydrological regime by decreasing the amount of spring runoff and its stages and by increasing winter flows. Even at a distance of 1800 km downstream of the dam, spring flows are now reduced by an estimated 3000 to 6000 m^3/sec (fig. 2) with a corresponding stage reduction of 2 to 4 m.

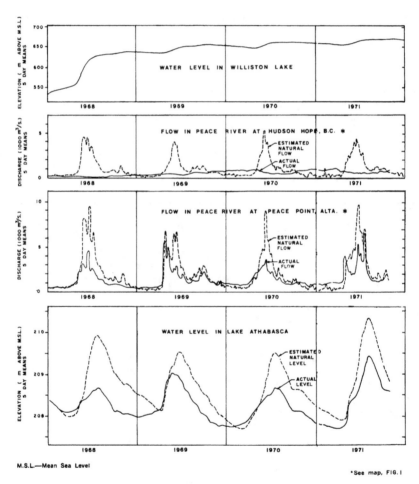

M.S.L.—Mean Sea Level

*See map, FIG. I

Figure 2.   *Lake levels, estimated natural Peace River flows, and actual Peace River flows during the filling of Williston Lake behind the W.A.C. Bennett Dam, 1968-1971 (adapted from Peace-Athabasca Delta Project Study Group, 1972).*

354

In the deltaic area at the western end of Lake Athabasca the lake outflow reaches the Peace River, and there is a complicated system of channels with reversible flow. Large lakes and wetlands in this area were in danger of re-verting to forest in response to the reduced inflow from the Peace River to Lake Athabasca, which resulted from attenuated stages of the Peace River during the spring runoff. A detailed study of the problem (Peace-Athabasca Delta Project Study Group, 1972) predicted an eventual reduction of 0.6 m in peak summer levels of the lake. To counteract this drop the Group recommen-ded the construction of a rock weir across the principal outlet channel. This has since been done. The weir has been reasonably successful but the final chapter remains to be written.

The twelfth and final generating unit has just been installed at the W.A.C. Bennett Dam and winter flow releases from the dam are only now reaching their design value of 1500 to 2500 m³/sec. This is an increase over natural winter flows by a factor of 10 at the dam and by a factor of 5 to 8 at Lake Athabasca. The increased winter flows lead to increased ice production, which may, in turn, lead to more severe ice jams during spring break-up. The northward flowing Peace River has always been subject to massive ice jams, but detailed records do not exist. Therefore it is difficult to establish conclusively any effect of the Bennett Dam. However, unusually large ice jams during the last few years have caused flooding of terraces that have not been flooded in human memory. Ice jams immediately downstream of the Lake Atha-basca outlet channels can divert a significant portion of the Peace River flow towards Lake Athabasca and can result in a rise of the water levels in the delta above the highest summer levels.

The morphological effects of ice jams are not well understood but they must be significant. Many valley flats along northern rivers are frequently flooded due to ice jams but are practically never flooded due to high flood flows. Each ice jam flooding event deposits a silt layer and some ice-rafted pebbles.

In the Peace-Athabasca Delta the effects of reduced summer flows and increased ice jamming in the Peace River and of weir construction in the lake outlet channel will combine and lead to gradual changes in the system of lake outlet channels. The final outcome is quite unpredictable.

The second effect considered due to the change in the hydrological re-gime after the construction of the W.A.C. Bennett Dam is the response of the tributaries of the Peace River below the dam. The reduced peak flows have reduced the capacity of the Peace River to transport its gravel bed material. Because the supply of gravel-sized material from the tributaries is not changed, the Peace River is aggrading in the vicinity of major con-fluences, and deltas are being built into the Peace River channel (Kellerhals and Gill, 1973) (fig. 3).

In simple situations, where bedload transport capacity of both the tributary and the Peace River can be calculated, estimates of the aggradation rate may be possible. Many of the tributaries are, unfortunately, steep, torrent-like streams with very coarse bed material. Their gravel bed-load transport is governed by the rate of supply rather than by channel hydraulics. As a result it is virtually impossible to compute the transport rate. Measurement of the bed load transport might be feasible but would be prohibitively expen-

Figure 3. *New delta forming at the confluence of Farrell Creek and the Peace River after the construction of the W.A.C. Bennett Dam located about 35 km upstream (photo: June, 1972).*

sive. Even in the Peace River itself, the bed material size is so highly variable (Church and Kellerhals, 1978) that a "typical" or "average" grain size distribution is not apparent. Lower stages in the main stem of the Peace River have induced the tributaries to degrade immediately upstream of the confluences. A bridge located on Farrell Creek (fig. 3) approximately 300 m upstream from the confluence with the Peace River has had its foundation exposed as a result of this effect (fig. 4). Concurrent with degradation, the increased tributary slope leads to increased rates of bank erosion and channel shifting during the spring freshet.

The Effects of Major Interbasin Diversions along Diversion Routes

As the scope of water resources projects increases, understanding the consequences of diverting relatively large flows from one natural channel to another becomes increasingly important. Kellerhals, et al. (1979) described 19 inter-basin river diversions in Canada, and gave post-construction data on 11 of them. The diversion routes were grouped into three geomorphological classes:

1. bedrock-controlled diversion routes
2. "steep" diversion routes in unconsolidated materials
3. alluvial diversion routes.

Figure 4.   Degradation on Farrell Creek showing the exposed bridge foundations at a site about 300 m above the confluence with the Peace River.

The main characteristics of these diversion routes are summarized in figure 5. The changes to be expected along the diversion routes and the procedures required for predicting them differ greatly between classes.

In a bedrock-controlled diversion route, the diversion channel cross section and slope are controlled by resistant bedrock outcrops. The main difficulty with such diversions is the marked change in water quality as the increased flows remove the overburden along the diversion route.

The Lake St. Joseph Diversion from the Albany River to the Root River in Ontario is an example of a bedrock-controlled diversion route. This diversion route is a small, rock-walled, structurally-controlled valley in the Canadian Shield. The valley previously contained a meandering stream with a narrow flood plain. The maximum diverted flow is about 200 m³/sec. This flow is two orders of magnitude greater than the natural, pre-diversion flows. The diversion flows have removed all traces of the pre-diversion channel and its flood plain, and the new channel virtually fills the bedrock valley bottom. Downstream water quality continues to be affected by the erosion of the fine-grained material remaining in the valley bottom.

In "steep" diversion routes in unconsolidated materials, the slopes along the diversions are such that the unconsolidated surface materials in the new channels are unable to withstand the erosive forces imposed by the diverted

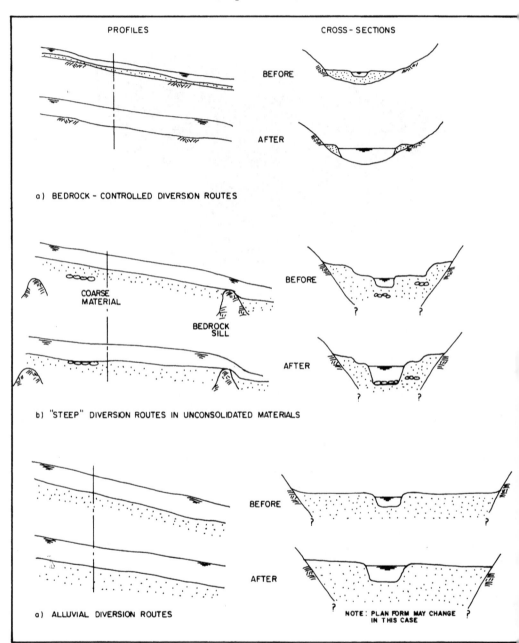

Figure 5. Conceptual sketches for classification of diversion routes.

flows. Estimates of stable channel dimensions can be made for such channels if they are composed of sand or gravel and if the location of resistant bedrock sills, large rivers, or other base levels are identifiable. Unfortunately, one can seldom estimate the vertical variation in the size of the bed materials, even in non-cohesive materials. If silts, cohesive materials, organics, or ice-rich permafrost are present, simple qualitative estimates of changes may be impossible.

One example of a "steep" diversion route in unconsolidated materials is the diversion from the Ogoki River to Lake Nipigon in Ontario. This route follows the natural channel of the Little Jackfish River, which ran on moraine and lacustrine silts. The pre-diversion mean flow in the Little Jackfish River is estimated to have been about 5 m^3/sec. The mean flow in the river after the diversion commenced in 1943 is approximately 110 m^3/sec. The diversion structure is designed for a maximum flow of 425 m^3/sec. Over the period of operation, the diversion route has undergone persistent degradation and bank erosion. The rates of these processes have been declining with time. Bridger and Day (1977) reported that approximately 23 x 10^6 m^3 of silt have been flushed into Ombabika Bay at a rate decreasing from 4 x 10^6 m^3/yr initially to 0.3 x 10^6 m^3/yr presently.

A particularly spectacular "steep" diversion is along Adam Creek in Northern Ontario (fig. 6). Ontario Hydro uses Adam Creek to pass spring freshet floods of the Mattagami River past a cascade of powerplants. The pre-diversion Adam Creek was a small, meandering stream incised somewhat into a 30 m-thick formation of unconsolidated Pleistocene sediments of the Hudson Bay Lowlands. Flows up to 2800 m^3/sec have been diverted through the creek and a canyon-like channel has been eroded into the clay and silt deposits. Because little gravel to boulder-size material is available to the flow, erosion is likely to continue downward to the Paleozoic sandstones which lie below the unconsolidated Pleistocene sediments.

In addition to the degradation, large scale slumping of the channel banks is occurring. R.J. Fulton (personal communication, 1978) indicated that the slumping banks are partly the result of modified groundwater flow in some of the Pleistocene beds. Modified groundwater flow is a response to the channel degradation. This diversion route will probably experience a long period of channel instability.

Tributaries along a "steep" diversion route must adjust to any degradation which occurs in the diversion channel as a result of the increased flows. The drainage system will respond to this change in base level until the process is arrested by some control such as resistant bedrock. Kellerhals et al. (1979) studied the Nechako - Cheslatta Diversion in British Columbia in some detail. A portion of this diversion route has entrenched with the consequent adjustment of the tributaries (fig. 7).

An alluvial diversion route directs flow into a river which has a genetic flood plain and is flowing in a self-formed channel. The diversion flows are not large enough to dissect the flood plain and the alluvial character of the channel-floodplain system is preserved, although there may be major changes in channel dimensions and channel morphology. Kellerhals et al. (1979) discussed the only Canadian case history of this class of diversion known to the

(a) *Degradation of the diversion channel showing the Kipling till overlying the more resistant Adam till.*

(b) *Low-flow channel cut into the Adam till.*

Figure 6. *Views of the extensive degradation along the Adam Creek diversion route, Ontario. Note lack of lag material in the eroded till (photos by R.G. Skinner, Geological Survey of Canada, August, 1969).*

*Figure 7.   Fan dissected by a tributary joining the degrading Cheslatta River,
B.C., where the Cheslatta River is a diversion route.*

authors.

Frenette and Caron (1976) presented information about the anticipated
effect of a proposed major diversion in Quebec where the mean flow of the
lower La Grande River is to be increased from 1760 m^3/sec to about 3310 m^3/
sec.   The increased flows in the La Grande River result from the diversion
of about 40 % of the normal flow of the Caniapiscau River and about 90 % of
the flow of the Eastmain and the Opinaca Rivers.   Frenette and Caron estimated the change to be expected in the width of some sand islands in the
estuarine part of the lower La Grande River.   They calculated that the width
of the waterway would increase by about 120 m, 10 % of the total effective
width of the channel, as a result of the diverted flows.   Because the lower
La Grande River is influenced by tides, it cannot easily be categorized into
one of the three classes of fluvial diversions presented by Kellerhals et al.
(1979), but it is essentially a channel with bedrock control.

Diversion channels which are deeply entrenched in unconsolidated material
may cause major bank slumping as the channel adjusts its width to the new
flow regime.   Bank slumping is another effect of the modified flow regime of
the lower La Grande River.

The Effects of Channel Straightening

Artificial cutoffs and channel straightening are often utilized in Canada to alleviate flooding upstream of the channel modifications, to facilitate the construction of bridges and highways, and to reduce the loss of agricultural land.

If a cutoff is constructed on an alluvial river, reach quantitative estimates can be made of the degradation associated with the cutoff if the properties of the material through which the channel is constructed are known. However, in many cases the bed of the new channel is controlled by a bedrock outcrop or becomes armoured as the channel cuts through the coarser material along the cutoff.

Bray and Cullen (1976) reported on the response of five artificial cutoffs on gravel-bed rivers in New Brunswick. The cutoff on the lower portion of the Coverdale River caused the greatest reduction in channel length and hence the greatest potential for damage due to degradation upstream of the cutoff. At the Coverdale River cutoff mean flow is 5.4 m³/sec and the mean annual flood is 55 m³/sec. The bed material in the natural channel has a median size of 47 mm. An air photo of the cutoff and a sketch of the longitudinal profiles before and after the cutoff was constructed are shown in figure 8. Computed degradation rates indicated that problems might occur with the bridge just upstream of the cutoff. However, no problems were encountered at the site, due to the bedrock control and some armouring of the cutoff section.

Parker and Andres (1976) presented the case of the East Prairie River in Alberta which was straightened and channelized over a distance of about 13 km. The river has a mean discharge (April-October) of 9.8 m³/sec. The bed of the natural channel consisted of a thin layer of medium to fine sand overlying a deep, high-density, highly plastic clay. Channelization took place between 1962 and 1966. About 11 km of the reach was straightened to double the slope of the natural channel. The lower portion of the reach was placed at a slope of about 0.00050 which corresponds to the approximate natural slope of the major length of the reach. The work on the lower portion commenced at the downstream end of the reach and due to an error, the channel was constructed 3.0 m too low. A knickpoint resulted from the mismatch of the two channelized portions of the reach. Over the period 1966 to 1972, the knickpoint essentially maintained its shape in the eroding cohesive clay making up the bed of the straightened channel. While doing so, the knickpoint has progressed upstream at an average rate of about 1.6 km/yr.

Severe degradation has taken place in the upper portion of the channelized section and in the river above the channelized section. Near two bridges located about 1.6 km upstream of the channelized section at least 4.5 m of degradation has occurred. One bridge had to be replaced. General degradation has resulted in a lowering of the bed by about 3.0 m at a distance of up to 9 km upstream of the channelized portion of the channel. If no control structure is built in the river, the process of degradation probably will not be arrested for some time because the eroding clay does not contain sufficient coarse material to armour the bed of the channel.

Aggradation is occurring in the lower part of the channelized section. The

362

*(a) Plan view of cutoff (N.B. Dept. of Natural Resources
Photo: 70022-98).*

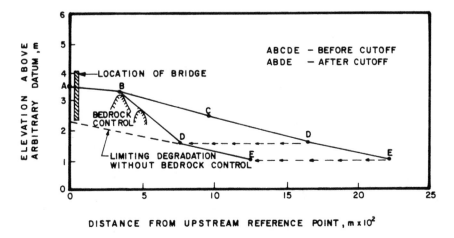

(b) Longitudinal profile before and after cutoff.

*Figure 8. Effects of a cutoff on the Coverdale River, N.B., where degradation
is controlled by bedrock outcrops.*

original excavated channel at the lowest reaches of the channelized section was
filled by 1972. Extensive overbank deposition takes place in these lower reaches of the channelized section during high discharges. A delta is forming at
the junctions of the East Prairie River and the West Prairie River and the
South Heart River, downstream of the channelized reach. Figure 9 shows the

general profile along the channelized portion of the channel and figure 10 shows the change in two channel cross sections over a 4 year period.

Based on their study of the East Prairie River, Parker and Andres (1976) presented a simplified model which can be used to make quantitive estimates of the response of a channel with a cohesive bed.

Extensive cutoffs on the Pembina River in Alberta were made in an attempt to reduce flooding on adjacent agricultural land and to reduce the effects of the lateral movement of the river in certain reaches adjacent to roadways (Nwachukwu and Quazi, 1977). The flood damage level in the river before modification corresponded to a discharge of 570 m^3/sec with an associated return period of about 15 years. During bankfull flow the natural channel in the reach had a water surface width of about 64 m, and the mean depth was 3.5 m. The sinuosity was about 2.3 and the average channel slope was about 0.00027 prior to the channel modifications.

Nine cutoffs were constructed over a 105 km reach of the Pembina River during the period 1954 to 1969. Two of the cutoffs on a typical portion of the reach are shown in the 1968 air photograph in figure 11. The upstream cutoff, G, was constructed in 1958 and the downstream cutoff, F_3, was constructed in 1966.

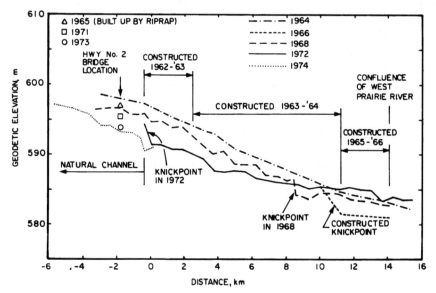

Figure 9.   Progress of degradation along the channelized reach of the East Prairie River, Alberta (adapted from Parker and Andres, 1976).

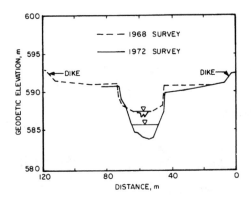

a) DEGRADATION AT SECTION AT 3 201 m D/S

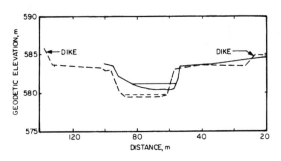

b) AGGRADATION AT SECTION AT 10 976 m D/S

*Figure 10. Cross sections showing degradation and aggradation over a four
year period on the East Prairie River, Alberta (adapted from
Parker and Andres, 1976).*

The cutoff program was expected to result in significant lowering of the
bed level over much of the 105 km reach of the Pembina River which was subject to flooding. Even after the passage of at least two major floods, the
cutoffs have not caused significant bed erosion except in the immediate loca-
lity of the cutoffs.

In almost all cases, the beds of the cutoff channels were in clay or were
on a shallow layer of sand underlain by clay. The cutoffs in clay developed
slowly and some have not developed fully even after 10 years. The cutoffs
did not function over the entire project reach as planned because the resistance of the clay beds to erosion was underrated. As a consequence, the
benefit of appreciable flood relief was confined to the immediate vicinity of
the cutoffs.

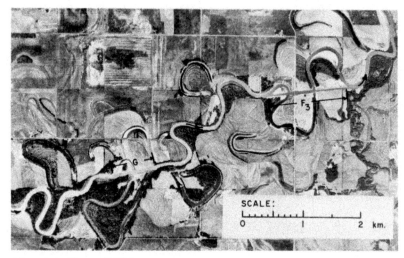

*Figure 11. Cutoff G (1958) and cutoff F_3 (1966) on the Pembina River, Al-
berta (Alberta Govt. Photo: C68 6397-6168-2640, YC 1552-5407-202:
Year 1968).*

The Effects of the Closure of a Tidal Estuary

The upper reaches of the Bay of Fundy experience the highest tides in
the world. Any natural or man-made change of the regime geometry of the
estuaries should result in relatively rapid responses in such a high energy
environment. Fluvial principles have been applied to the study of such changes, but as Kestner (1966) indicated, fundamental differences between rivers
and estuaries in their response to engineering works must be understood.
Engineering works in rivers which cause a reduction in cross section may have
effects on the local geometry of the river, but the discharge of the river is
essentially unchanged assuming that there is no major storage upstream of the
constriction. In an estuary, a reduction in the extent of the tidal flats by
dyking, a reduction of the cross section, or a closure such as that resulting
from a causeway will reduce the volume of the tidal prism and as a consequence the magnitude of the tidal discharge in the estuary. Thus, engineering
works in estuaries may result in widespread and often irreversible consequences.

For example, along the lower Petitcodiac River in New Brunswick, an
estuary of the.Bay of Fundy, the tidal prism was shortened by about 14 km
due to the construction of a causeway. The causeway was constructed to
provide a highway link, to establish a recreational freshwater lake, and to
avoid rehabilitation of 20 km of dykes and a great number of aboiteaux. The
causeway and associated control structure were built between 1966 and 1968.

The estuary has adjusted rapidly to the changed conditions (fig. 12). A cross section located 4.8 km downstream of the causeway (C. Desplanque, 1979 personal communication) has undergone substantial aggradation (fig. 13). The 1960 cross section in this figure is assumed to be the stable cross section before closure of the upper part of the estuary. Between 1966 and 1971, about 14×10^6 m^3 of silt has accumulated between this cross section and the causeway.

The infilling of the upper estuary has not resulted in any significant physical damage; however, it has reduced the magnitude of the famous tidal bore on the Petitcodiac River. Furthermore, small tankers and other ships can no longer navigate the river, because the narrowed width does not allow adequate room for maneuvering the vessels. The newly deposited mud flats are now being colonized by salt-tolerant plants (compare fig. 12b and fig. 12c).

A relatively simple semi-quantitative estimate of the response of the estuary can be made by moving the surface width-distance relationship of the pre-causeway tidal prism downstream to the causeway control structure (McCrea, 1975) (fig. 14). With the assumption that the mean annual flood in the fluvial reach of the river was the formative discharge the upstream width of the estuary was determined by the Lacey (1958) width equation for regime channels. Assuming that an adequate sediment source is available, the ultimate shape of the new estuary can be estimated by means of figure 14. This figure also shows the probable width-distance relationships for various times as the estuary gradually approaches the ultimate regime shape for the modified conditions. The ultimate regime bed level at the causeway should be approximately equal to the bed level at the head of tide for the pre-causeway estuary.

The Effects of Road Construction Along a Northern River in Permafrost

As more development takes place in northern Canada, it is found that the behavior of rivers in permafrost regions differs in some respects from that of rivers in the south. The thermal regime of the river banks becomes an important consideration when planning civil engineering works on or near the banks of such rivers. Removal of the vegetal mat from the top of a river bank may result in accelerated bank erosion due to an increase in the thickness of the active layer of the material making up the river bank.

Cooper and Hollingshead (1973) presented a case study of accelerated bank erosion on a section of the Liard River, near Watson Lake, Yukon (latitude 60°10'). The banks of the river at the site are about 5.5 m above low water and were located in a region of discontinuous permafrost before any man-made changes were made. Evidence from air photographs indicated that the banks were stable for a period of 20 yr before the vegetation was removed from a portion of the valley flat near the river to facilitate the construction of a logging road. This road was constructed on the bank of the river with a southern exposure. The material making up the bank consisted of fine to medium sand.

After construction of the logging road bank erosion increased during each annual flood in the section from which the bank vegetation was removed. Field evidence indicates that the banks on which the vegetation was not removed have remained stable. The probable processes involved in this case

(a) 1963 flight, before the construction of the causeway
(N.B. Dept. of Natural Resources Photo: 6320-162).

(b) 1970 flight, two years after completion of the causeway
(N.B. Dept. of Natural Resources Photos: 70021-36, 37).

(c) 1978 flight, ten years after completion of the causeway.
Note colonization of newly deposited mud flats. (N.B.
Dept. of Natural Resources, Photos: 78502-6, 7, 8).

Figure 12.   Air photographs of the Petitcodiac River at Moncton, N.B., showing the response of the tidal estuary due to the construction of
a causeway.

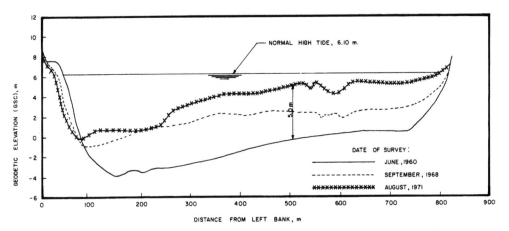

Figure 13. Changes at cross section on the Petitcodiac River, N.B., located 4.8 km downstream of the causeway (after C. Desplanque, 1979, personal communication).

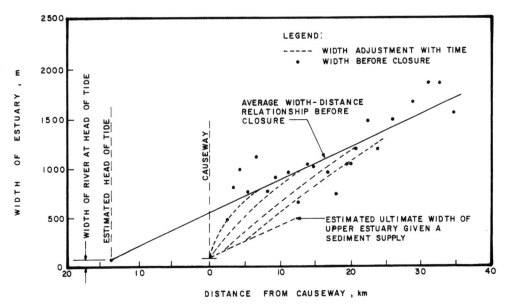

Figure 14. Simplified evaluation of the response of the Petitcodiac River estuary, N.B., to closure (adapted from McCrea, 1975).

are outlined in figure 15.

SUMMARY AND CONCLUSIONS

The case studies presented in this paper demonstrate that it is essential to have an understanding of the fluvial processes and the nature of the geological controls operating in a river system before attempting to predict the response of the river system to a change in the hydrological and sediment regime as the result of the works of man. Quantitative prediction of the response of river systems to such changes is restricted mainly by the complexity of geological constraints and the diversity of materials affecting a river. The basic physical processes are reasonably well understood, but predictive attempts are forever frustrated by the lack of knowledge of the basic input parameters. Furthermore, due to the nature of the channels of the majority

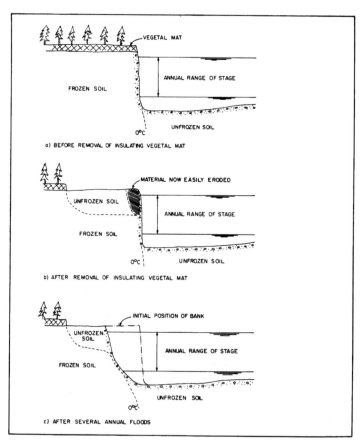

Figure 15. Schematic diagram to indicate probable processes influencing bank erosion in regions of permafrost.

of rivers in Canada, there is little opportunity to apply directly the extensive laboratory and field research related to sand-bed channels. The best estimates of the change in fluvial systems due to human interference are made by utilizing a combination of the qualitative geomorphological approach with the more quantitative approach which is generally established from two-dimensional steady uniform flow concepts. Carefully documented case studies will continue to play a major role in prediction even though they are seldom directly transferable.

ACKNOWLEDGEMENTS

The writers wish to express their appreciation for the assistance and comments provided by the following persons: Dr. M. Church, Dept. of Geography, University of B.C.; Mr. D. Demerchant, A.D.I. Limited; Mr. C. Desplanque, Maritime Resource Management Service; Dr. M. Frenette, Dept. of Civil Engineering, Laval University, Mr. M. Quazi and Mr. B. Nwachakwu, Alberta Dept. of the Environment.

REFERENCES

Bray, D.I., and Cullen, A.J., 1976, Study of artificial cutoffs on gravel-bed rivers: American Society of Civil Engineers, Fort Collins, Proceedings of Third Annual Symposium Waterways, Harbors, and Coastal Engineering Division, p. 1399-1417.

Bridger, K.C., and Day, T.C., 1977, Ogoki River diversion effects on commercial and sports fisheries: University of Regina, Canadian Association of Geographers, Abstracts of Annual Meetings, p. 60-67.

Church, M., and Kellerhals, R., 1978, On the statistics of grain size variation along a gravel river: Canadian Journal of Earth Sciences, v. 15, p. 1151-1160.

Cooper, R.H., and Hollingshead, A.B., 1973, River bank erosion in regions of permafrost, in Fluvial Processes and sedimentation: Edmonton, University of Alberta, Hydrology Symposium Proceedings No. 9, p. 272-283.

Frenette, M., and Caron, O., 1976, Predictions of the hydraulic and morphological changes of "La Grande Riviere" lower reach system: American Society of Civil Engineers, Fort Collins, Proceedings of Third Annual Symposium of the Waterways, Harbors, and Coastal Engineering Division, p. 1229-1247.

Gregory, K.J., and Walling, D.E., 1973, Drainage basin form and process: London, Edward Arnold, 458 p.

Kellerhals, R., and Gill, D., 1973, Observed and potential downstream effects of large storage projects in Northern Canada: Madrid, Proceedings of 11th Congress of the International Commission on Large Dams, p. 731-754.

Kellerhals, R., Church, M., and Bray, D.I., 1976, Classification and analysis of river processes: American Society of Civil Engineers Proceedings, Journal of Hydraulics Division, v. 102, HY7, p. 813-829.

Kellerhals, R., Church, M., and Davies, L.B., 1979 (in press), Morphological effects of interbasin river diversions: Canadian Journal of Civil Engineering, v. 6.

Kerr, J.A., 1973, Physical consequences of human interference with rivers, in Fluvial Processes and sedimentation: Edmonton, University of Alberta, Hydrology Symposium Proceedings No. 9, p. 664-696.

Kestner, F.J.T., 1966, The effects of engineering works in tidal estuaries; in Thorn, R.B., ed., River engineering and water conservation works: London, Butterworths, p. 226-238.

Kuiper, E., 1965, Water resources development: London, Butterworths, 483 p.

Lacey, G., 1958, Flow in alluvial channels with sandy mobile beds: Institute of Civil Engineers, Proceedings, v. 9, p. 145-164.

Lane, E.W., 1955, The importance of fluvial morphology in hydraulic engineering: American Society of Civil Engineers Proceedings, Journal of Hydraulics Division, v. 81, paper 745, 17 p.

McCrea, J.H., 1975, Study of Petitcodiac River estuary [B.Sc. thesis] : Fredericton, University of New Brunswick, 69 p.

Neill, C.R. and Galay, V.J., 1967, Systematic evaluation of river regime: American Society of Civil Engineers Proceedings, Journal of Waterways and Harbor Division, v. 93, WW1, p. 25-53.

Nwachukwu, B.A., and Quazi, M.E., 1977, Assessment of Pembina River cutoffs: Edmonton, Alberta Department of the Environment, Report of the Technical Services Division, 22 p.

Parker, G., and Andres, D., 1976, Detrimental effects of river channelization: American Society of Civil Engineers, Fort Collins, Proceedings of Third Annual Symposium of the Waterways, Harbors, and Coastal Engineering Division, p. 1248-1266.

Peace-Athabasca Delta Project Study Group, 1972, The Peace-Athabasca Delta - a Canadian resource: Ottawa, Information Canada, 144 p.

9 780367 460587